AF291331

Synthesis Lectures on Mechanical Engineering

This series publishes short books in mechanical engineering (ME), the engineering branch that combines engineering, physics and mathematics principles with materials science to design, analyze, manufacture, and maintain mechanical systems. It involves the production and usage of heat and mechanical power for the design, production and operation of machines and tools. This series publishes within all areas of ME and follows the ASME technical division categories.

George Manolis · Christos Panagiotopoulos

Vibrations of Structural Systems

 Springer

George Manolis
Department of Civil Engineering
Aristotle University of Thessaloniki
Thessaloniki, Greece

Christos Panagiotopoulos
Department of Civil Engineering
Aristotle University of Thessaloniki
Thessaloniki, Greece

SIMPLEGMA P.C.
Thessaloniki, Greece

ISSN 2573-3168   ISSN 2573-3176   (electronic)
Synthesis Lectures on Mechanical Engineering
ISBN 978-3-032-12278-0   ISBN 978-3-032-12279-7   (eBook)
https://doi.org/10.1007/978-3-032-12279-7

This Springer imprint is published by the registered company Springer Nature Switzerland AG
The registered company address is: Gewerbestrasse 11, 6330 Cham, Switzerland

If disposing of this product, please recycle the paper.

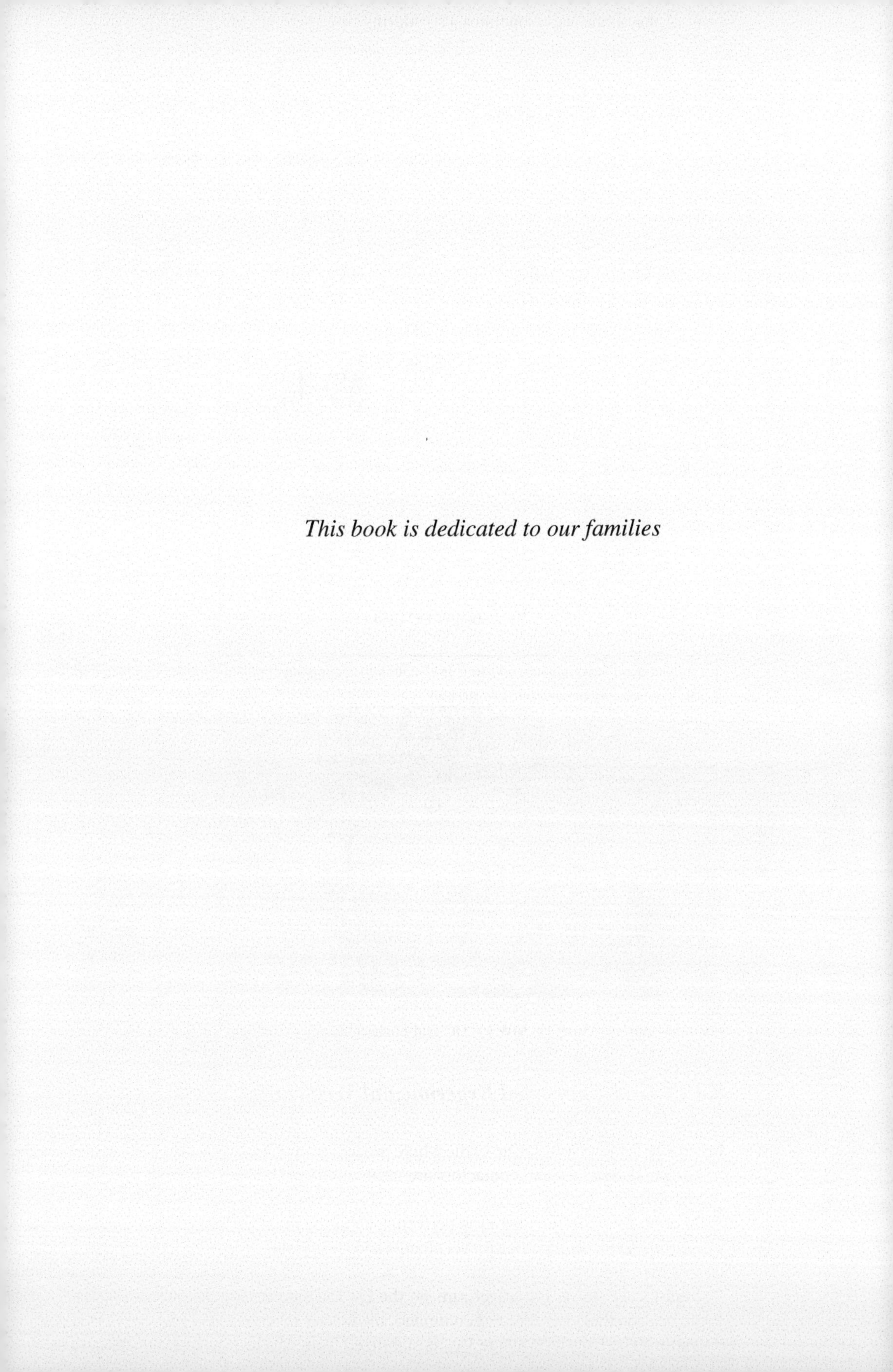

This book is dedicated to our families

Foreword

The field of structural dynamics and vibrations has exploded over the last few decades, as it finds applications to a variety of problems in civil, mechanical and aerospace engineering. Starting in the mid-nineteenth century with the problem posed by the vibrations of metallic bridges due to the passage of locomotives, when the methods of solution were analytical, a tremendous impulse came to the field during the mid-twentieth century with the development of the finite element method in parallel with the emergence of the electronic computer. This powerful numerical tool, coupled with the use of computers, was employed to solve all types of structures subjected to a variety of external loads. Nowadays, much attention is paid to issues such as structural health monitoring, passive and active protection systems, retrofit and rehabilitation strategies, all of which require structural dynamics as the background. The present book starts with the basic structural dynamic models, namely the single degree-of freedom system, the multiple degree of freedom system and the continuous system, while discussing in parallel the mathematical methods used for formulating their governing equations of motion. Next, the Rayleigh-Ritz method that was popular in the past, plus the contemporary finite element and boundary element methods are presented. This is followed by material on structural control, structural identification and finally earthquake engineering, which includes coupled-field problems. Furthermore, an internet-based computational platform labelled as the 'Symplegma Development Environment' (SDE) is discussed and explained as the means for the numerical solution of small-scale problems in structural dynamics. In sum, the present book is a useful addition to the current scientific literature in that it fills a gap between the more mathematical treatment of the subject of vibrations, favoured in mechanical and aerospace engineering, and the more applied treatment of problems in structural dynamics and earthquake engineering, favoured in civil engineering. It should also be added that the novelty of this work is in linking text material with the SDE internet-based platform for the solution of exercises and small-scale problems. In closing, it was a pleasure for me to peruse through the book and communicate my comments to the publisher. I am familiar with the research work of both authors, since Prof. Manolis completed his doctoral dissertation under my supervision at the University of Minnesota in Minneapolis, USA, on the use of the boundary elements for examining the dynamic response of underground tunnels, while Dr. Panagiotopoulos completed his dissertation on boundary element formulations for transient problems at the Aristotle University in Thessaloniki, Greece, under the supervision of Prof. Manolis. Thus, both authors are well placed to produce work in structural dynamics, and I congratulate them for their efforts to present this demanding subject to a mixed audience comprising undergraduate students in their senior year, graduate students specializing in structural dynamics, researchers and finally practising engineers.

Shanghai, China
Summer 2025

Dimitri E. Beskos
Chair Professor, Civil Engineering Department
Tongji University
Shanghai, China

Emeritus Professor, Civil Engineering Department
University of Patras
Patras, Greece

Preface

The present book addresses three interconnected fields, namely Structural Dynamics and Vibrations, Mathematical Modelling and Numerical Methods and finally Computer Programming using the Java Virtual Machine. The target audience includes undergraduate students at the senior level studying civil engineering and specializing in structural mechanics, graduate students embarking in Masters and/or Doctoral programs in civil and mechanical engineering, plus researchers and practitioners wishing to upgrade their skills and knowledge in the field of structural vibrations.

This book is the culmination of a continuous cooperation between the two authors for over two decades in teaching structural dynamics at both graduate and undergraduate levels, while conducting research in elastodynamics. The impetus for converting loose teaching notes into book form for an English-speaking audience came from a common initiative to develop web-based, free access software for the solution of elementary problems in structural dynamics. It was felt that using calculators and spreadsheets to compute the time and/or frequency evolution of the kinematic and force variables in a structure was somewhat outdated, while the use of commercially available finite element software for the same purpose seemed redundant. Thus, a middle path was taken, whereby special-purpose, Java-based code was developed and made available through the Aristotle University's servers to users who were students and were issued e-mail addresses of the type 'name@civil.auth.gr'. Alternatively, it could be accessed by anyone using computers with an Aristotle University IP address. Later on, this restriction was relaxed and the two sites that were successively developed from the late 1990's onwards (namely edusoft.civil. auth.gr and dynasoft.civil.auth.gr) were made freely accessible.

At about the same time, work began on the development of a more ambitious project, the 'Symplegma Dynamic Environment' (or SDE) site, with capabilities that go beyond the solution of elementary structural dynamic problems. In essence, the material presented in this book is centred around the use of the SDE site, which works in a Java programming environment. This way, it is possible to solve both elementary and non-trivial problems in structural dynamics in parallel with the reading material so as to reinforce both the underlying theory and the methodology. The difference between transient/steady-state problems as contrasted with static analyses is that all results (e.g., displacements and tractions) are now dependent on either the time or the frequency variables, resulting in voluminous data and the need for plotting it.

In closing, we wish to thank our teachers from whom we learned structural dynamics, as well as our students to whom we taught this subject. This interaction between teachers and students made our academic experience all the more interesting and valuable. Thanks are also due to researchers with whom we collaborated on a number of projects that had to do with the dynamic response of both discrete and continuous media.

Thessaloniki, Greece

June 2025

George Manolis

Professor Honorarius

Member of Academia Europaea

Laboratory for Experimental Strength of Materials and Structures

Department of Civil Engineering

Aristotle University

Thessaloniki, Greece

gdm@civil.auth.gr

https://strength.civil.auth.gr/gdm

Christos Panagiotopoulos

Department of Civil Engineering

Aristotle University of Thessaloniki

Thessaloniki, Greece

cpanagia@civil.auth.gr

SIMPLEGMA P.C.

Thessaloniki, Greece

christos.panagiotopoulos@simplegma.com

https://www.simplegma.com

Acknowledgements We wish to acknowledge our colleagues and graduate students who worked with us on various research projects over the years at the universities and institutes, where we were employed. Most of these projects involved, directly or indirectly, elements of structural dynamics and this further served as the impetus for writing this book. Thanks are also due to the students who attended our undergraduate classes on structural dynamics; it was a privilege to teach and interact with them.

Contents

Acronyms

BEM	Boundary element method
DOF	Degree-of freedom
EMA	Experimental modal analysis
FEM	Finite element method
FSI	Fluid-structure-interaction
FT	Fourier transform
IFT	Inverse Fourier transform
MDOF	Multiple-degree-of freedom
OMA	Operational modal analysis
PCA	Principal component analysis
SDE	Symplegma Development Environment
SDOF	Single-degree-of freedom
SHM	Structural health monitoring
SSI	Soil-structure-interaction

Introduction

1

1.1 Definitions

Dynamics

Dynamics constitutes a special branch of mechanics focusing on the kinematics of bodies under the influence of external forces, but talking into account at the same time the inertia forces which develop. Time is manifested in both the resulting strain and stress fields that develop in the components of the structure in question. The starting point for analysis is a mathematical model of the dynamic system and the derivation of the equations of motion that describe this system. This is done using Newtonian mechanics, as expressed by the three basic laws that describe the relationship between the motion of an object and the forces acting upon it. These laws can be paraphrased as follows: (i) A body remains at rest, or in motion at a constant speed along a straight line, unless acted upon by a force; (ii) when a body is acted upon by a force, the time rate of change of its momentum equals that force and (iii) if two bodies exert forces upon each other, these forces have the same magnitude but opposite directions. Furthermore, it is possible to use energy formulations for deriving the equations of motion of complex systems. More specifically, Hamilton's principle states that the dynamic response of a system is determined by a variational problem involving a functional based on a single function, namely the Lagrangian, which contains all information relevant to that system and to the forces acting upon it. This variational problem is equivalent to dynamic equilibrium and allows for an alternative derivation of the differential equations of motion of the system. In reference to their mechanical behavior, dynamic systems can be classified as rigid versus deformable, as linear versus nonlinear and as discrete versus continuous [1].

© The Author(s), under exclusive license to Springer Nature Switzerland AG 2026
G. Manolis and C. Panagiotopoulos, *Vibrations of Structural Systems*, Synthesis Lectures
on Mechanical Engineering, https://doi.org/10.1007/978-3-032-12279-7_1

Vibrations

The term 'vibration' refers to the repeated motion of a body about a mean value, which is referred to as the equilibrium position. By introducing the term 'vibration field', we can delineate both qualitative as well as quantitative aspects regarding the analysis of dynamic systems. These systems can be classified as either continuous, for which closed form solutions are recoverable, or discrete, leading to numerical methods of solution. The first step is the formulation of a mathematical model and the derivation of the equation(s) of motion of the dynamic system. These are differential equations that may be categorized as either ordinary or partial. Their solution entails distinguishing between free and forced vibrations, with the former case corresponding to the homogeneous solution and the later yielding a particular solution. Furthermore, solutions can be recovered in either the time or the frequency domains, or possibly in any other transformed domain. When the vibration response of a dynamic system is measured, either in the field or in the laboratory, it is essential that the transient signals recorded be processed for the purposes of identifying the properties of the system. This procedure can be carried out in either the time domain or the frequency domain. It should be noted here that the term 'operational modal analysis' refers to the effort behind recovering the dynamic characteristics of the dynamic system, and specifically its natural frequencies (eigenfrequencies) and their associated modal shapes (eigenvectors).

Structural Systems

The terms *system* and *structure* are often used interchangeably in civil and mechanical engineering, meaning much the same thing. A more general term would be *structural system*. This allows for a mathematical definition of what a structure is, namely a mapping P that connects an input function $w(t))$ to an output function as $z(t) = P[w(t)]$. This way, structural analysis implies that the system is known along with the input, so one can solve for the output. Structural control is a step further, where we wish to ascertain that the output falls within certain acceptable (or prescribed) bounds, implying that some sort of modification must be made to the system for the duration of a given input. The final possibility comprises inverse problems, where both input and output are known, and the task is to determine the structural system properties from this information. Redefining structural control more carefully implies a design process encompassing a set of actions applied to the system so that it will respond in a certain, predictable way to external forces. Thus, control implies the existence of an additional substructure K, appropriately connected to structure P, which will satisfy a predetermined criterion for the desirable response of the combined system $P+K$. Another issue that arises is that of structural optimization, which implies that from a set of possible and/or acceptable designs pertaining to a specific structural project, only one is best in terms of certain constraints. A typical such constraint is minimum weight design, but it is possible to specify alternative performance criteria (including economic considerations) such as a cap on the level of the stresses or on the level of the displacements in order to keep the structure operating within acceptable bounds.

1.2   Mathematical Modelling

Modern design codes invariably state that the purpose of structural dynamic analysis is to limit the vibrations resulting from the action of external forces, to a threshold that ascertains both the safety of the occupants and continuous operation of the mechanical systems housed in the structure, all within certain pre-described limit states. At the same time, complete functionality must be allowed for as the design loads act upon the structure. In order to conduct the analysis, mathematical models for the mechanical behavior of structural systems must be developed, as well as methods for describing the various loads. These mathematical models depend on various simplified assumptions regarding the structure and its true behavior, inasmuch as the latter can be measured. As a consequence, there is divergence between results produced by the mathematical models and those recorded from the actual structure, and it is these incompatibilities that are labelled 'modelling errors'. The analysis of structural systems is carried out in three steps, as can be seen in Fig. 1.1. The first step has to do with the modelling of the mechanical behavior of the components comprising the structure. The second step is the formulation of the governing equations for this model, which are based on the laws of motion, the kinematics and the constitutive laws. The third step is the solution of the equations of motion, either analytically or numerically, and the subsequent interpretation of the results.

1.3   Programming Languages and Numerical Tools

At present, there is a plethora of programming languages, application packages and general software that can all be used in conjunction with the teaching of structural dynamics, as well as all other topics in engineering science. As examples, one can mention Matlab [2], Simulink [3] and LS-Dyna, which are commercially available software packages addressing many fields in the physical sciences. At the same time, there are many open source software packages (or freeware) that are available. e.g., Octave [4], OpenModelica [5], Salome-Meca, etc. More specifically, Simulink amd OpenModelica are programming languages focusing

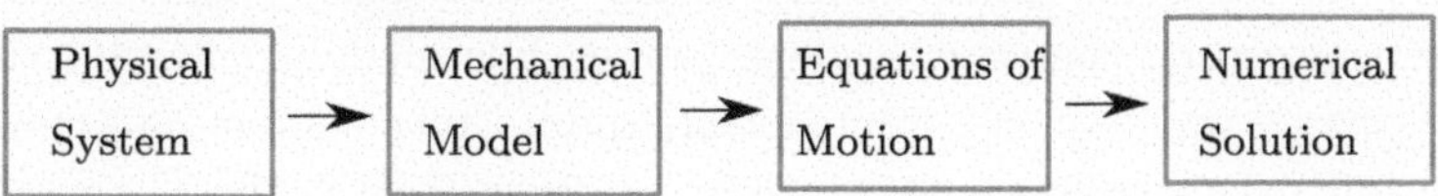

Fig. 1.1 Modeling and solution procedure for problems in structural dynamics

on technical applications and operate in a dedicated workspace environment. Next, Matlab and Octave are actually application software for the analysis and design of engineering systems, while LS-DYNA and Salome-Meca are integrated finite element method (FEM) programs that can be used, among other applications, for problems in structural dynamics and mechanical vibration. For instance, Salome-Meca is actually the fusion of the FEM solver Code-Aster with the pre- and post-processing workspace platform Salome. Furthermore, one can find programming languages which have scripting capabilities, along with dynamic space allocation of the variables that can be used for solving problems such as those of interest in structural vibrations and structural control. Well-known examples of these languages are Python and the more recent Julia. Finally, one should mention what is probably the most widely used programming language, namely Java, which is suitable for the development of a vast number of applications. The material presented above is by no means an exhaustive list of contemporary software. However, for our purposes, we have chosen Java as the programming language for our applications, with the Java Virtual Machine (JVM) environment used for carrying out the input/output process.

Groovy Programming Language

More specifically for the purposes of the present work, the 'Groovy' programming language has been selected for setting up the computational platform. In terms of classification, Groovy can be viewed as a higher order programming language containing Java as a subset. In particular, Groovy is a dynamic, object-oriented programming language similar to languages such as Python, Ruby, Perl and Smalltalk. It runs on the JVM platform and communicates with all software packages that have been developed for Java. Furthermore, Groovy can be used for scripting purposes. As with Java, the hierarchy of the programming structure is as follows: *class*, *object*, *method*, *instance*. For more information, one should consult the dedicated site http://groovy-lang.org.

Application Dynasoft

The educational software package Dynasoft (http://dynasoft.civil.auth.gr) was developed using the Java programming language for use within the context of coursework in structural dynamics (comprising two undergraduate and one graduate level courses) offered by the Civil Engineering Department at Aristotle University of Thessaloniki, Greece since the academic year 2014–2015. It is an improved version focusing on structural dynamics and derives from the earlier, more general-purpose educational package Edusoft (http://edusoft.civil.auth.gr). This latter package was developed from a research project running from 2008 onward by the department of civil engineering at Aristotle University in Thessaloniki, Greece in an effort to produce user-friendly, specialized software for educational purposes. Dynasoft has also been introduced in the School of Production Engineering and Management at the Technical University of Crete in Chania, Greece for teaching an undergraduate course in Dynamics, Vibrations and Structural Control since 2016. Dynasoft is a simple, easy to use application comprising five separate modules, namely:

- Single degree-of-freedom oscillator (`sdof`)
- Three-dimensional, single story frame (`storey`)
- Three degree-of-freedom, three story plane frame (`frame`)
- Response spectrum calculation (`specalc`)
- Artificially-generated accelerograms (`ground`).

The SDE Workspace and the Climax Library

A more advanced software package has been subsequently developed for addressing a wider range of problems in vibrations and structural dynamics, leading up to structural control applications. This open source software is based on existing, freely available modulus written for the Java programming language. Specifically, a software library labelled 'Climax' was compiled for numerical solutions going beyond the programming of closed-form solutions that derive from the theory of structural vibrations, to address modern computational techniques such as finite elements, boundary elements, finite differences and hybrid methods. The interface seen by the user is labelled 'SDE' standing for 'Symplegma Development Environment', which makes available all software modules from the Climax software library. In addition, the SDE can serve as an interface with the programming language Groovy for further developments. This software is simply labeled as 'Symplegma' for short, and is accessible through the web page http://symplegma.civil.auth.gr/. It was first made available for a course on Computational Mechanics (2016) as well as for a course on Vibrations and Dynamics, Vibration and Control of Structures (2017) both taught at the School of Production Engineering and Management, Technical University of Crete, Chania, Greece. Subsequently, it was used for a seminar on Applied Acoustics offered at the Department of Music Technologies and Acoustics of the School of Music and Optoacoustic Technologies of the Hellenic Mediterranean University, Rethymno, Greece.

1.4   Brief Bibliography

This book derives in a large part from material written by the authors as notes when teaching structural dynamics at an undergraduate level [6] and also from notes with instructions for incorporating the SDE software package as a tool for solving exercises and small-scale problems. As was previously mentioned, the SDE package derives from earlier software platforms and is under continuous development by the authors.

A survey of the vast literature in the fields of vibrations and structural dynamics is purposely kept brief here by restricting it to a listing of references that were primarily consulted in developing lecture notes for an undergraduate audience, as well as for first year master-level students attending speciality programs in structural dynamics and earthquake engineering.

We begin with the subject of mechanical vibrations and point out the early work by [7], followed by that of [8] on vibrations of structures under moving loads plus that of [9] on the vibrations of continuous systems. A comprehensive account of the theoretical background behind vibrations can be found in [10], while the propagation of elastic waves in structural elements, i.e., the topic of elastodynamics, is developed in [11]. Material on the control of vibrations can be found in [1, 12], while the use of numerical methods for the analysis of structural vibrations under moving loads is presented in [13]. Continuing with structural dynamics, as perceived relevant to civil engineering, where beams, columns and frames are the dominant elements, we have the early works by [14] and by [15]. Material relevant to earthquake engineering was gradually introduced in structural dynamics, e.g., see the books by [16] and [17], where the use of response spectra for the analysis of structures to seismic motions became standard practice and are now included in contemporary code regulations [18]. Structural dynamics, as a discipline more relevant to aerospace structures is treated in [19],while more advanced topics such as the consideration of continuous (versus discrete structural systems) are presented in [20, 21]. Finally, the development of stochastic mechanics should be mentioned [22] and its use in the analysis of structures to ground induced loads can be found in [23].

In reference to topics that are relevant to computations associated with structural dynamics and vibrations, we note the development of numerical methods that are well suited for the solution of both transient and steady-state problems over the last fifty years. In this context, we mention the book on finite elements [24], followed by books on boundary element methods, e.g., [25] and [26]. In parallel to these numerical methods, we have the emergence of software packages (subroutines) readily available for inclusion in special-purpose computer programs that can be used for a variety of methods of analysis, including transform methods, eigenvalue extraction and time-stepping solutions, see the book by [27] that lists programs in both Fortran and C++.

Moving on to contemporary topics in structural dynamics, there has been much work on signal analysis, i.e., the extraction of information regarding the structure in question from measurements conducted either in situ or under laboratory conditions, see for instance the books by [28, 29]. This type of work can be seen within the context of structural health monitoring and of damage detection in important infrastructure such as bridges, see the article by [30]. Along these lines we mention the use of damping systems for absorbing energy from vibrating structures and thus furnishing better protection from environmentally-induced loads [31]. Furthermore, for a better interpretation of the data contained in recorded transient signals, it is slowly becoming standard practice to resort to artificial intelligence (AI), and particularly to machine learning techniques such as neural networks [32]. Finally, web-based software is becoming available for the study of structural dynamics for the benefit of students, as well as for practising engineers and researchers [33].

In the non-inclusive list of references that was presented above, we felt important to add some books that are used for teaching in civil engineering departments [6, 34–36] and in mechanical engineering departments [37, 38] of the national Greek universities. These books are obviously written in Greek, with their corresponding titles translated in English in this bibliography.

References

1. Inman DJ, Singh RC (1994) Engineering vibration, vol 3. Prentice Hall Englewood Cliffs, NJ
2. The MathWorks Inc (2022) Matlab version: 9.13.0 (z2022b)
3. Simulink Documentation (2020) Simulation and model-based design
4. Eaton JW, Bateman D, Hauberg S, Wehbring R (2025) GNU Octave version 10.1.0 manual: a high-level interactive language for numerical computations
5. Fritzson P (2011) Introduction to modeling and simulation of technical and physical systems with modelica. Wiley-IEEE Press
6. Manolis G, Koliopoulos P, Panagiotopoulos C (2015) Dynamics of structures. Hellenic Academic Libraries Link, 2015. Kallipos electronic book project, National Technical University of Athens (in Greek)
7. Den Hartog JP (1985) Mechanical vibrations. Courier Corporation
8. Frỳba L (2013) Vibration of solids and structures under moving loads, vol 1. Springer Science & Business Media
9. Leissa A, Qatu MS (2011) Vibrations of continuous systems. (No Title)
10. Meirovitch L (2010) Fundamentals of vibrations. Waveland Press
11. Graff KF (1975) Wave motion in elastic solids. Oxford University Press
12. Inman DJ (2017) Vibration with control. Wiley, New York
13. Bajer CI, Dyniewicz B (2012) Numerical analysis of vibrations of structures under moving inertial load, vol 65. Springer Science & Business Media
14. Hurty WC, Rubinstein MF (1964) Dynamics of structures
15. Biggs JM (1964) Introduction to structural dynamics. McGraw-Hill
16. Clough RW, Penzien J (1975) Dynamics of structures. McGraw-Hill
17. Chopra AK (1995) Dynamics of structures: theory and applications to earthquake engineering. Prentice-Hall (1995)
18. European Standard EN 1998-1 (2004) Eurocode 8: Design of structures for earthquake resistance—part 1: General rules, seismic actions and rules for buildings
19. Craig Jr RR, Kurdila AJ (2006) *Fundamentals of structural dynamics*. Wiley, New York
20. Kausel E (2017) *Advanced structural dynamics*. Cambridge University Press
21. Rao SS (2007) Vibration of Continuous Systemsg. Wiley, New York
22. Nigam NC, Saunders H (1986) Introduction to random vibration
23. Manolis GD, Koliopoulos PK (2001) Stochastic structural dynamics in earthquake engineering. (No Title)
24. Bathe K-J, Wilson E (1976) Numerical methods in finite element analysis
25. Manolis GD, Beskos DE (1988) Boundary element methods in elastodynamics. Taylor & Francis

26. Dominguez J (1993) Boundary elements in dynamics. WIT Press
27. Hadjian AH (2002) Fundamental period and mode shape of layered soil profiles. Soil Dyn Earthq Eng 22(9–12):885–891
28. Brandt A (2023) Noise and vibration analysis: signal analysis and experimental procedures. Wiley, New York
29. Brincker R, Ventura C (2015) Introduction to operational modal analysis. Wiley, New York
30. Farrar CR, Worden K (2007) An introduction to structural health monitoring. Philos Trans R Soc A: Math Phys Eng Sci 365(1851):303–315
31. Dadoulis GI, Manolis GD (2024) A comparative study on the effectiveness of a moving versus a fixed passive damper in beam vibration mitigation. Acta Mech 235(10):6403–6412
32. Brunton SL, Kutz JN (2022) Data-driven science and engineering: machine learning, dynamical systems, and control. Cambridge University Press
33. Panagiotopoulos CG, Manolis GD (2016) A web-based educational software for structural dynamics. Comput Appl Eng Educ 24(4):599–614
34. Anastasiadis K (1986) Dynamics of structures, vol I. Ziti
35. Anastasiadis K (1986) Dynamics of structures, vol II. Ziti
36. Katsikadelis JT (2020) Dynamic analysis of structures. Academic
37. Antoniadis I, Kanarahos A (1998) Dynamics of machines. Papasotiriou Publishers
38. Natsiavas S (2001) Vibrations of mechanical systems. Zitis Publishers

Structural Dynamics of Elementary Systems 2

2.1 The Single-Degree-of-Freedom System

Figure 2.1 depicts a simple structural system comprising a rigid body of mass m, which is attached to a base by a spring of stiffness k plus a damper with damping coefficient c and is subjected to an external transient force $f(t)$.

The equation of motion of this single degree-of-freedom (SDOF) system can be formulated using d'Alembert's principle, which states that the time rate of change of an oscillating mass is equal to the sum of external forces acting upon it. In essence, this is nothing more than a statement of dynamic equilibrium, where the inertia force f_I in the body, the restoring force in the spring f_S and the viscous force in the damper f_D, all balance the externally applied force f. More specifically, the inertia force according to Newton's second law equals the mass of the SDOF system times its acceleration, while from Hooke's law, the force in the spring is the product of its stiffness times the relative movement of its two ends. Finally, the viscous force in the damper, which is a linearized approximation of the energy lost in a system as it vibrates, is equal to the velocity of the mass times the damping coefficient. From kinematics, the velocity and the acceleration of the SDOF system are equal to the first and second time derivatives of its displacement $u(t)$, respectively. In here, overdots are used to symbolize time t derivatives. Consequently, the equation of motion is an ordinary differential equation (ODE) of second order, i.e.,

$$f_I(t) + f_D(t) + f_S(t) = f(t) \rightarrow$$
$$m\ddot{u}(t) + c\dot{u}(t) + ku(t) = f(t) \tag{2.1}$$

© The Author(s), under exclusive license to Springer Nature Switzerland AG 2026   9
G. Manolis and C. Panagiotopoulos, *Vibrations of Structural Systems*, Synthesis Lectures
on Mechanical Engineering, https://doi.org/10.1007/978-3-032-12279-7_2

Fig. 2.1 The single
degree-of-freedom system

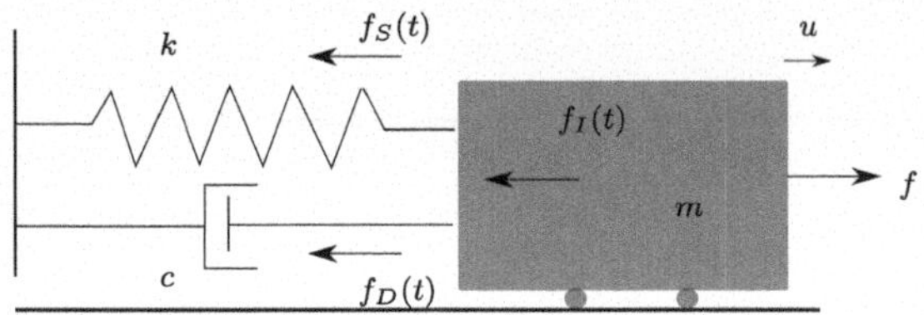

2.2   Free Vibrations

In the differential equation of motion, Eq. (2.1), it is necessary to consider the initial conditions that describe the state of the system at time $t = 0$ at which vibrations commence. Usually, these conditions are zero if the system is at rest. Specifically, initial conditions apply to the dependent variable and its time derivatives up to one less than the highest order time derivative appearing in the equation of motion. In our case, these are the zeroth derivative (i.e., the initial displacement) and the first derivative (i.e., the initial velocity). These two terms are respectively labelled as $u(t_0) = u_0$ and $\dot{u}(t_0) = \dot{u}_0$. If for some reason it is necessary to evaluate the initial acceleration, this can be done by using the equation of motion as follows:

$$\ddot{u}(t_0) = \ddot{u}_0 = \frac{f_0 - c\dot{u}_0 - ku_0}{m} \tag{2.2}$$

In the above, $f_0 = f(t_0)$ is the external force at the onset of time. It is observed that in order to initiate motion in the system, at least one of the initial displacement, velocity and external force variables must have a finite value at $t = 0$. In the absence of an external force, the motion of the system is labelled as free vibration and can be evaluated over time by solving for the homogeneous part of the equation of motion subject to the initial conditions $u_0, \dot{u}_0$.

We now define the most important parameter for studying SDOF systems, namely the natural frequency (or eigenfrequency) $\omega_0 = \sqrt{k/m}$ and its inverse, the natural period (or fundamental period) $T_0 = 2\pi/\omega_0$. The natural frequency is measured in (rad/s), and if divided by 2π corresponds to the number of vibration cycles per unit of time, i.e., $F_0 = \omega_0/2\pi$ (in Hz). Correspondingly, the natural period $T_0 = 1/F_0$ is the time (in s) required for the SDOF system to execute a full cycle of vibration. Depending on the numerical value of the damping coefficient c, the free vibration of the SDOF system is either undamped if $c = 0$ and damped if $c \neq 0$. Furthermore, damped vibrations are classified as either *sub-critically damped (underdamped)*, *critically damped* or *super-critically damped (overdamped)*. These three categories result when the damping coefficient is less than, equal to or greater than the value for critical damping, namely $c_{cr} = 2m\omega_0$. This last value determines the amount of damping in the system that results in a motion that is no longer vibratory, meaning the displacement will not change sign about the equilibrium position of the SDOF system. The

Table 2.1 Free vibrations of SDOF systems

Undamped vibrations	$\xi = 0$	$u(t) = u_0 \cos(\omega_0 t) + \dot{u}_0 \dfrac{\sin(\omega_0 t)}{\omega_0}$
		$= p \cos(\omega_0 t + \theta) \qquad (2.4)$
Subcritical vibrations	$\xi < 1$	$u(t) = \left(u_0 \cos(\omega_d t) + \dfrac{\dot{u}_0 + u_0 \xi \omega_0}{\omega_d} \sin(\omega_d t) \right) e^{-\xi \omega_0 t}$
		$= p \cos(\omega_d t + \theta) e^{-\xi \omega_0 t} \qquad (2.5)$
Critical vibrations	$\xi = 1$	$u(t) = \left(u_0(1 + \omega_0 t) + \dot{u}_0 t \right) e^{-\omega_0 t} \qquad (2.6)$
Supercritical vibrations	$\xi > 1$	$u(t) =$ $\left(u_0 \cosh(\omega_d t) + \dfrac{\dot{u}_0 + u_0 \xi \omega_0}{\omega_d} \sinh(\omega_0 t) \right) e^{-\xi \omega_0 t} \qquad (2.7)$

Comments: (i) We have that ω_d is the damped natural frequency of an SDOF and is equal to $\omega_d = \omega_0 \sqrt{1 - \xi^2}$ for subcritical vibrations and $\omega_d = \omega_0 \sqrt{\xi^2 - 1}$ for supercritical vibrations. (ii) The amplitude of vibration in Eq. (2.4) is equal to $|u| = \sqrt{u_0^2 + \dfrac{\dot{u}_0^2}{\omega_0^2}}$, while the phase angle is equal to $\theta = \tan^{-1}(-\dfrac{\dot{u}_0}{\omega_0 u_0})$. (iii) Similarly, in Eq. (2.5) the amplitude of vibration is $|u| = \sqrt{u_0^2 + \dfrac{(\dot{u}_0 + u_0 \xi \omega_0)^2}{\omega_0^2}}$ and the phase angle is $\theta = - \tan^{-1}(\dfrac{\dot{u}_0 + u_0 \xi \omega_0}{\omega_0 u_0})$

damping ratio ξ is defined in terms of critical damping as follows:

$$\xi = \frac{c}{c_{cr}} = \frac{c}{2 m \omega_0} \qquad (2.3)$$

Listed below in Table 2.1 are the free vibration solutions for the transient displacement $u(t)$, which correspond to the homogeneous solution for the equation of motion of the SDOF system, Eq. (2.1), and for all damping cases. Next, Fig. 2.2 plots the displacement amplitude versus time for an undamped SDOF system with a natural period $T = 2.507s$ under the influence of a unit initial displacement u_0 at $t = 0$. Also, Fig. 2.3 plots the damped vibrations of this SDOF system as functions of the damping ratio ξ ranging from 0.0 to 1.0. Note that damping does not appreciably change the natural period of an SDOF system, so the assumption that $T_d \to T$ is often made. Furthermore, Fig. 2.4 depicts the exponential decay in the free vibrations of the aforementioned SDOF system in the presence of 5 percent of critical damping, i.e., for $\xi = 0.05$. Finally, Fig. 2.5 is similar to Fig. 2.3 and is a parametric plot in terms of damping of the free vibrations of the SDOF system under a unit initial velocity of $\dot{u}_0 = 1.0$.

In reference to damped subcritical vibrations, we define the logarithmic decrement δ_n as the natural logarithm of the ratio of two vibration amplitudes separated by an integer number n of natural periods, see Table 2.1 and Fig. 2.4:

$$\delta_n = \ln \frac{u(t_i)}{u(t_i + n T_d)} = \ln \frac{u_i}{u_{i+n}} = n \frac{2\pi \xi}{\sqrt{1 - \xi^2}} \xrightarrow{\xi \ll 1} \delta_n = n 2\pi \xi \qquad (2.8)$$

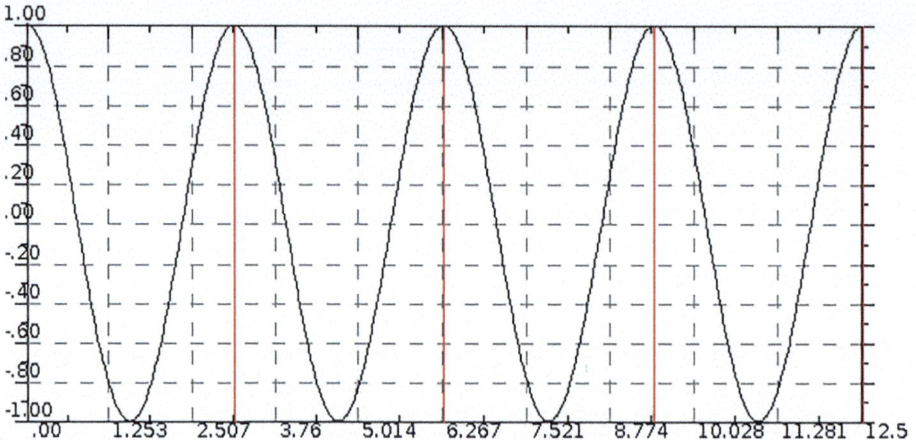

Fig. 2.2 Undamped free vibration of an SDOF system with $T = 2.507s$ activated by a unit initial displacement $u_0 = 1.0$

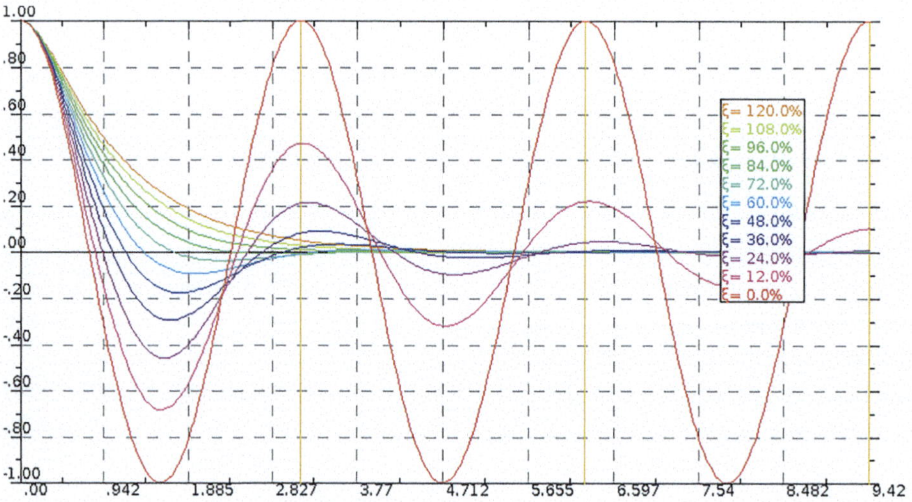

Fig. 2.3 Damped free vibrations of an SDOF oscillator with $T = 2.507s$ under a unit initial displacement and parametric in the damping ratio $\xi = c/c_{cr}$

In the above equation, the damped period of vibration T_d is defined as $T_d = 2\pi/\omega_d$. In computing this logarithmic decrement, attention must be paid not to choose times during which the amplitude of vibration is nearly zero, but instead use two consecutive maximum amplitude values. It should be noted that from now on, we will focus on underdamped vibrations, as most structures exhibit viscous damping which is quite small as compared to their critical damping coefficient c_{cr}.

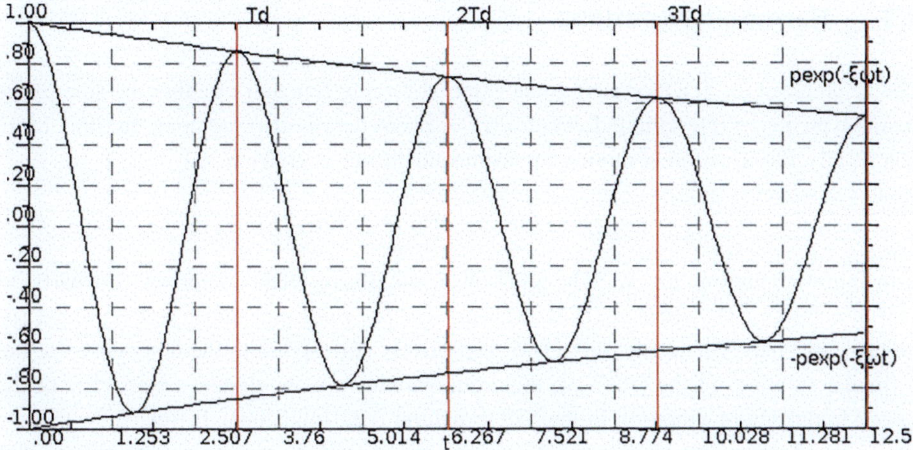

Fig. 2.4 Free vibration response for an underdamped SDOF system with $T = 2.507s$ and damping ratio $\xi = 0.05$ under a unit initial displacement of $u_0 = 1.0$

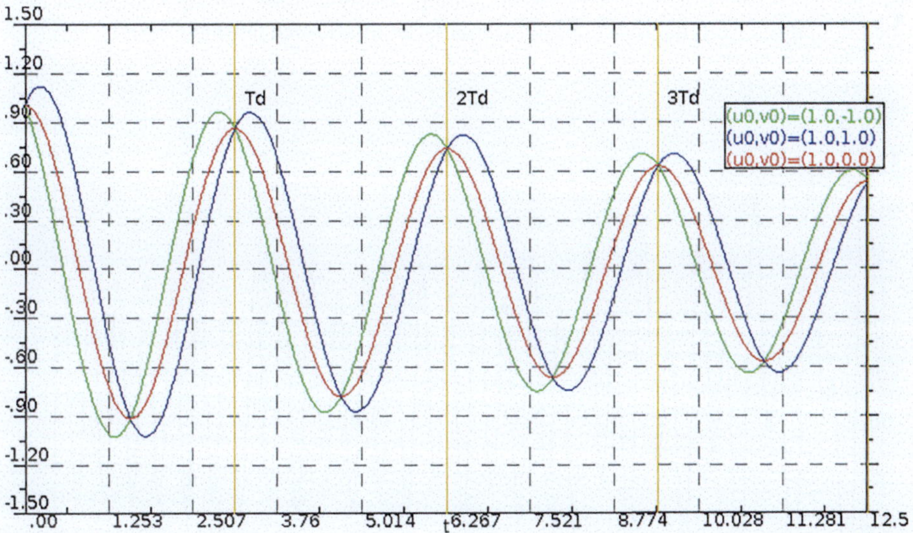

Fig. 2.5 Damped free vibrations of an SDOF oscillator with $T = 2.507s$ under a unit initial velocity and parametric in the damping ratio $\xi = c/c_{cr}$

2.3   Harmonic Vibrations

The response of an SDOF system to an external forcing function of either the sine or the cosine type, i.e., $f(t) = f_0 \sin(\omega t)$ or $f(t) = f_0 \cos(\omega t)$ is also harmonic in time. More specifically, the equation of motion for a sinusoidal force is given as

$$m\ddot{u}(t) + c\dot{u}(t) + ku(t) = f_0 \sin(\omega t) \tag{2.9}$$

with initial conditions u_0, $\dot{u}_0$. The solution to the above ODE comprises two parts, as $u(t) = u_c(t) + u_p(t)$, where $u_c(t)$ is the *transient (or free)* part and $u_p(t)$ is *forced (or steady-state)* part. Obviously, these two respectively correspond to the homogeneous and particular solutions of the ODE. An important parameter is the dimensionless ratio of the forced frequency of vibration to the SDOF natural (i.e., fundamental) frequency, namely $\beta = \omega/\omega_0$. A plot of the displacement response of an SDOF system with a natural period of $T = 0.785s$ and for a damping ratio of $\xi = 0.05$ is shown in Fig. 2.6, where the steady-state response gradually evolves with time after the transient response is damped out. Thus, the response of an SDOF system to harmonic input is harmonic, but emerges only after some time has elapsed from the onset of the externally imposed force at time $t = 0s$.

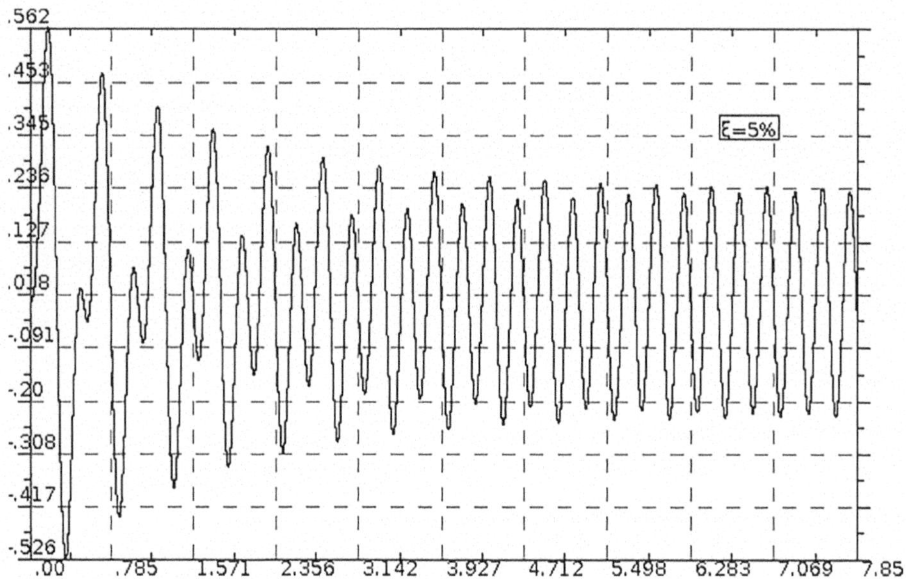

Fig. 2.6 Harmonic vibration of an SDOF system with natural period $T = 0.785s$ and a damping ratio of $\xi = 0.05$ for a unit sinusoidal force input. Note the gradual dominance of the forced vibration part as the free vibration part is damped out with time t

Table 2.2 Harmonic vibrations of SDOF systems

$u(t) = u_c(t) + u_p(t)$ (2.10)	
Free (or transient) vibration regime	$u_c(t) = (C_1 \sin(\omega_d t) + C_2 \cos \omega_d t)\, e^{-\xi \omega_0 t}$ (2.11) Note: Constants C_1, C_2 depend on the initial conditions
Forced (or steady-state) vibration regime	$u_p(t) =$ $\dfrac{f_0}{k} \dfrac{1}{(1-\beta^2)^2+(2\xi\beta)^2}\left((1-\beta^2)\sin(\omega t) - 2\xi\beta \cos(\omega t)\right)$ $= \dfrac{f_0}{k} \dfrac{1}{\sqrt{(1-\beta^2)^2+(2\xi\beta)^2}} \sin(\omega t - \theta)$ $= \|u\| \sin(\omega t - \theta)$ (2.12) where $\|u\| = \dfrac{f_0}{k} \dfrac{1}{\sqrt{(1-\beta^2)^2+(2\xi\beta)^2}}$ is the vibration amplitude and $\theta = \tan^{-1}\dfrac{2\xi\beta}{1-\beta^2}$ is the phase angle

The dynamic load factor (DLF), which appears in Eq. (2.12) of Table 2.2, is labelled as D and defines the amount by which the maximum value observed in the SDOF system forced vibrations exceeds (or is smaller than) the correspond static displacement of this system resulting from the magnitude of the external force. It is given by the following equation:

$$D = \frac{\max(u(t))}{u_{st}} = \frac{1}{\sqrt{(1-\beta^2)^2 + (2\xi\beta)^2}} \qquad (2.13)$$

The value of β which yields a maximum value for $\xi = 0$ is $\beta = 1$, resulting in $D(\beta, \xi) = D(1, 0) = \infty$. This is the definition of resonance, since $\beta = 1 \rightarrow \omega/\omega_0 = 1 \rightarrow \omega = \omega_0$. In the case when $\xi \neq 0$, it can be shown that the maximum value for the dynamic load factor materializes when $\beta = \sqrt{1 - 2\xi^2}$ and is equal to $D_{max} = \dfrac{1}{2\xi\sqrt{1-\xi^2}}$. Finally, the resonance frequency is given as $\omega_{res} = \omega_0\sqrt{1 - 2\xi^2}$. This information is summarized in Fig. 2.7, which plots the SDOF system's DLF versus β and is parametric in terms of the damping ratio ξ.

2.3.1 Critical Damping in the Resonance Region

Values of β in the neighborhood of $\beta = \sqrt{1 - 2\xi^2} \approx 1$ result in a maximum for the DLF $D(\beta, \xi)$. Therefore, the corresponding frequency range is referred to as the resonance region. The width of this region is delineated by the following lower and upper values of the frequency ratio, β_1 and β_2, outside which the DLF drops from its maximum value D_{max} to the value $D_{max}/\sqrt{2}$, as shown in Fig. 2.8. Numerical values for β_1 and β_2 are recovered from the solution of the following equation:

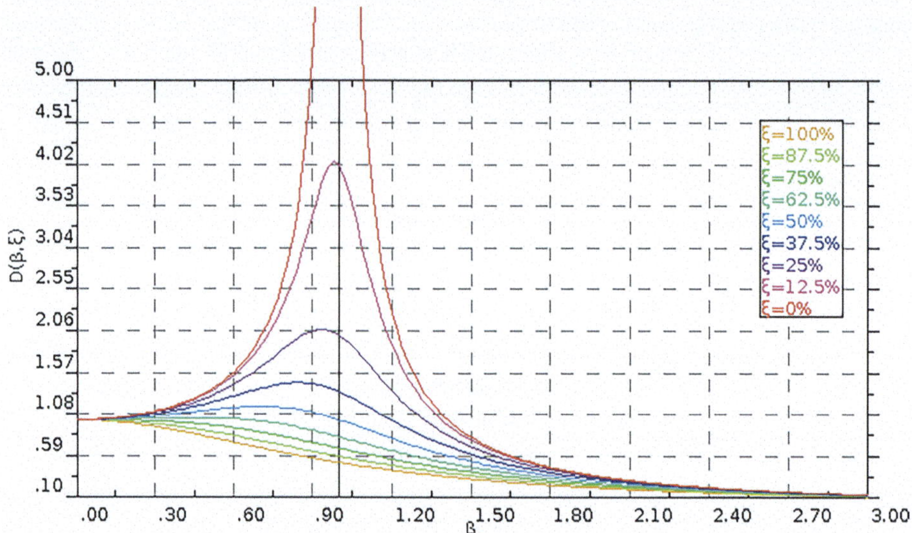

Fig. 2.7 Dynamic load factor $D(\beta, \xi)$ for an SDOF system plotted against frequency ratio $\beta = \omega/\omega_0$ and for different values of the damping ratio ξ

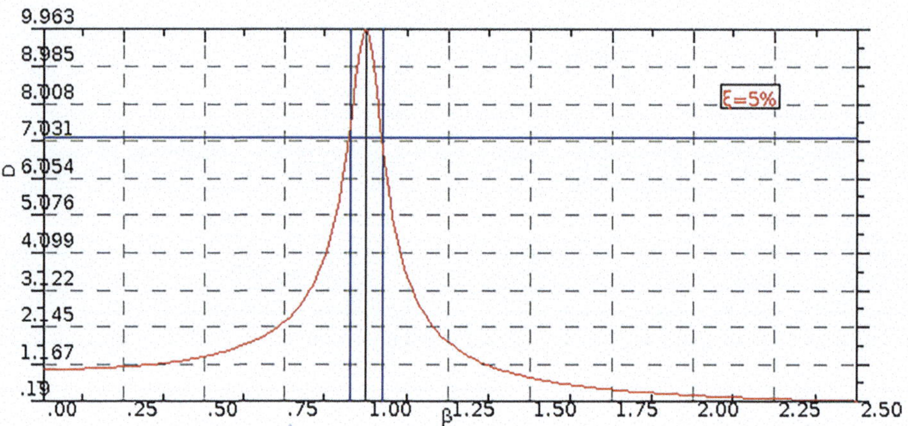

Fig. 2.8 DLF spectrum $D(\beta, \xi)$ versus frequency ratio β for a fixed damping ratio of $\xi = 0.05$. *Note* The horizontal line caps the value $D_{max}/\sqrt{2}$, while the two vertical lines bracket the values $\beta_{1,2} = 1 \pm \xi$

$$D(\beta, \xi) = D_{max}/\sqrt{2} \to \frac{1}{\sqrt{(1 - \beta^2)^2 + (2\xi\beta)^2}} = \frac{1}{2\sqrt{2}\xi\sqrt{1 - \xi^2}} \tag{2.14}$$

This yields a quadratic equation when $\xi \ll 1$, whose roots are

$$\beta_{1,2} = 1 \pm \xi \tag{2.15}$$

while the width of the resonance region is $\beta_2 - \beta_1 = 2\xi$, i.e., the corresponding frequency range is $\omega_2 - \omega_1 = 2\omega_0$. In case the frequency response spectrum of the SDOF system is available from experimental measurements under time harmonic conditions (see Fig. 2.7), the values β_1 and β_2 (or correspondingly the values ω_1 and ω_2) can be directly measured and used to compute an accurate value for the SDOF system damping ratio as follows:

$$\xi = \frac{\beta_2 - \beta_1}{2} = \frac{\omega_2 - \omega_1}{2\omega_0}. \tag{2.16}$$

2.4   General Solution for an SDOF System: Duhamel's Integral

We start with the general case of of an SDOF system under an external load $f(t)$ and initial conditions for the displacement u_0 and the velocity $\dot{u}_0$ at time $t_0 = 0$. The complete solution is the superposition of the homogeneous (free vibration because of the initial conditions) and particular (forced vibration due to the load) solutions of the equation of motion as follows:

$$u(t) = \left(u_0 \cos(\omega_d t) + \frac{\dot{u}_0 + u_0 \xi \omega_0}{\omega_d} \sin(\omega_d t) \right) e^{-\xi \omega_0 t}$$

$$+ \frac{1}{m\omega_d} \int_0^t f(\tau) \sin(\omega_d(t - \tau)) e^{-\xi \omega_0(t - \tau)} \, d\tau \tag{2.17}$$

Actually, it is the second term in Eq. (2.17) that is the Duhamel integral, which is valid for any transient forcing function. For the special case of undamped vibration, it is sufficient to substitute in Eq. (2.17) the values $\xi = 0$ and $\omega_d = \omega_0$ to derive a simpler form. We note that Duhamel's integral may be evaluated either analytically or numerically using Simpson's rule (i.e., the trapezoidal rule) for integration. Regarding the derivation of Duhamel's integral, it is defined as the time convolution of the external force with the unit impulse response of an SDOF system. This unit impulse response is actually a fundamental solution for the ODE of motion, i.e.,

$$h(t - \tau) = \frac{e^{-\xi \omega_0(t - \tau)}}{m\omega_d} \sin(\omega_d(t - \tau)) \tag{2.18}$$

A fundamental solution derives from a special type of external force, namely the impulse $\delta(t - \tau)/m$ under zero initial conditions, where the generalized function δ is Dirac's delta function that is used to represent a point load in time τ. As before, for the undamped vibration case, both fundamental solution, Eq. (2.18) and Duhamel's integral, Eq. (2.17), are recovered by simple substitution for $\xi = 0$. Next, for the case of critical damping, when $\xi = 1$, the fundamental solution is

$$h(t - \tau) = \frac{1}{m}(t - \tau)e^{-\omega(t-\tau)} \tag{2.19}$$

while for overdamped systems, when $\xi > 1$, we have

$$h(t - \tau) = \frac{e^{-\xi\omega_0(t-\tau)}}{m\omega\sqrt{\xi^2 - 1}} \sinh\left(\omega\sqrt{\xi^2 - 1}\,(t - \tau)\right) \tag{2.20}$$

At the same time, the corresponding response due to the initial conditions has to be modified accordingly in Eq. (2.17), as previously mentioned. Finally, it should be stressed that the SDOF system response to the unit impulse (or fundamental solution) $h(t)$ can be used to yield the full response $u(t)$ for any type of external forcing function through time convolution as follows:

$$u(t) = f(t) * h(t) + \text{IC} \tag{2.21}$$

and by adding the influence of the initial conditions (IC). In the literature, $h(t)$ is also referred to as a transfer function in the time domain.

2.4.1   Numerical Evaluation of Duhamel's Integral

Computation of the transient response of an SDOF system by Duhamel's integral is valid, provided the system's material behavior remains in the linear range and the material parameters (mass, damper, spring) are time-independent. It should be kept in mind that Duhamel's integral is a time convolution integral, see Eq. (2.21), and it is possible in certain cases to evaluate it analytically. In here we will present an algorithm for its numerical evaluation based on Simpson's rule, as given below:

$$\int_{x_1}^{x_2} f(x)dx = \frac{x_2 - x_1}{6}\left(f(x_1) + 4f(\frac{x_1 + x_2}{2}) + f(x_2)\right). \tag{2.22}$$

The solution algorithm is presented step-by-step in Table 2.3 for the case of sub-critical damping. It is possible, following a few modifications, to tailor the algorithm for the other two cases of critical and super-critical damping.

Class SDOF

We now introduce software package `courses.structuraldynamics` that refers to the SDOF oscillator through the class `sdof`, see Listing 2.1. In order to use the class `sdof`, we must first activate software package, see line 1 below. Next, line 3 defines an object that describes the external load applied to the SDOF system, which in turn is

Table 2.3 Simpson's rule for the numerical evaluation of Duhamel's integral for under-damped SDOF systems

Step	Procedure
Input	SDOF system parameters: k, m, c
Initial computations	Evaluation of SDOF system natural frequencies: $\omega_0 = \sqrt{\frac{k}{m}}$, $\omega_d = \omega_0\sqrt{1-\xi^2}$ Initialization: $p_1 = f_0$, $g_1 = 0$, $A_0 = 0$, $B_0 = 0$, $t = \Delta t$
Recursive computations at each time step $i = 1, .., N$	1. Computations at steps $j = 2, 3$ $\tau_j = t + (j-1)\Delta t/2$ $p_j = f(\tau_j)e^{\xi\omega_0\tau_j}\cos(\omega_d\tau_k)$ $g_j = f(\tau_j)e^{\xi\omega_0\tau_j}\sin(\omega_d\tau_k)$ 2. Computation of Simpson's rule parameters $A_t = \frac{\Delta t}{6}(p_1 + 4p_2 + p_3) + A_{t-\Delta t}$ $B_t = \frac{\Delta t}{6}(g_1 + 4g_2 + g_3) + B_{t-\Delta t}$ 3. Computation of SDOF displacement, velocity and acceleration response $u(i\Delta t) =$ $e^{\xi\omega_0 t}(A_t\sin(\omega_d t) - B_t\cos(\omega_d t))/(m\omega_d)$ $\dot{u}(i\Delta t) = e^{\xi\omega_0 t}(A_t\cos(\omega_d t) + B_t\sin(\omega_d t))/m - \xi\omega_0 u(i\Delta t)$ $\ddot{u}(i\Delta t) =$ $-2\xi\omega_0\dot{u}(i\Delta t) - \omega_0^2 u(i\Delta t) + f(i\Delta t)/m$

defined in line 5 using five real number variables, namely `sdof(double k, double m, double c, double u0, double vo)`. These respectively are the stiffness, the mass, the damper, the initial displacement and the initial velocity. We continue with line 6 where the external force is prescribed, plus with line 7 requesting the solution by defining the method `duhamel()`. Once all variables have been specified along with the solution, line 9 calls the methods `Disp()`, `Velc()` and `Accl()` from the class `sdof` to respectively recover the time histories of the displacement, velocity and acceleration of the SDOF system.

```
1  import courses.structuraldynamics.*
2
3  f={sin(it*sqrt(100.0)*0.8)}
4
5  theSDOF=new sdof(100.0, 1.0, 0.1, 0.0, 0.0)
6  theSDOF.setRHS(f as DoubleFunction)
7  theSDOF.duhamel()
8
9  u=theSDOF.Disp(); v=theSDOF.Velc(); a=theSDOF.Accl()
```

Listing 2.1 Brief description of the single degree of freedom class `sdof`

2.5   SDOF System Response to Impact

Impact is a loading of small duration in comparison with the natural period of an SDOF system. Because of its small duration, the maximum response of the oscillator materializes fast, before the damping mechanisms can develop. As a consequence, damping is often ignored when studying impact loads. The response comprises two phases, the first coinciding with the duration of the impact load. In this case, the transient displacement can be computed from Duhamel's integral, usually assuming zero initial conditions. The second phase commences after the removal of the external load and is a free vibration regime. The initial conditions for this phase are computed from the final conditions of the first phase, resulting in a displacement and velocity at time $t = t_\delta$, i.e., the duration of the first phase. A useful quantity for impact loading is the dynamic load factor $D = u_{max}/u_{st}$, where u_{st} is the equivalent static displacement generated by the magnitude of force f_0 of the impact load, resulting in $u_{st} = f_0/k$.

2.6   Base Excitation

There are cases where no external forces are applied to the SDOF structural system, but instead motion is initiated because of base motion. This could be the result of ground seismicity, of ambient vibrations due to nearby traffic, of elastic wave motion initiated by underground blasts, etc., see Fig. 2.9.

In order to derive the appropriate form of the equation of motion for the SDOF system, it is necessary to differentiate between the total motion of its mass u_t and that of its base u_g. The SDOF spring element is activated by the relative motion of its two ends; likewise,

Fig. 2.9 SDOF system under base motion

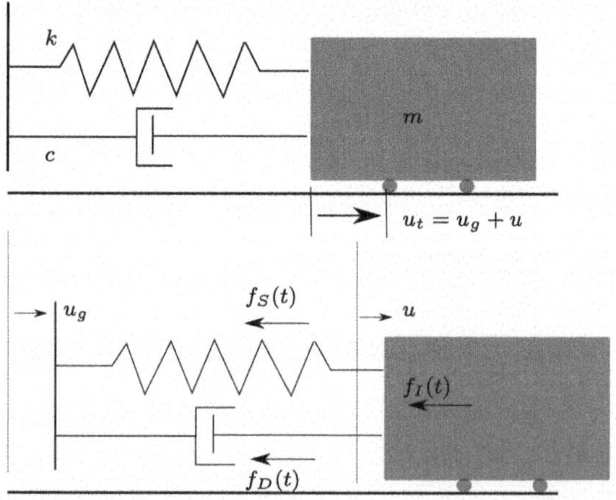

the damper is activated by the relative velocity of its two ends. The SDOF mass, however, is activated by the total acceleration which it experiences. As can be seen in Fig. 2.9, it is necessary to define the relative motion between that of the top mass u_t and that of the base u_g as u, i.e.,

$$u(t) = u_t(t) - u_g(t) \tag{2.23}$$

The same definition holds true for the velocity and the acceleration. Therefore, the equation of motion is

$$f_I(t) + f_D(t) + f_S(t) = 0 \rightarrow$$
$$m\ddot{u}_t(t) + c\dot{u}(t) + ku(t) = 0 \rightarrow$$
$$m(\ddot{u}_g(t) + \ddot{u}(t)) + c\dot{u}(t) + ku(t) = 0 \rightarrow$$
$$m\ddot{u}(t) + c\dot{u}(t) + ku(t) = -m\ddot{u}_g(t) \tag{2.24}$$

In sum, the equivalent external force is now $f(t) = -m\ddot{u}_g(t)$. It is obvious that the equation of motion can be formulated in terms of the total displacement u_t, in which case the equivalent external force is $f(t) = ku_g(t) + c\dot{u}_g(t)$.

2.7   Gravity Loads

Within a gravity field, the SDOF system first assumes a deformed position due to its own weight plus any other dead loads, i.e., u_μ, see Fig. 2.10. After that, it will vibrate about its equilibrium position as $\bar{u}(t)$ if transient loads $f(t)$ are present. The total displacement about the original, undeformed position is the sum of the aforementioned two displacements, $u_\mu + \bar{u}(t)$. Next, the equation of motion becomes

$$m\ddot{u}(t) + c\dot{u}(t) + ku(t) = f(t) + w \tag{2.25}$$

The spring deformation due to the weight is $u_\mu = w/k$, while its first and second time derivatives of the transient displacement are $\dot{u} = \dot{\bar{u}}$ and $\ddot{u} = \ddot{\bar{u}}$ respectively. Substituting in Eq. (2.25) yields

$$m\ddot{\bar{u}}(t) + c\dot{\bar{u}}(t) + k\bar{u}(t) + ku_\mu = f(t) + w \rightarrow$$
$$m\ddot{\bar{u}}(t) + c\dot{\bar{u}}(t) + k\bar{u}(t) = f(t) \tag{2.26}$$

Thus, when computing the dynamic response of an SDOF system, the gravity loads can be ignored provided the transient displacement is measured about its deformed shape, and after these have acted upon the structure. The cumulative effect of dead and live loads can therefore be evaluated through superposition of these two states.

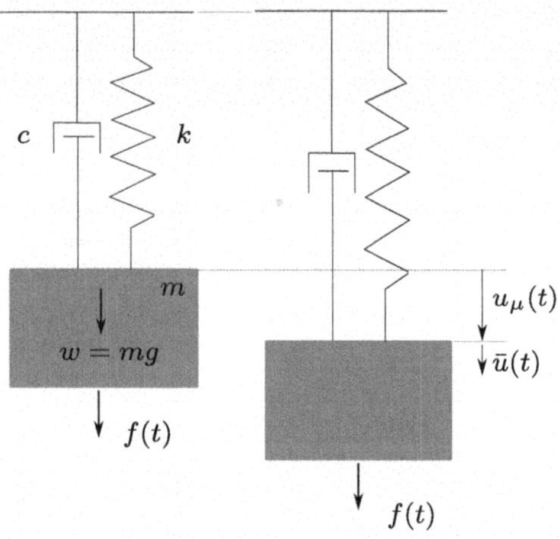

Fig. 2.10 SDOF system with gravity loads included

2.8   Vibration Isolation

Vibration isolation is a two-fold problem, aiming at (a) minimizing the transmission of the dynamically-induced forces generated in the SDOF system to the surrounding space (supports, ground, neighboring structural system) and (b) minimizing the externally-induced ground motion input to the SDOF system itself. The usual approach for this subject is to consider harmonic motions, as these may cause resonance phenomena.

2.8.1   Force Transmission to the Surrounding Area

The harmonic vibration environment previously mentioned is important for an additional reason, namely the fact that a transient motion can be reconstructed from a series of harmonics using Fourier synthesis, as will be discussed in Sect. 2.10. The forced motion of an SDOF oscillator under the influence of an external harmonic motion of the type $f(t) = f_0 \sin \omega t$ is given in Eq. 2.12 of Table 2.2. As can be seen in Fig. 2.11, the transmitted force is the reaction of the base to the sum of forces that develop in the spring and damper elements, i.e.,

$$\begin{aligned}
f(t) &= f_S(t) + f_D(t) \\
&= ku(t) + c\dot{u}(t) \\
&= f_T \sin(\omega t - \theta - \phi)
\end{aligned} \tag{2.27}$$

where $\phi = tan^{-1}(2\xi\beta)$ is a phase angle and the transmitted force magnitude is

Fig. 2.11 Transmission of forces from the SDOF system to its base

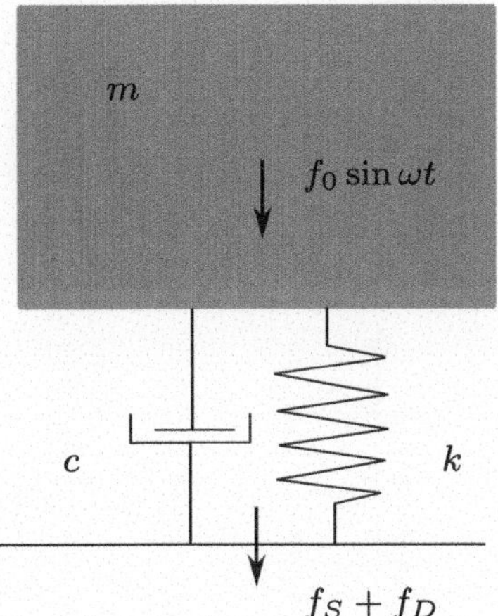

$$f_T = f_0 \sqrt{\frac{1 + (2\xi\beta)^2}{(1 - \beta^2)^2 + (2\xi\beta)^2}} \tag{2.28}$$

We define $D_f(\beta, \xi) = f_T / f_0$ as the dimensionless dynamic load factor (DLF) for the transmitted force. As can be seen in Fig. 2.12, there is a reduction of the transmitted force magnitude in the frequency range $\beta > \sqrt{2}$. Furthermore, higher values of the damping ratio ξ result in lower values for the dynamic load factor D_f. In the resonance region, the controlling factor for a reduction in the transmitted force defined by D_f is only through the addition of large damping ratio values ξ.

2.8.2   Ground Motion Input

Mathematically speaking, this is the same problem as the one discussed in Sect. 2.6 and is defined by Eq. 2.24. In order to give an example, the ground motion is assumed to be a harmonic function $u_g(t) = u_0 \sin \omega t$ and the resulting dynamic load factor $D_u(\beta, \xi)$ for the transmission of base motion to the SDOF system is given by the expression

$$D_u(\beta, \xi) = \frac{(\ddot{u}_t)_{max}}{(\ddot{u}_g)_{max}} = \frac{(u_t)_{max}}{(u_g)_{max}} = D_f(\beta, \xi) \tag{2.29}$$

which turns out to be the same as $D_f(\beta, \xi)$.

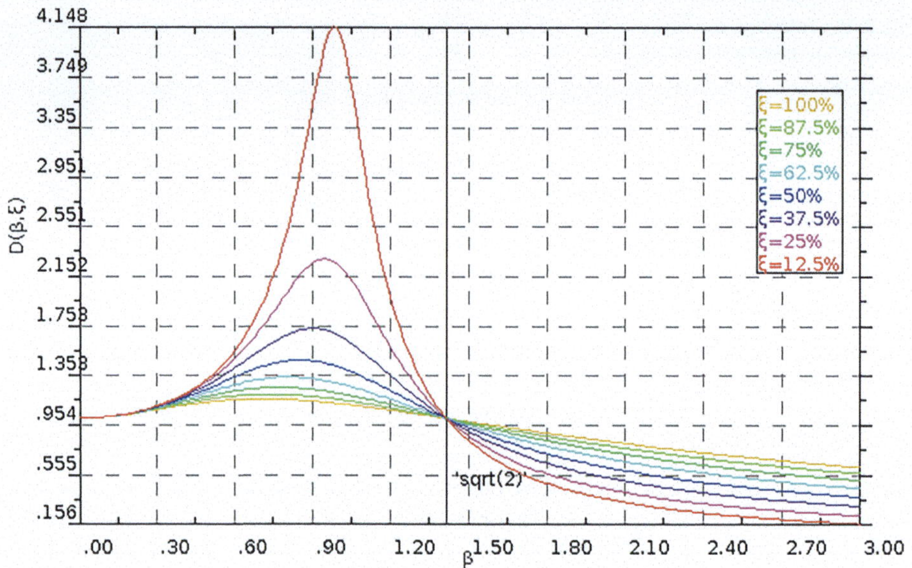

Fig. 2.12 SDOF dynamic load factor $D_f(\beta, \xi)$ for the transmission of force versus frequency ratio $\beta = \omega/\omega_0$ as function of the damping ratio ξ

2.9   Mechanical Energy

The mechanical energy $\mathcal{E}(t)$ of a SDOF system is the sum of its kinetic $\mathcal{T}(t) = \frac{1}{2}m\dot{u}^2$ plus potential $\mathcal{V}(t) = \frac{1}{2}mu^2$ energies,

$$\mathcal{E}(t) = \mathcal{T}(t) + \mathcal{V}(t) = \frac{1}{2}m\dot{u}^2 + \frac{1}{2}mu^2 \tag{2.30}$$

By integrating the equation of motion (2.1) over time and after it has been multiplied by the velocity $\dot{u}$,

$$\int_0^t (m\ddot{u} + c\dot{u} + ku)\,\dot{u}\,dt = \int_0^t f\dot{u}\,dt \tag{2.31}$$

we recover the following relation:

$$\int_0^t (m\ddot{u} + ku)\,\dot{u}\,dt = \mathcal{E}(t) - \mathcal{E}(0) = \int_0^t (f - c\dot{u})\,\dot{u}\,dt \tag{2.32}$$

This last relation shows that the change in the mechanical energy over the time interval $[0, t]$ is equal to the work of the external force $f(t)$ minus the energy lost due to damping $c\dot{u}(t)$.

The time derivative of Eq. (2.32) is the rate of change of the mechanical energy, i.e., the power:

$$\frac{d\mathcal{E}(t)}{dt} = (m\ddot{u} + ku)\,\dot{u} = -c\dot{u}^2 + f\dot{u}. \tag{2.33}$$

2.10   Frequency Domain Analysis

It is possible to formulate and solve dynamic problems in either the time domain or in a transformed domain, such as the Fourier (or frequency) and the Laplace domains. In here, we will present the former domain in detail.

2.10.1   Fourier Series Expansion of a Periodic Function

Any periodic function $f(t)$, see Fig. 2.13, can be represented by a trigonometric series with infinite terms as follows:

$$f(t) = a_0 + \sum_{n=1}^{\infty} (a_n \cos n\omega_p t + b_n \sin n\omega_p t) \tag{2.34}$$

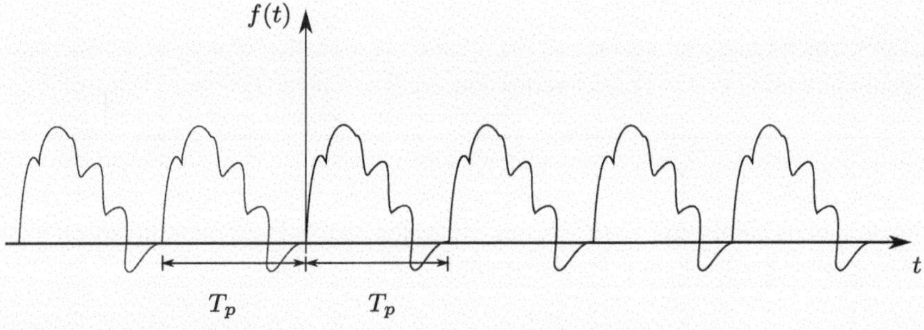

Fig. 2.13 Periodic function $f(t)$ defined from ($-\infty$ to ∞) with period T_p

In the above, a_n, b_n are coefficients to be determined and $\omega_p = T_p/2\pi$ is the cyclic frequency (in Hz) of function $f(t)$. Using the orthogonality property of the trigonometric functions, allows for the computation of the expansion coefficients a_0, a_n and b_n as follows:

$$a_0 = \frac{1}{T_p} \int_{-T_p/2}^{T_p/2} f(t)\, dt \tag{2.35a}$$

$$a_n = \frac{2}{T_p} \int_{-T_p/2}^{T_p/2} f(t) \cos n\omega_p t\, dt \tag{2.35b}$$

$$b_n = \frac{2}{T_p} \int_{-T_p/2}^{T_p/2} f(t) \sin n\omega_p t\, dt \tag{2.35c}$$

This expansion converges to $f(t)$ as the number of terms $n \to \infty$, provided the function obeys the Dirichlet conditions:

(a) The number of singular points of $f(t)$ within one period is finite.
(b) The number of maxima/minima within one period is finite.
(c) The function is integrable within one period, i.e.,

$$\int_{-T_p/2}^{T_p/2} |f(t)|\, dt = k < \infty.$$

Functions which satisfy points (a) and (b) is called sectionally continuous. Furthermore, at a discontinuity point t_d, the Fourier series converges to a mean (average) value of

$$\frac{1}{2}\left(f(t_d^-) + f(t_d^+)\right)$$

At discontinuous points, the Fourier series expansion results in an error, irrespective of the number of terms used, which is known as Gibb's phenemenon.

2.10.2 Steady-State Vibrations to a Periodic Input

Following some time after application of a periodic external force at $t = 0$, the response of an SDOF system stabilizes since its transient part is damped out and becomes periodic as well. This steady-state response comprises three parts, one from each of the expansion terms shown in Eq. (2.34). Specifically, there is a static-like response of the SDOF system to the constant term a_0 as follows:

$$u_{a_0}(t) = \frac{a_0}{k}. \tag{2.36}$$

For the sinusoidal part of the periodic expansion of the load, the response is given by Eq. 2.12, i.e.,

$$u_{b_n}(t) = \frac{b_n}{k} \frac{1}{(1-\beta_n^2)^2 + (2\xi\beta_n)^2} \left((1-\beta_n^2)\sin(n\omega_p t) - 2\xi\beta_n \cos(n\omega_p t)\right), \qquad (2.37)$$

where the ratio $\beta_n = \frac{n\omega_p}{\omega_0}$. Correspondingly, the response to the cosinusoidal part of the expansion is given as

$$u_{a_n}(t) = \frac{a_n}{k} \frac{1}{(1-\beta_n^2)^2 + (2\xi\beta_n)^2} \left(2\xi\beta_n \sin(n\omega_p t) + (1-\beta_n^2)\cos(n\omega_p t)\right). \qquad (2.38)$$

By superimposing the above parts resulting from the loading terms in Eq. (2.34), the response of the SDOF system, following some simplifications, can be written as follows:

$$u(t) = \frac{a_0}{k} + \sum_{n=1}^{\infty} \frac{1}{k} \frac{1}{(1-\beta_n^2)^2 + (2\xi\beta_n)^2} \left(\left(a_n 2\xi\beta_n + b_n(1-\beta_n^2)\right)\sin(n\omega_p t)\right.$$
$$\left. + \left(a_n(1-\beta_n^2) - b_n 2\xi\beta_n\right)\cos(n\omega_p t)\right). \qquad (2.39)$$

2.10.3 Complex Fourier Series

We now introduce complex number formalism in the SDOF system analysis. At first, we list Euler's formulas relating the complex exponential function with the trigonometric functions, i.e.,

$$\cos\theta = \frac{1}{2}\left(e^{i\theta} + e^{-i\theta}\right) \qquad (2.40a)$$

$$\sin\theta = -\frac{i}{2}\left(e^{i\theta} - e^{-i\theta}\right) \qquad (2.40b)$$

plus the inverse relations,

$$e^{i\theta} = \cos\theta + i\sin\theta \qquad (2.41a)$$

$$e^{-i\theta} = \cos\theta - i\sin\theta \qquad (2.41b)$$

The previously mentioned sine and cosine Fourier series expansions of Eq. (2.34) can now be expressed using Eq. (2.40) as a unified complex series

$$f(t) = \sum_{n=-\infty}^{\infty} c_n e^{in\omega_p t} \qquad (2.42)$$

where the expansion coefficients are computed as follows:

$$c_n = \frac{1}{T_p} \int_{-T_p/2}^{T_p/2} f(t) e^{-in\omega_p t}\, dt \tag{2.43a}$$

$$c_{-n} = \frac{1}{T_p} \int_{-T_p/2}^{T_p/2} f(t) e^{in\omega_p t}\, dt \tag{2.43b}$$

$$c_0 = \frac{1}{T_p} \int_{-T_p/2}^{T_p/2} f(t)\, dt = a_0. \tag{2.43c}$$

The paradox of expanding a real function $f(t)$ in terms of complex series in Eq. (2.42) can be answered by observing that for each term $c_n e^{in\omega_p t}$, there exists a complex conjugate term $c_{-n} e^{-in\omega_p t}$ term, so that their sum yields a real-valued result. Next, we present the forced response of an SDOF system to a periodic function which has been expanded using the series with complex-valued exponential terms. The equation of motion is modified for a complex-valued input, i.e.,

$$m\ddot{u}(t) + c\dot{u}(t) + ku(t) = f_0 e^{i\omega t} \tag{2.44}$$

In order to solve the above equation, we assume a solution in the form $Ce^{i\omega t}$. By substituting in the equation of motion, the solution is now

$$u(t) = H(\omega) f_0 e^{i\omega t} \tag{2.45}$$

where $H(\omega)$ is the complex-valued, *frequency response function*:

$$H(\omega) = \frac{1}{-m\omega^2 + ic\omega + k} = \frac{1}{k(1 - \beta^2 + 2i\xi\beta)} \tag{2.46}$$

If we consider that $f_0 = c_n$ because of Eq. (2.43a) and $\omega = n\omega_p$, superposition of terms allows for a synthesis of the solution of Eq. (2.42) as follows:

$$u(t) = \sum_{n=-\infty}^{\infty} c_n H(\omega_p) e^{in\omega_p t}. \tag{2.47}$$

2.10.4 Fourier Integral for Non-periodic Functions

By expanding Eq. (2.42), employing Eq. (2.43a) for coefficients c_n and setting $\frac{1}{T_p} = \frac{\omega_p}{2\pi}$, we have that

$$f(t) = \frac{1}{2\pi} \sum_{n=-\infty}^{\infty} \left(\int_{-T_p/2}^{T_p/2} f(t)e^{-in\omega_p t}\, dt \right) e^{in\omega_p t} \omega_p \tag{2.48}$$

In order to write the Fourier integral of a non-periodic function, we must assume that the period of an original periodic function $f(t)$ becomes infinite, i.e., $T_p \to \infty$ and $\omega_p \to d\omega$. At the same time we convert the discrete period $n\omega_p$ to a continuous variable ω. By moving to the limit in Eq. (2.48), the infinite series summation of frequency ω_p terms becomes an integral in the differential $d\omega$,

$$f(t) = \frac{1}{2\pi} \int_{-\infty}^{\infty} F(\omega)e^{i\omega t}\, d\omega \tag{2.49}$$

The inverse relation is

$$F(\omega) = \int_{-\infty}^{\infty} f(t)e^{-i\omega t}\, dt \tag{2.50}$$

Thus, the above two relations define a *Fourier transform pair*. More specifically, we have the direct Fourier transform $\mathcal{F}[f(t)] = F(\omega)$ of a function of time $f(t)$ using Eq. (2.50). At the same time, given the transformed function $F(\omega)$, the original function $f(t)$ can be reconstituted from Eq. (2.49), which is the inverse Fourier transform and symbolized as $\mathcal{F}^{-1}[f(\omega)] = f(t)$.

An important property of Fourier transforms is the transformation of the time derivatives, given below as

$$\mathcal{F}\left[\frac{d^n f}{dt^n}(t) \right] = (i\omega)^n F(\omega). \tag{2.51}$$

In order to compute the response of an SDOF system to an arbitrary external forcing function using the Fourier transform, which implies solving the problem in the frequency domain instead of the original time domain, we start with the equation of motion, Eq. (2.1), multiply both sides with the term $e^{-i\omega t}$ and integrate over time from $-\infty$ to ∞:

$$m\mathcal{F}\left[\frac{d^2 u}{dt^2}(t) \right] + c\mathcal{F}\left[\frac{du}{dt}(t) \right] + k\mathcal{F}[u(t)] = \mathcal{F}[f(t)] \to$$

$$m(i\omega)^2 U(\omega) + ic\omega U(\omega) + kU(\omega) = F(\omega) \to$$

$$(-m\omega^2 + ic\omega + k)U(\omega) = F(\omega) \to$$

$$U(\omega) = F(\omega)\frac{1}{(k - m\omega^2 + ic\omega)}$$

Thus, we recover the equivalent to the time convolution integral given by Eq. (2.21) as follows:

$$U(\omega) = H(\omega)F(\omega) \tag{2.52}$$

Observe that the time convolution becomes a simple product in the frequency domain. Specifically, function $H(\omega)$ is the transfer function which converts the input to the SDOF system into an output. Furthermore, the transfer function is the Fourier transform of the impulse response function $h(t)$, i.e.,

$$H(\omega) = \frac{1}{k\left(1 - \frac{\omega^2}{\omega_0^2} + i2\xi\frac{\omega}{\omega_0}\right)}. \tag{2.53}$$

In Listing 2.2 given below we compute the complex-valued, transfer function for a linear, SDOF oscillator and plot its amplitude over a given frequency range ω.

```
k=100; m=1; c=3.5
om0=sqrt(k/m)
UnitReal=new Complex(1.0, 0.0)

H={
   C=new Complex(k-m*(it*om0)**2, c*(it*om0))
   return ((UnitReal/C).abs())
}
thePlot.clear()
thePlot.addFunction(new plotfunction(linspace(0.0,
     201, 0.01),H as DoubleFunction))
thePlot.show()
```

Listing 2.2 Computation of real part of complex transfer function $H(\omega)$

Once the transfer function $H(\omega)$ is available, we can use Eq. (2.50) to solve for the SDOF system response in the frequency domain. What is left then is to transform the solution to the time domain using the inverse Fourier transformation of Eq. (2.49)) as follows:

$$u(t) = \frac{1}{2\pi} \int_{-\infty}^{\infty} U(\omega)e^{i\omega t}\, d\omega$$

$$= \frac{1}{2\pi} \int_{-\infty}^{\infty} F(\omega)H(\omega)e^{i\omega t}\, d\omega. \tag{2.54}$$

It should be noted that the inverse transformation is not a simple analytical procedure. In most cases, it is performed numerically by using the Fast Fourier Transform (FFT) algorithm. In order to use this algorithm, however, reference must first be made to the conversion of a continuous signal into a discrete one.

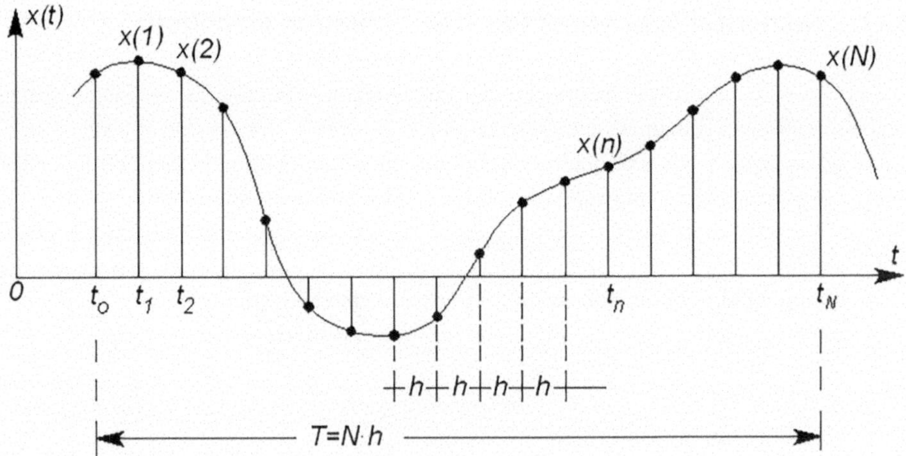

Fig. 2.14 Sampling a random continuous time signal

In order to convert a continuous (analog) signal $x(t)$ into a discrete (digital) $f(n)$ one, most data points along the time axis must be omitted. This process is known as *sampling* and it is important that the sampled points retain with sufficient accuracy the original phenomenon that was recorded. Consider the random signal $x(t)$ shown in Fig. 2.14. The first step is to chose N time intervals with equal spacing h. Assuming that a reliable record time length is T_{tot}, with starting time t_0, then $T_{tot} = Nh$. Despite the fact that random signals are not necessarily periodic, we can consider that the total recording time T_{tot} is the period of this signal. This assumption implies that the minimum possible number of repetitive cycles is $f_{min} = 1/T_{tot}$. Thus, the choice of step h determines the maximum signal frequency f_{max} and the corresponding minimum period T_{min} for sampling a continuous signal. This is known as the Nyquist frequency and is defined as follows:

$$f_{max} = \frac{1}{T_{min}} = \frac{1}{2\,h}. \tag{2.55}$$

The FFT Algorithm in SDE

The Climax software has been programmed to perform the discrete Fourier transform using the FFT algorithm available in the Apache Commons Mathematics library. In a subsequent section, we will give application examples, along the the necessary code to help familiarize the reader with the Symplegma Development Environment (SDE) work environment.

2.11 Numerical Solution of the SDOF Equation of Motion

In a previous section it was mentioned that the equation of motion for an SDOF system can be solved using Duhamel's integral. Despite its generality, this closed-form solution has two disadvantages: (a) It is difficult to extend it to the solution of multi-degree-of-freedom (MDOF) systems; (b) it is not valid for non-linear systems, as it is based of the principle of superposition. For these cases and for the sake of generality, specialized numerical integration methods for the equations of motion have been developed over the years that are labelled as time-stepping algorithms. Regarding this group of methods, the solution procedure for the equation of motion (Eq. 2.1) using time integration is based on two premises:

- satisfaction of Eq. (2.1), or of the corresponding system of equations for MDOF systems, at a series of discrete points in time $t_n = n\Delta t$
- a priori estimation of the time variation of the dependent variables and of the forcing function within a given time step Δt.

Following these two basic premises, any time function $g(t)$ is replaced by the sequence of points $g_n = g(t_n)$, while the original differential equation becomes a difference equation.

In general, there are two ways for classifying time stepping methods: (a) those based on algorithms which address the solution of first -order differential equations versus those that solve second-order differential equations; (b) those based on algorithms that use explicit numerical integration schemes versus those that use implicit numerical integration schemes. In the first case, the solution for the next time step is based on information already computed and coming from the previous time steps, while in the second case, assumptions have to be made regarding the variation of the solution for an immediately future time step, with the possibility of correcting these assumptions later. Critical issues regarding reliability and efficiency of a given time stepping algorithm is its (a) numerical accuracy and (b) numerical stability. The latter is essential for convergence of the computations to finite (but not necessarily accurate) values. If the convergence criterion requires restrictions on the fundamental parameter for time discretization, namely the time step Δt, then the algorithm is known as conditionally stable. Otherwise, it is known as unconditionally stable.

2.11.1 Numerical Integration of Second-Order Ordinary Differential Equations

In what follows, we will present the two most basic time-stepping algorithms currently used for the numerical integration of the equation of motion of an SDOF system, which is classified as a second-order, ordinary differential equation in the time variable. These algorithms can also integrate systems of equations of motion resulting from MDOF representations of structures. In this case, the material parameters (stiffness k, mass m, damper c) are no longer scalars but matrices, while the dependent variables (displacement u, velocity v, acceleration a and external force f) are now vectors.

Central Difference Method

As the name suggests, the central difference method is based on expressing the velocity as a finite difference scheme involving displacement values symmetrically placed about the current time interval, i.e.,

$$\dot{u}(t) \approx \frac{u(t + \Delta t) - u(t - \Delta t)}{2\Delta t} \rightarrow \dot{u}_n = v_n = \frac{u_{n+1} - u_{n-1}}{2\Delta t},$$

As far as the acceleration is concerned, it is a central difference expansion of the velocity resulting in

$$\ddot{u}(t) \approx \frac{u(t + \Delta t) - 2u(t) + u(t - \Delta t)}{\Delta t^2} \rightarrow \ddot{u}_n = \dot{v}_n = a_n = \frac{u_{n+1} - 2u_n + u_{n-1}}{\Delta t^2}.$$

All that remains is to substitute these expansions in the equation of motion and re-arrange terms so as to produce a time-integration algorithm. The necessary steps defining the central difference method for computer implementation are listed in Table 2.4. It is important to note that the method is conditionally stable, with the stability criterion imposed on the time step Δt, which is a function of the natural period of the SDOF oscillator:

$$\Delta t \leq \Delta t_{cr} = \frac{T_0\sqrt{1 - \xi^2}}{\pi} = \frac{2\sqrt{1 - \xi^2}}{\omega_0} \tag{2.56}$$

Despite this being a rather restrictive condition, the method is used for the solution of wave propagation problems which involve high frequencies, resulting in a very small time step Δt in order to secure accuracy. The advantage of central differences is their simplicity and fast execution. Thus, the method is desirable for use with MDOF systems, especially if the resulting system matrix can be diagonalized, as would be the case with the use of modal analysis that will be discussed in a following section.

The Newmark-Beta Family of Algorithms

These algorithms are the most commonly used for the numerical solution of problems in structural dynamics. They are also simple to program, but most important, they yield unconditionally stable algorithms with the appropriate choice of parameters, namely the coefficients β and γ. These two coefficients control how the initial and final acceleration values, as registered in a given time interval, represent the actual variation of the displacement and

Table 2.4 The central difference method

Step	Procedure
I. Input data and preliminary computations	1. SDOF system parameters and forcing function input: $k,\ m,\ c,\ f$
	2. Initial displacement and velocity input: $u_0,\ v_0$
	3. Initial acceleration computed from the equation of motion: $a_0 = m^{-1}(f - ku_0 - cv_0)$
	4. Time step input: Δt
	5. Integration constants: $w_0 = \frac{1}{\Delta t^2},\ w_1 = \frac{1}{2\Delta t},\ w_2 = 2w_0,\ w_3 = \frac{1}{w_2}$
	6. Negative step displacement setup: $u_{-1} = u_0 - \Delta t v_0 + w_3 a_0$
	7. Effective stiffness matrix setup: $\hat{k} = w_0 m + w_1 c$
II. Recursive computations for time step $i = 1, .., N$	1. Effective forcing function at step i $\hat{f}_i = f_i - (k - w_2 m)u_{i-1} - (w_0 m - w_1 c)u_{i-2}$
	2. Solution for the displacement at time step i $u_i = \hat{k}^{-1}\hat{f}_i$
	3. Recovery of velocity and acceleration at time step i $v_i = \frac{2}{\Delta t}(u_i - u_{i-1}) - v_{i-1}$ $a_i = \frac{4}{\Delta t^2}(u_i - u_{i-1} - \Delta t v_i) - a_{i-1}$

velocity within that time interval. More specifically, the step-wise variation of the velocity is given as

$$\dot{u}_{n+1} = \dot{u}_n + (1 - \gamma)\Delta t\, \ddot{u}_n + \gamma\Delta t\, \ddot{u}_{n+1} \tag{2.57}$$

while for the displacement it is

$$u_{n+1} = u_n + \Delta t\, \dot{u}_n + \left(\frac{1}{2} - \beta\right)\Delta t^2 \ddot{u}_n + \beta\Delta t^2 \ddot{u}_{n+1} \tag{2.58}$$

By substituting $\ddot{u}_{n+1}$ from the second of the above equations into the first one, we can derive the velocity and acceleration at the new time step $n + 1$ (parametric in the coefficients β and γ) as functions of the acceleration, velocity and displacement at the immediately previous (i.e., the current) time step n as well as of the displacement at time step $n + 1$. By replacing all these expressions in the equation of motion written for the new time step $n + 1$, we have an equilibrium statement in the form $\hat{k}u_{n+1} = \hat{f}_{n+1}$. Upon solution, we obtain u_{n+1} and the process is repeated until the end of time interval of interest is reached. In Table 2.5 which follows, Newmark's method is presented in algorithmic form for use in programming on a computer. The β-Newmark family of algorithms is unconditionally stable for the following choice of parameters:

$$0.5 \leq \gamma \leq 2\beta$$

For any other choice of these parameters, the stability criterion expressed in terms of the time step is

Table 2.5 The Newmark-β method

Steps	Procedure
I. Input of data and preliminary computations	1. SDOF system parameters and the external load input: k, m, c, f
	2. Determination of the initial displacement and velocity: u_0, v_0
	3. Computation of the acceleration from the equation of motion: $a_0 = m^{-1}(f - ku_0 - cv_0)$
	4. Time step input: Δt
	5. Computation of the integration constants: $w_0 = \frac{1}{\beta \Delta t^2}, w_1 = \frac{\gamma}{\beta \Delta t}, w_2 = \frac{1}{\beta \Delta t},$ $w_3 = \frac{1}{2\beta} - 1,$ $w_4 = \frac{\gamma}{\beta} - 1, w_5 = \frac{\Delta t}{2}\left(\frac{\gamma}{\beta} - 2\right),$ $w_6 = \Delta t(1 - \gamma), w_7 = \Delta t\gamma$
	6. Formation of the effective stiffness matrix: $\hat{k} = k + w_0 m + w_1 c$
II. Recursive computations for a given time step $i = 1, ..., N$	1. Computation of the effective external force i $\hat{f}_i = f_i + m(w_0 u_{i-1} + w_2 v_{i-1} + w_3 a_{i-1}) + c(w_1 u_{i-1}n + w_4 v_{i-1} + w_5 a_{i-1})$
	2. Solution for the displacement at time step i $u_i = \hat{k}^{-1}\hat{f}_i$
	3. Recovery of the acceleration and velocity at time step i $a_i = w_0(u_i - u_{i-1}) - w_2 v_{i-1} - w_3 a_{i-1}$ $v_i = v_{i-1} + w_6 a_{i-1} + w_7 a_i$

$$\Delta t \leq \Delta t_{cr} = \frac{1}{\omega_0} \frac{\xi(\gamma - 0.5) + \sqrt{0.5\gamma - \beta + \xi^2(\gamma - 0.5)^2}}{0.5\gamma - \beta}. \tag{2.59}$$

Two common choices for the parameters that result in popular variants are

(a) the constant acceleration method, $\gamma = \frac{1}{2}$ and $\beta = \frac{1}{4}$
(b) the linear acceleration method, $\gamma = \frac{1}{2}$ and $\beta = \frac{1}{6}$.

The first of the above methods is unconditionally stable, but of lower accuracy compared to the second one, which however is conditionally stable. The stability condition in this case for the time step Δt is determined by Eq. (2.59) upon substitution of the following parameter values, $\gamma = \frac{1}{2}$ and $\beta = \frac{1}{6}$, thus yielding

$$\Delta t_{cr} = \frac{2\sqrt{3}}{\omega_0}, \tag{2.60}$$

Note that the above criterion is independent of the system damping ξ.

2.11.2 Numerical Integration of First-Order Ordinary Differential Equations

Numerical integration of the equations of motion for dynamic systems commenced with the development of algorithms for various sub-categories of first-order differential equations. From a mathematical viewpoint, the literature is rich in the development of such algorithms for initial value, first-order differential equation systems. In here, we will present methods most commonly used for structural dynamics and for structural control. At first, the velocity $v = \frac{du}{dt} = \dot{u}$ appearing in the equation of motion (2.1) can be expressed as follows:

$$\dot{v} = \frac{f}{m} - \frac{c}{m}v - \frac{k}{m}u \tag{2.61}$$

Next, we form the vector of the two dependent variables as $y = \{u \quad v\}^T$, (superscript T denotes the transpose operator) allowing for the following form of the second-order equation of motion of an SDOF system:

$$\dot{u}(t) = v(t) \tag{2.62}$$

$$\dot{v}(t) = \frac{f(t)}{m} - \frac{c}{m}v(t) - \frac{k}{m}u(t) \tag{2.63}$$

These can be condensed as follows:

$$\dot{y} = \tilde{f}(y, t) \tag{2.64}$$

Also, the vector of the initial conditions is

$$y(t_0) = y_0 = \begin{bmatrix} u(t_0) \\ v(t_0) \end{bmatrix} = \begin{bmatrix} u(t_0) \\ \dot{u}(t_0) \end{bmatrix} \tag{2.65}$$

Note that the mathematical space defined by the displacement and velocity, namely the two dependent variables of the problem, $y = \{u \quad v\}^T$, is labelled as the *state space* and finds extensive use in the field of structural control. Similarly, in applied mechanics we have a representation of the dependent variables defined as the *phase space*, while in non-linear systems we have the notation *configuration space*.

Euler's Method

By direct numerical integration of Eq. (2.64), we have the following simple recursive relation

$$y(t_{n+1}) = y(t_n) + \Delta t \, \tilde{f}(y_n, t_n) \tag{2.66}$$

This relation is known the *forward Euler method* and derives after considering the first derivative (the velocity) as a forward time difference of the displacement:

$$\dot{y}(t_n) = \frac{y(t_{n+1}) - y(t_n)}{\Delta t} \tag{2.67}$$

It is evident that this is an explicit integration algorithm. In a similar manner, by using a backward in time difference scheme for the velocity, i.e.,

$$\dot{y}(t_n) = \frac{y(t_n) - y(t_{n-1})}{\Delta t} \tag{2.68}$$

we derive the *backward Euler method*

$$y(t_{n+1}) = y(t_n) + \Delta t \, \tilde{f}(y_{n+1}, t_{n+1}) \tag{2.69}$$

which turns out to be an implicit algorithm.

The Runge-Kutta Method

Despite being the most commonly used method for integrating differential equations of all types, the Runge-Kutta method is seldom used in structural dynamics. This is due to the following reasons: (i) The complexity inherent in formulating this method for MDOF systems and (ii) the lack of a well-defined stability criterion. The method, however, has a sound mathematical foundation and is capable of producing high-accuracy results. In the *second-order Runge-Kutta method*, starting at every time step n, an intermediate solution is first computed as y_{n+1}^*, which in turn is used to form the actual solution y_{n+1}. The algorithm can be formulated in these two steps as follows:

$$y_{n+1}^* = y_n + \Delta t\,\tilde{f}(y_n, t_n)$$

$$y_{n+1} = y_n + \frac{\Delta t}{2}\left[\tilde{f}(y_n, t_n) + \tilde{f}(y_{n+1}^*, t_{n+1})\right] \tag{2.70}$$

A more accurate version is the *fourth-order Runge-Kutta method*, which at every time step n requires the computation of four ancillary quantities that are subsequently used to synthesize the solution at the following time step $n + 1$:

$$k_1 = \Delta t\,\tilde{f}(y_n, t_n)$$

$$k_2 = \Delta t\,\tilde{f}(y_n + k_1/2, t_{n+1/2})$$

$$k_3 = \Delta t\,\tilde{f}(y_n + k_2/2, t_{n+1/2})$$

$$k_4 = \Delta t\,\tilde{f}(y_n + k_3, t_{n+1})$$

$$y_{n+1} = y_n + \frac{\Delta t}{6}\,[k_1 + 2k_2 + 2k_3 + k_4] \tag{2.71}$$

This fourth-order Runge-Kutta method has been programmed in object `solve` of class `sdof` (see Sect. 2.13.1) for the numerical solution of an SDOF system to an arbitrary forcing function.

2.12   Non-linear SDOF Systems

Up to this point, we have assumed that the mechanical parameters of the SDOF oscillator (mass, stiffness, damper) remain constant in time. Thus, the corresponding forces which develop (inertia force, restoring force, damping force) are linear functions of the corresponding kinematic variables (acceleration, displacement, velocity), see Eq. (2.1). As a consequence, the principle of superposition holds, which is crucial in solving linear systems. More specifically, the principle of superposition used here states that if an external force f_{I} acts on an SDOF system, then the system responds as u_{I}; if a second external force f_{II} is applied consecutively and yields response u_{II}, then total system response to the sum of these forces $f_{\mathrm{I}} + f_{\mathrm{II}}$ is $u_{\mathrm{I}} + u_{\mathrm{II}}$. Furthermore, the *principle of homogeneity* holds for for linear systems, which states that if the system response is u to an external force f, then the response to force αf is αu, where α is a scalar multiplier. More specifically, there are two basic types of nonlinearities encountered in structural and mechanical systems, namely (i) the geometric nonlinearity and (ii) the material nonlinearity. In the former case, stability phenomena result and this subject will not be pursed further. In the latter case, the resulting equation of motion for an SDOF system is

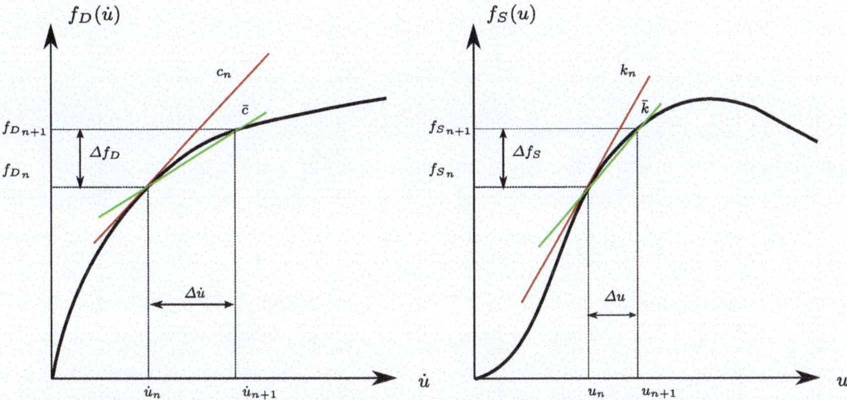

Fig. 2.15 Non-linear damping and restoring forces developing in an SDOF oscillator

$$m(t)\ddot{u}(t) + f_D(u, \dot{u}, t) + f_S(u, \dot{u}, t) = f(t) \tag{2.72}$$

Quite often we encounter a special case to the above equation as follows:

$$m\ddot{u}(t) + f_D(\dot{u}) + f_S(u) = f(t) \tag{2.73}$$

When considering non-linear systems, the best way to proceed is by writing the dynamic equilibrium equation in incremental form, see Fig. 2.15. We therefore assume that between two consecutive time steps t and $t+\Delta t$, a force increment $\Delta f = f(t+\Delta t) - f(t)$ is applied and the instantaneous equation of motion is

$$m\,\Delta\ddot{u} + c(t)\,\Delta\dot{u} + k(t)\,\Delta u = \Delta f \tag{2.74}$$

In the above, material parameters $c(t)$ and $k(t)$ represent averaged values within time step Δt. In practice, these averaged values (i.e., $\bar{c}_n$ and $\bar{k}_n$, see Fig. 2.15), can be computed using an iterative procedure, since the displacement and velocity at the end of the time interval $t+\Delta t$ are still unknown. The simplest approach is to use the original slope of the restoring force and of the damping force diagrams at time t. Specifically, if t_n is the begining of the time interval in question and t_{n+1} is the end, and after time increment Δt elapses, we have

$$c(t) \doteq \left.\frac{df_D}{dt}\right|_n = c_n \qquad k(t) \doteq \left.\frac{df_S}{dt}\right|_n = k_n \tag{2.75}$$

By substitution in the incremental form of the equation of motion,

$$m\,\Delta\ddot{u} + c_n\,\Delta\dot{u} + k_n\,\Delta u = \Delta f \tag{2.76}$$

the increments in the kinematic variables are

$$\Delta u = u_{n+1} - u_n,$$

$$\Delta \dot{u} = \dot{u}_{n+1} - \dot{u}_n,$$

$$\Delta \ddot{u} = \ddot{u}_{n+1} - \ddot{u}_n \tag{2.77}$$

The incremental form of the the equation of motion, Eq. (2.76), can be solved by any of the methods that were presented in Sect. 2.11.1, while a numerical example is presented in Sect. 2.14.8. Furthermore, the incremental method for nonlinear problems can be improved if within a given time step, the error committed by simply using the tangential forms of the restoring and damping forces in Eq. (2.75), is minimized. This can be achieved by use of the well-known Newton-Raphson method. For the case where the equations of motion are cast as a system of first-order differential equations, the solution algorithms that were presented in Sect. 2.11.2 remain valuable numerical tools and can be used for nonlinear systems as well. Relevant examples are given in the corresponding paragraphs of the section on numerical applications.

2.12.1 The Simple Pendulum

The equation of motion for the simple pendulum shown in Fig. 2.16 above is derived by equating the kinetic with the potential energy of its mass, i.e.,

$$\frac{d^2\theta}{dt^2} + \frac{g}{l}\sin\theta = 0 \tag{2.78}$$

We have that θ is the kinematic variable describing the oscillator (i.e., the angular displacement) expressed as the angle traced with respect to the vertical axis at a given time t. Furthermore, g is the acceleration of gravity and l the undeformed length of the rod supporting the mass of the pendulum. For small angles $\theta \ll 1$, we have the approximation $\sin\theta \approx \theta$, so that the equation of motion is linearized and becomes

Fig. 2.16 The simple pendulum

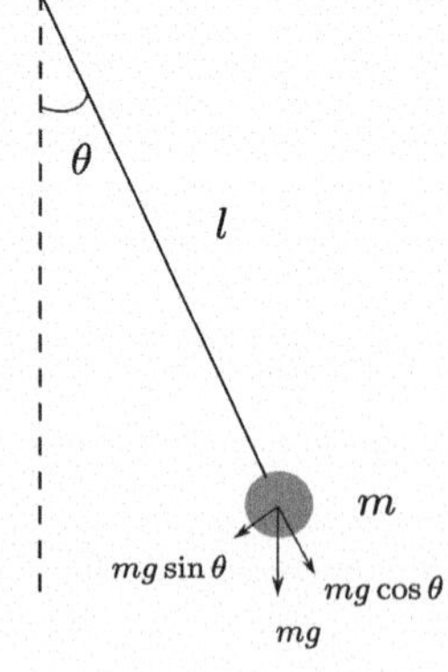

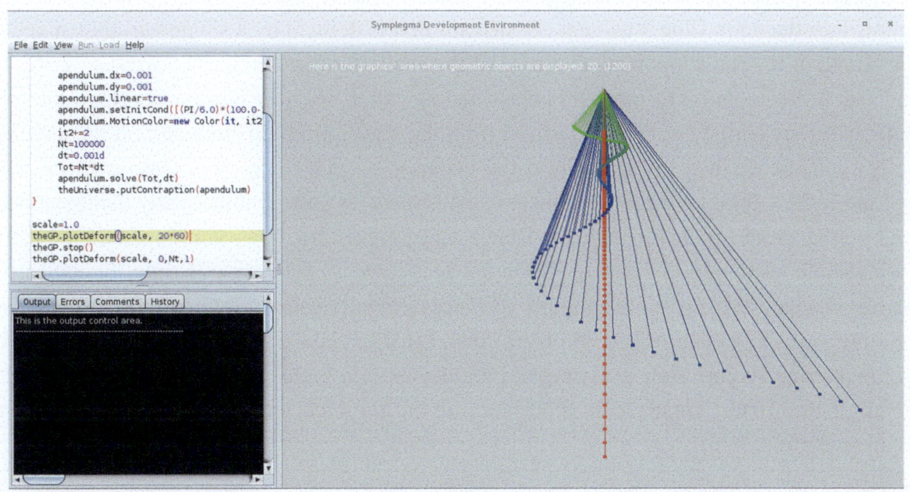

Fig. 2.17 Snapshot of simulationof the experiment on a wave sequence of independent pendulums

$$\frac{d^2\theta}{dt^2} + \frac{g}{l}\theta = 0 \tag{2.79}$$

From the above equation it becomes obvious that the mass of the simple pendulum plays no role in its equation of motion. In order to fully define the problem, initial conditions on the pendulum's angular displacement and velocity must be given. These are defined as follows:

$$\theta(t_0) = \theta_0 \quad \text{and} \quad \dot{\theta}(t_0) = \dot{\theta}_0 \tag{2.80}$$

In sum, the simple pendulum is the most elementary non-linear system in structural dynamics. For this reason, there is a special class in the educational package `courses.structuraldynamics` that focuses on both SDOF and MDOF nonlinear systems. Furthermore, Fig. 2.17 reproduces a snapshot of the motion from a classical experiment in mechanics known as the wave sequence of independent pendulums. The response produced from use of object `pendulum` is derived from the numerical solution of the differential equation of a pendulum, either in its original non-linear form or the its linearized form.

Class Pendulum

This is a first reference to object textttcourses.structuraldynamics that refers to SDOF systems and materializes through class textttpendulum. All material presented here is ancillary to source code [2.3] given below. In order to use class textttpendulum, we must first import `structuraldynamics` from `courses` package, see Line 1 in Listing 2.3. The pendulum is introduced in Line 3 using a constructor that is defined by a single variable, namely the length of the rod supporting the mass of the oscillator, `pendulum(double leng)`. In Line 4, the equation of motion is classified as either linear or non-linear. Next, the initial conditions on the angle θ_0 measured from the equilibrium position and on the angular velocity $\dot{\theta}_0$ are defined in Line 5 by using instance `setInitCond` of class `pendulum`. In Line 7, the equation of motion (linear or non-linear) is solved numerically using method `solve` that employs the fourth-order Runge-Kutta algorithm. Upon solution, the kinematic variables about the equilibrium position are recovered as functions of time. Specifically, we use instance `Theta` for the angular displacement and instance `DTheta` for the angular velocity, as can be seen in Line 8 of the code. Finally, class `pendulum` is an application of the interface/connection `contraption` and as such yields a graphical representation. Finally, Lines 10 to 13 manage this interface plus the graphical representation of the response as an animation.

```
 1  import courses.structuraldynamics.*
 2
 3  apendulum = new pendulum(1.0)
 4  apendulum.linear=true
 5  apendulum.setInitCond([PI/8.0] as double[], [0.0] as
        double[])
 6  Nt=1000; dt=0.01d ; Tot=Nt*dt
 7  apendulum.solve(Tot,dt)
 8  th=apendulum.Theta(); dth=apendulum.DTheta()
 9
10  theGP.setIsoScale(true)
11  theUniverse.putContraption(apendulum)
12  scale=1.0
13  theGP.plotDeform(1.0, 0,Nt,10)
```

Listing 2.3 Object `pendulum`

2.12.2 The Duffing Oscillator

The well-known Hooke's law of elasticity defines the relation between the restoring force and the lengthening of a spring as a linear one of the type $f_s = ku$. There exist, of course, springs that do not obey Hooke's law, while most springs obey it only for small values of the displacement between their two ends. A more general version of this law results if a third order polynomial function of the displacement is assumed for the restoring force, i.e., $f_s = ku + \mu u^3$. If parameter $\mu > 0$, the resulting stiffness in the spring decreases with increasing displacement, while if $\mu < 0$, the opposite is true. Systems comprising of such types of springs are known as having either a negative or positive stiffness, respectively. Thus, the equation of motion of a SDOF system augmented by the non-linear stiffness reads as

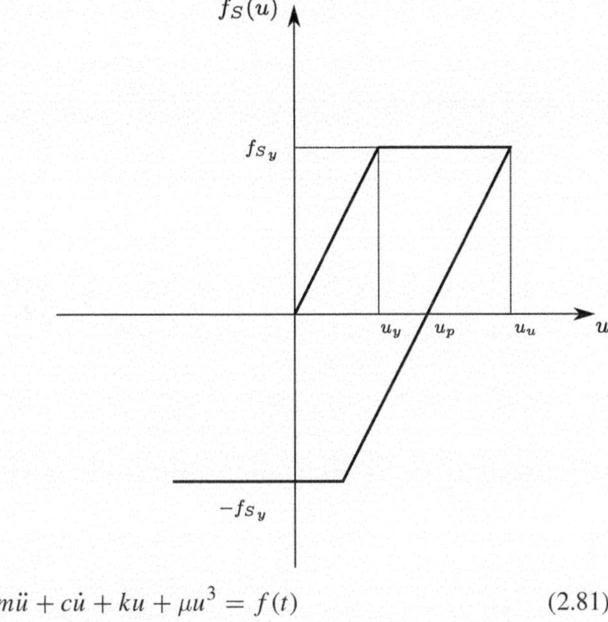

Fig. 2.18 Constitutive law for an elasto-plastic system relating the restoring force to the displacment

$$m\ddot{u} + c\dot{u} + ku + \mu u^3 = f(t) \tag{2.81}$$

The above is known as the Duffing equation and was formulated in relation to the analysis of mechanical systems. This nonlinear differential equation is of second order and can be written as a system of two first order differential equations as follows:

$$\dot{u} = v$$
$$\dot{v} = \frac{f(t)}{m} - \frac{c}{m}v - \frac{k}{m}u - \frac{\mu}{m}u^3 \tag{2.82}$$

This system of first-order differential equations can easily be treated numerically, as would be the case using the Runge-Kutta family of methods.

2.12.3 Elastic-Perfectly Plastic SDOF Systems

Another example of a non-linear SDOF system is the case of an elastic-perfectly plastic spring element, as can be seen in Fig. 2.18. The spring response depends on the displacement level, starting with a constant stiffness k, until the value for the restoring force and of the corresponding displacement reach limit values of (f_{Sy}) and (u_y), respectively. Once the latter value has been exceeded, the spring material (e.g., steel) is yielding, there is no further resistance available and the restoring force is flat, i.e., $(k_h = 0)$. Upon unloading, the spring will revert back to the elastic range and its stiffness will be k, but there will remain a residual displacement u_p when the external force has been completely removed, i.e., when $f_S = 0$.

2.13 Entities in Class `courses.structuraldynamics`

2.13.1 Class `sdof`

Description: This class is addresses the solution of an SDOF system under arbitrary initial conditions and forcing function.

Instance
A constructor for an instance of an object comprising three versions. The first defines five numerical parameters, in double precision, which correspond to the stiffness, mass and damper of the oscillator plus the initial displacement and the initial velocity. In subsequent calls to the constructor, variables not originally defined are considered to be zero, but can be redefined during the course of the solution.

- `sdof(double k, double m, double c, double u0, double v0)`

- `sdof(double k, double m, double c)`

- `sdof(double k, double m)`

Methods
The naming of the methods is self-explanatory.

    void setCritDampRatio(double xsi)

    void setInitCond(double u0, double v0)

    void setRHS(DoubleFunction df)

    double getCritDampRatio()

    double getDamping()

    double getNaturalFrequency()

    double getNaturalPeriod()

continues …

Class `sdof` (...continued)

Methods

- double maxDisp()
- double maxVelc()
  double minDisp()
- double minVelc()

- double[ ] Disp()
- double[ ] Velc()
- double[ ] Accl()

- void duhamel(double tot,   double dt)

- void duhamel(int N)

- void duhamel(int N, double dt)

- void duhamel(int N, int M)

- double dt()
- Complex TransferFunction(double omega)

A similar usage to the one employing Duhamel's closed-form solution, which is actually integrated numerically, is the use of **solve**, which numerical solves the equation of motion of an SDOF system by using the Runge-Kutta method. Furthermore for a unconditionally stable procedure there is the choice of the **bNewmark** method an implementation of β-Newmark algorithm.

continued ...

Class `sdof` (…continued)

The object `sdof` applies the interface/interconnect methods (`interface`) available in Climax that are labelled as `contraption`. The interface `contraption` helps with a graphical representation of a given entity. In order to realize a mapping in this environment after a snapshot of a certain object `sdof` has been created, this object must be inserted in the 'universe', namely `theUniverse`, residing in environment SDE, using the command `putContraption`.

Methods:

- void setOrigin(double x0, double y0)

- void setLength(double elen)

- void setWidth(double w)

- void setHeight(double h)

2.14 Numerical Examples

2.14.1 Response Spectrum for an Orthogonal Pulse Input

Consider an SDOF system without damping under the action of an orthogonal pulse, as shown in Fig. 2.19. In order to reconstitute the response of the system, it is necessary to consider two time intervals, the first when the oscillator executes forced vibrations and the next one when it executes free vibrations. Thus, the first phase ($t \leq t_1$) is determined by the time during which the pulse is active, while the second phase ($t > t_1$) commences as soon as the pulse stops. The free vibrations are activated by initial conditions that are equal to the displacement and velocity registered by the oscillator at the instant the pulse is removed. The maximum response of the SDOF system, which is of interest to the design engineer, may occur in either the first or the second phase, and depends on the ratio of the time duration of the pulse (t_1) to the natural period of the oscillator ($T_0 = 2\pi/\omega_o$).

Fig. 2.19 External forcing function in the form of an orthogonal pulse, see Sect. 2.14.1

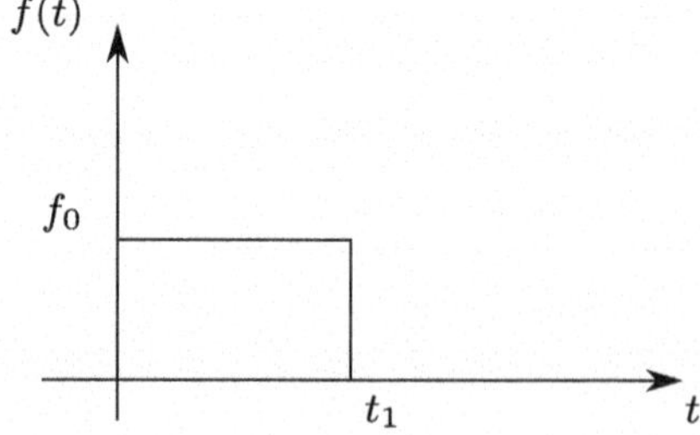

During the first phase (I), defined for times $0 < t \leq t_1$, the external forcing function remains constant $f(t) = f_o$, giving the following form of the equilibrium equation:

$$m\ddot{u}(t) + ku(t) = f_0$$

The closed form solution defined by Duhamel's integral for a constant in time load and for zero initial conditions is given by Eq. (2.17):

$$u^I(t) = \frac{f_0}{m\omega_0} \int_0^t \sin(\omega_0(t - \tau))\, d\tau = \frac{f_0}{m\omega_0} \left[\frac{\cos(\omega_0(t - \tau))}{\omega_0} \right]_0^t$$

$$= \frac{f_0}{k}(1 - \cos(\omega_0 t))$$

where the equivalent static displacement is $u_s t = f_0/k$.

During the second phase (II) where $t > t_1$, the SDOF oscillator will execute free vibrations for which the initial conditions must be known. Specifically, from the end of phase I we have that $u_0^{II} = u^I(t_1)$ for the initial displacement and $\dot{u}_0^{II} = \dot{u}^I(t_1)$ for the initial velocity. The velocity in phase I is computed by taking the time derivative of the displacement as follows:

$$\dot{u}^I(t) = \frac{f_0}{k}\sin(\omega_0 t)$$

Now we have the initial conditions for the second phase, i.e.,

$$u_0^{II} = u^I(t_1) = \frac{f_0}{k}(1 - \cos(\omega_0 t_1)) \quad \text{and} \quad \dot{u}_0^{II} = \dot{u}^I(t_1) = \frac{f_0}{k}\sin(\omega_0 t_1)$$

The solution is synthesized in reference to Eq. (2.4), where the time delay $\tilde{t} = t - t_1$ must be taken into account. Finally, we have the solution for $t > t_1$ as

$$u^{II}(t - t_1) = u_0^{II}\cos(\omega_0(t - t_1)) + \dot{u}_0^{II}\frac{\sin(\omega_0(t - t_1))}{\omega_0} \rightarrow$$

$$u^{II}(\tilde{t}) = \frac{f_0}{k}(1 - \cos(\omega_0 t_1))\cos(\omega_0 \tilde{t}) + \frac{f_0}{k}\sin(\omega_0 t_1)\frac{\sin(\omega_0 \tilde{t})}{\omega_0}$$

In order to compute the dynamic load factor D, we first assume that the maximum displacement occurs during phase (I). This maximum is determined by setting the first time derivative of the displacement (i.e., the velocity $\dot{u}^I(t)$) equal to zero, and solving for the time that maximizes the displacement:

$$\dot{u}^I(t) = \frac{f_0}{k}\sin(\omega_0 t) = 0 \rightarrow \omega_0 t = \pi \rightarrow t = \frac{\pi}{\omega_0} = \frac{T_0}{2}$$

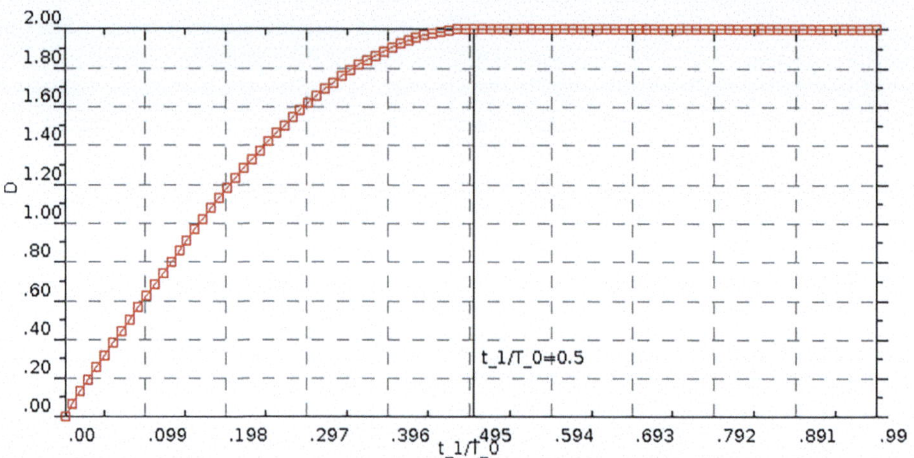

Fig. 2.20 Response spectrum for an orthogonal in time pulse, see Sect. 2.14.1

This implies that the maximum displacement will occur during phase (I) only if $t_1 \geq T_0/2$, in which case teh DLF be equal to $D = 2$. If we assume that $t_1 < T_0/2$, then this maximum amplitude can be found from Table 2.1,

$$u_{max} = p = \sqrt{\left(u_0^{II}\right)^2 + \left(\frac{\dot{u}_0^{II}}{\omega_0}\right)^2} = \frac{2f_0}{k}\sin\left(\pi\frac{t_1}{T_0}\right)$$

In this case, the DLF is equal to

$$D = 2\sin\left(\pi\frac{t_1}{T_0}\right)$$

It is worth noting that the dynamic load factor depends only on the ratio t_1/T_0, where t_1 is the duration of the pulse and $T_0 = 2\pi/\omega_0$ is the natural period of the oscillator. If the dynamic load factor D is plotted against the natural period of the oscillator or against a derived variable (e.g., the eigenfrequency), then it is labeled as a response spectrum (RS), see Fig. 2.20. This is an important graph as it allows for determining the maximum displacement without recourse to solving the underlying differential equation.

2.14.2 SDOF System Response to a Linear Pulse

In this case, an SDOF oscillator is subjected to a pulse that commences at time t_1 with magnitude f_1 and increases linearly until time t_2, reaching a terminal value of f_2, past which it disappears. We will compute the response of the SDOF oscillator for this cases using object `sdof` from class `courses.structuraldynamics` in the SDE. The loading is shown

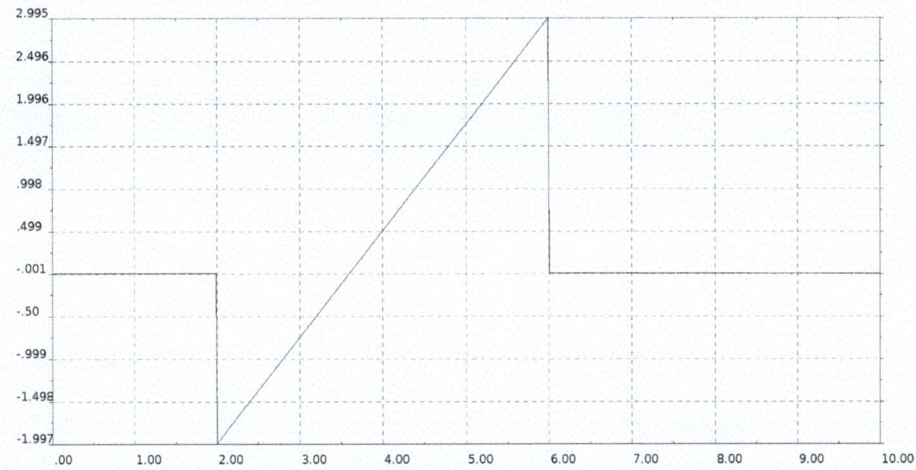

Fig. 2.21 Linear pulse input to an SDOF, see Sect. 2.14.2

in Fig. 2.21, while for the solution to the SDOF system, code of Listing 2.4 was implemented. Next, the response of the SDOF oscillator is shown in Fig. 2.22 in terms of the displacement. Regarding the time rate of change of the external work and of the energy dissipation due to damping, this is shown in Fig. 2.23 with calculations performed using code of Listing 2.5.

```
import courses.structuraldynamics.*
thePlot.clear()
t1=2.0;  f1=-2.0
t2=6.0;  f2=3.0

f={double t ->
    val=0.0d
    if(t>=t1 && t<=t2){val=f1+(f2-f1)/(t2-t1)*(t-t1)}
    return val
}
thePlot.addFunction(new plotfunction(linspace
    (0.0,10.0,1000) as double[],f as DoubleFunction))
thePlot.show()

k=4.0; m=1.0; c=1.0
theSDOF=new sdof(4.0, 1.0, 1.0)

theSDOF.setRHS(f as DoubleFunction)
theSDOF.duhamel(20.0, 0.001)

resPlot=new PlotFrame()
pf=new plotfunction(theSDOF.dt(),theSDOF.Disp())
pf.setName("xsi="+theSDOF.getCritDampRatio())
resPlot.addFunction(pf)
```

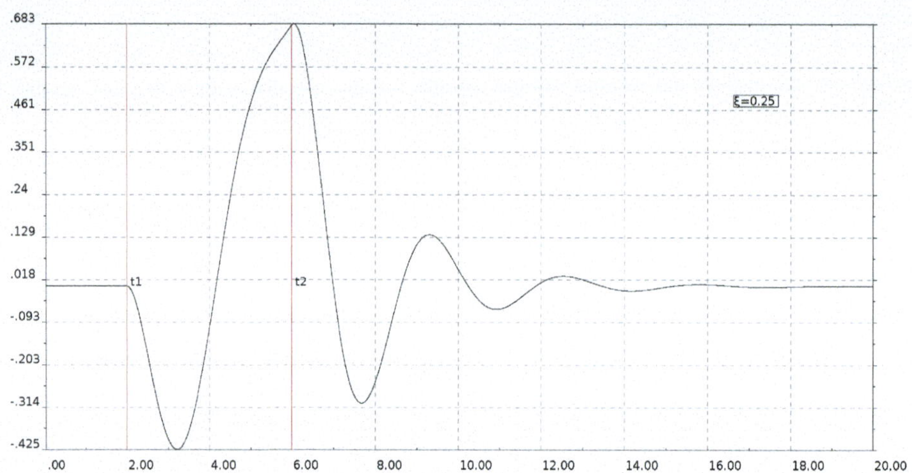

Fig. 2.22 Displacement response of the SDOF oscillator, see Sect. 2.14.2

```
24  println("xsi="+theSDOF.getCritDampRatio())
25
26  resPlot.makeLegend(true)
27  resPlot.text(t1+0.1,0.001,"t1")
28  resPlot.text(t2+0.1,0.001,"t2")
29  resPlot.vline(t1,Color.red)
30  resPlot.vline(t2,Color.red)
31  resPlot.show()
```

Listing 2.4 Solution using implementation `duhamel` for the SDOF oscillator object `sdof` under a linear pulse input and construction of diagrams

```
33  nrgPlot=new PlotFrame()
34  dissNRG= new double[theSDOF.Velc().length]
35  excfNRG= new double[theSDOF.Velc().length]
36  (0..<theSDOF.Velc().length).each{
37      dissNRG[it]=c*theSDOF.Velc()[it]*theSDOF.Velc()[
        it]
38      excfNRG[it]=f(it*theSDOF.dt())*theSDOF.Velc()[it]
39  }
40
41  pf=new plotfunction(theSDOF.dt(),dissNRG)
42  pf.setName("dissipation rate")
43  nrgPlot.addFunction(pf)
44
45  pf=new plotfunction(theSDOF.dt(),excfNRG)
46  pf.setName("external work rate")
47  nrgPlot.addFunction(pf)
48
```

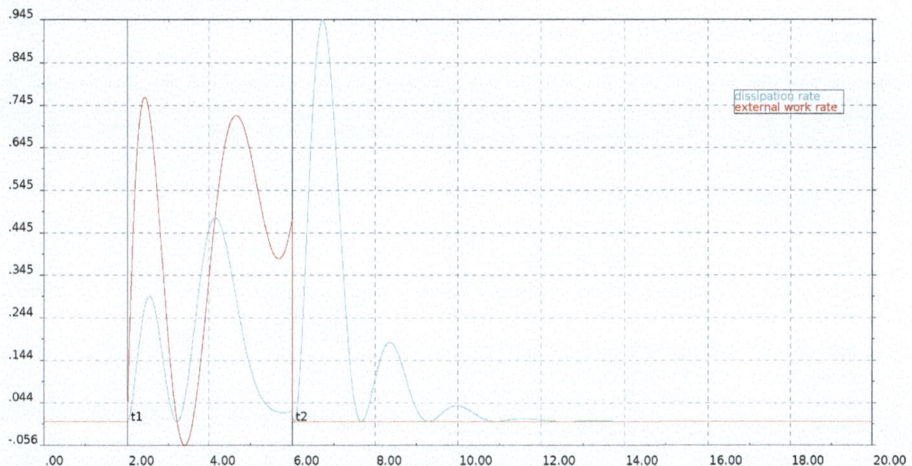

Fig. 2.23 Rate of change of the internal energy and of the external work in the SDOF system, see Sect. 2.14.2

```
49  nrgPlot.setAutoColor(true)
50  nrgPlot.makeLegend(true)
51  nrgPlot.text(t1+0.1,0.001,"t1")
52  nrgPlot.text(t2+0.1,0.001,"t2")
53  nrgPlot.vline(t1)
54  nrgPlot.vline(t2)
55  nrgPlot.show()
```

Listing 2.5 Complementary code for the creation of diagrams showing the rate of change of work in the SDOF system

2.14.3 Free Vibration Class sdof Using the Environment SDE

We start with a series of SDOF oscillators exhibiting a gradual increase in their damping ratio ξ, starting from 0.0 up to to 1.0. By using class sdof embedded in the SDE, inserting it with the help of implementation theUniverse that is based on the method putContraption, and finally solving the problem in the SDE environment, we first see the graphical environment given in Fig. 2.24. Subsequently, the response computed for a series of SDOF oscillators with the same undamped natural period of $T = 10s$ and differentiated only by their damping ratio ξ is plotted in a series of diagrams, as depicted in Fig. 2.25.

```
1  import courses.structuraldynamics.*
2  theUniverse.cls()
3  thePlot.clear()
```

```
4  Nt=2000
5  dt=0.001 as double
6  Tot=Nt*dt as double
7  (0..6).each{
8     theSDOF = new sdof(100.0,1.0,0.0, 0.1,0.0)
9     xsi=1.2*it/6
10    println("sdof= "+it+", xsi: "+xsi)
11    theSDOF.setCritDampRatio(xsi)
12    x0=(it-1)*10.0; y0=0.0
13    theSDOF.setOrigin(x0,y0)
14    theSDOF.elen=20
15    theSDOF.solve(Tot, dt)
16    theUniverse.putContraption(theSDOF)
17
18    pf = new plotfunction(dt, theSDOF.Disp())
19    pf.setName("xsi="+xsi)
20    thePlot.addFunction(pf)
21 }
22 thePlot.setAutoColor(true)
23 thePlot.makeLegend(true)
24 thePlot.show()
25
26 scale=100.0
27 theGP.plotDeform(scale, 0,Nt,1)
```

Listing 2.6 Code for computation of an SDOF oscillator response with the graphical representations

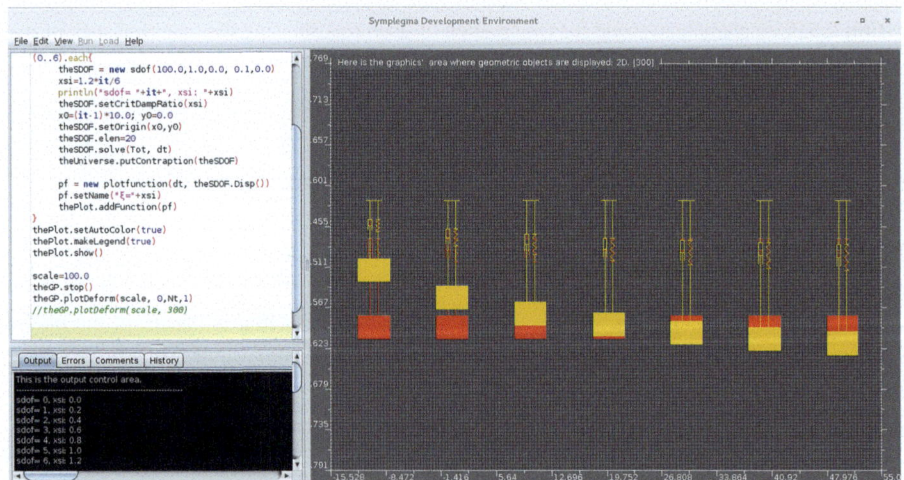

Fig. 2.24 Solution window plus graphical representation for a series of SDOF oscillators, see Sect. 2.14.3

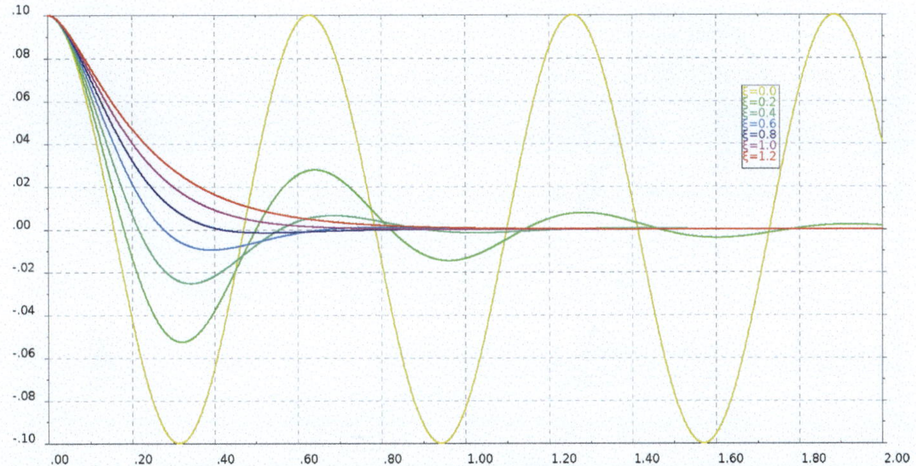

Fig. 2.25 Response of a series of SDOF oscillators with fixed undamped natural period of $T = 10s$ and generated by using different damping ratios ξ, see Sect. 2.14.3

2.14.4 Response Spectrum for a Triangular Pulse Input

Although this task could be done analytically by superimposing three time regimes (linearly increasing load, linearly decreasing load and free vibrations), we will produce a solution by using class `sdof` that numerically solves the closed form solution to the differential equation of motion, e.g., `duhamel`.

The triangular load is shown in Fig. 2.26, defined for the following input data: Load magnitude $f_0 = 96.6$ N and half-time duration $t_1 = 0.025$ s. The mass and stiffness of the SDOF oscillator are $m = 3$ kg and $k = 2700$ N/m, respectively. The solution is parametric in the damping coefficient ξ so as to observe how motion dampens out as damping increases. The pertinent results are shown in Fig. 2.27.

Fig. 2.26 External triangular pulse, see Sect. 2.14.4

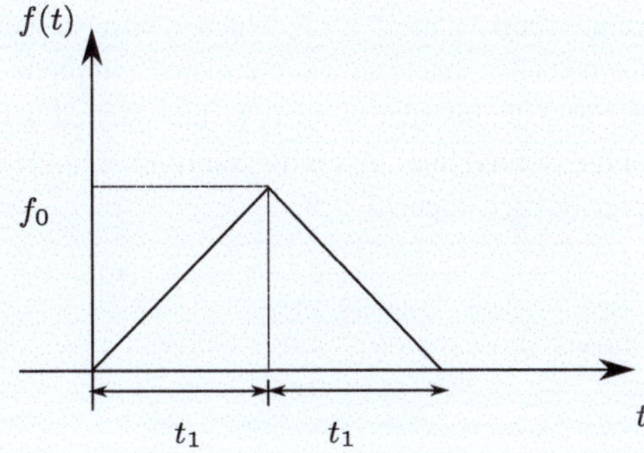

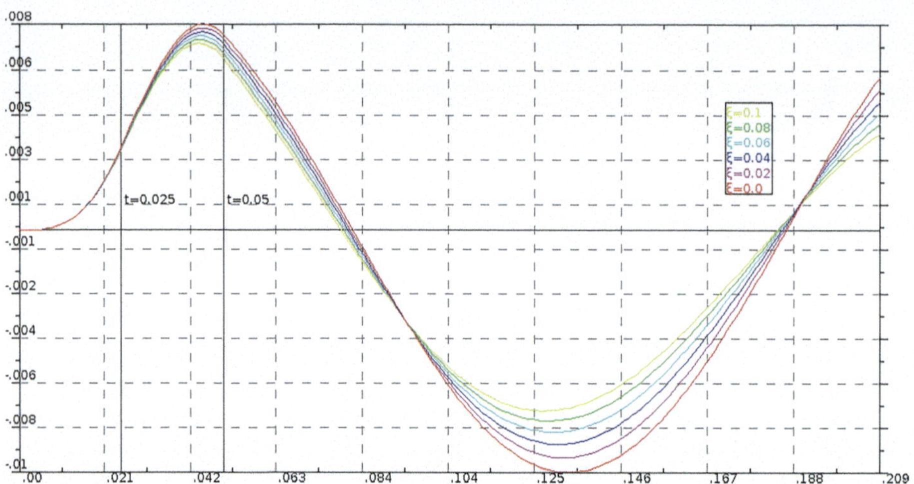

Fig. 2.27 SDOF oscillator displacement to a triangular pulse for different damping ratios ξ, see Sect. 2.14.4

Specifically, the ratio of the time duration of the increasing branch to the decreasing branch of the external load, as compared to the natural period of the undamped SDOF oscillator, is $t_1/T \approx 0.12$. Furthermore, the ratio of the maximum dynamic displacement u_{max} to the equivalent static displacement ($u_{st} = f_0/k$) is approximately $u_{max}/u_{st} \approx 0.715$. When damping ($\xi = 2 - 10\%$) is taken into account in the computations, the ratio u_{max}/u_{st} assumes values in the range 0.694 to 0.617.

We next consider the construction of a response spectrum by numerically solving the equation of motion for the undamped oscillator. At first, we define the dynamic load factor D in the time interval $t_1/T \leq q = 2.0$. In terms of the SDOF system stiffness, this inequality results in $k \leq \frac{4\pi^2 q^2 m}{t_1^2}$. Thus, we compute the SDOF response for a stiffness range $[0, \frac{4\pi^2 q^2 m}{t_1^2}]$, and for every stiffness value we pick the largest absolute value of the dynamic displacement u_{max} normalized by its equivalent static value u_{st}. This procedure generates a string of values for the pair $D, t_1/T$, which in turn is used to plot the response spectrum. Specifically, the response spectrum was computed and plotted by using code of Listing 2.7 and is shown in Fig. 2.28. In order to verify results derived in the first part of this Section, the intersection of the two red lines in this diagram correspond to the pair $D = 0.715, t1/T = 0.12$ that was previously computed.

```
import courses.structuraldynamics.*
thePlot.clear()
tx=[]; Dv=[]
t1=0.025; f0=96.9; m=3.0
f={double t ->
    val=0.0d
```

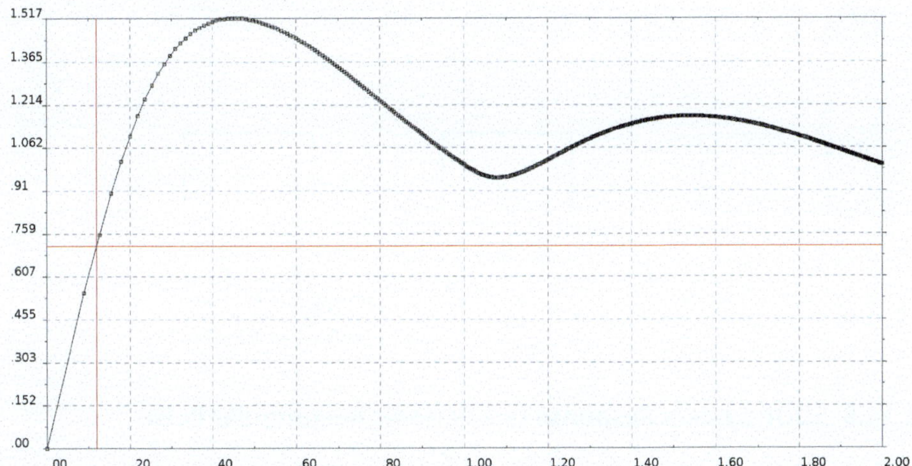

Fig. 2.28 Response spectrum for a triangular pulse load. Note that the ratio $\frac{t_0}{T}$ is the abscissa and the dynamic load factor D is the ordinate, see Sect. 2.14.4

```
 7      if(t<=t1){val=1.0d*t/t1}else if(t<=2*t1){val=1.0d
       -(t-t1)/t1}
 8      return val*f0
 9  }
10
11  Nm=500
12  (0..Nm).each{
13      q=2.0 // max t1/T
14      k=(q*q*4*PI*PI*m/(t1*t1))*it/Nm
15      theSDOF=new sdof(k, m)
16
17      theSDOF.setRHS(f as DoubleFunction)
18      theSDOF.duhamel(5*t1, t1/100.0)
19      tx.add(t1/theSDOF.getNaturalPeriod())
20      dval=max(theSDOF.maxDisp(),abs(theSDOF.minDisp())
       )/(f0/k)
21      Dv.add(dval)
22  }
23
24  thePlot=new PlotFrame()
25  thePlot.addFunction(new plotfunction(tx,Dv))
26  thePlot.setMarker(true)
27  thePlot.vline(0.12,Color.red)
28  thePlot.hline(0.715,Color.red)
29  thePlot.show()
```

Listing 2.7 Solution method `duhamel` for the SDOF oscillator class `sdof` to a traingular pulse load including computation of the resposne spectrum and the plotting of graphs

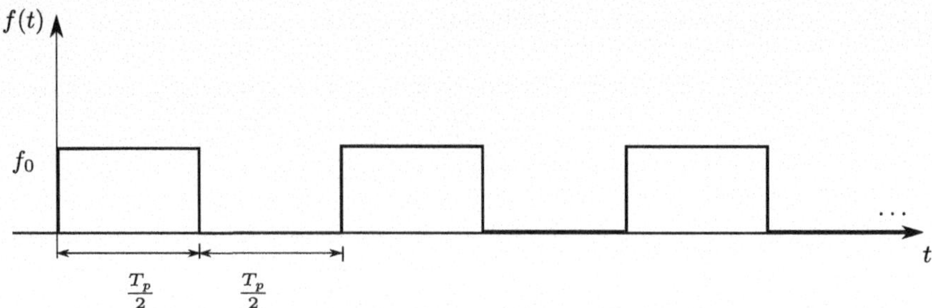

Fig. 2.29 Periodic orthogonal pulse load, see Sect. 2.14.5

2.14.5 SDOF System Response to a Periodic Rectangular Pulse

This periodic pulse load is shown in Fig. 2.29 and its time variation is described by the following equation:

$$f(t) = \begin{cases} f_0 & \text{for } 0 \le t \le (T_p/2)^- \\ 0 & \text{for } (T_p/2)^+ \le t \le T_p \end{cases}$$

This type of load will be expanded as a time-harmonic Fourier series with coefficients computed according to Eq. (2.35), starting with the zeroth term a_0,

$$a_0 = \frac{1}{T_p} \int_0^{T_p} f(t)\, dt = \frac{1}{T_p} \int_0^{T_p/2} f(t)\, dt + \frac{1}{T_p} \int_{T_p/2}^{T_p} f(t)\, dt$$

$$= \frac{1}{T_p} \int_0^{T_p/2} f_0\, dt = \frac{f_0}{2}.$$

Next, the coefficients multiplying the cosine function expansion terms, a_n, are computed as follows:

$$a_n = \frac{2}{T_p} \int_0^{T_p} f(t) \cos n\omega_p t\, dt = \frac{2}{T_p} \int_0^{T_p/2} f_0 \cos n\omega_p t\, dt = \frac{2}{T_p} f_0 \left[\frac{\sin n\omega_p t}{n\omega_p} \right]_0^{T_p/2}$$

$$= \frac{2}{T_p} \frac{f_0}{n\omega_p} \left(\sin\left(n\omega_p \frac{T_p}{2} \right) - \sin 0 \right) = \frac{f_0}{n\pi} \sin(n\pi) = 0.$$

Finally, the coefficients b_n for the sine expansion terms are

$$\beta_n = \frac{2}{T_p} \int_0^{T_p} f(t) \sin n\omega_p t \, dt = \frac{2}{T_p} \int_0^{T_p/2} f_0 \sin n\omega_p t \, dt = \frac{2}{T_p} f_0 \left[-\frac{\cos n\omega_p t}{n\omega_p} \right]_0^{T_p/2}$$

$$= \frac{2}{T_p} \frac{f_0}{n\omega_p} \left(-\cos\left(n\omega_p \frac{T_p}{2} \right) + \cos 0 \right) = \frac{f_0}{n\pi} \left(1 - \cos(n\pi) \right)$$

$$= \begin{cases} \frac{2f_0}{n\pi} & \text{for } n \text{ odd,} \\ 0 & \text{for } n \text{ even} \end{cases}$$

By introducing these Fourier coefficients, the periodic function $f(t)$ can be expanded using Eq. (2.34) as follows:

$$f(t) = \frac{f_0}{2} + \sum_{n=1,3,5,\ldots}^{\infty} \frac{2f_0}{n\pi} \sin(n\omega_p t)$$

We now use the platform SDE to construct and plot the above periodic function for values of $f_0 = 10$ and $T_p = 1.0$, as the number of expansion terms increases. More specifically, Fig. 2.30 shows the representation for 1, 2, 4 and 8 terms, while in Fig. 2.31, 100 terms were employed. Specifically, these coefficients were computed and the function representations plotted by using code of Listing 2.8. Note the manifestation of Gibb's phenomenon in these expansions at the corners of the function, which are points with slope discontinuity.

```
1  Tp=1.0 // period
2  Nt=1000 // discrete steps of time for computation
3  dt=Tp/(Nt-1) // time step
4
5  f0=10.0
6  omp=2.0*PI/Tp
7
8  nPeriods=10 // number of periods to be calculated and
       plotted
9  TimeHistoryPlot = new PlotFrame()
10 TimeHistoryPlot.setTitle("SDE Figure: Fourier series"
      )
11
12 Niter=10 // keep it small
13 Nterms=1
14 (1..Niter).each{
15    f=[]
16    for(int q=0;q<nPeriods;q++) {
17       for(int k=0;k<Nt;k++) {
18          if(q==0) {
19             val=f0/2.0
20             for(int n=1;n<=Nterms;n=n+2) {
21                val+=2.0*f0*sin(n*omp*k*dt)/(n*PI)
22             }
```

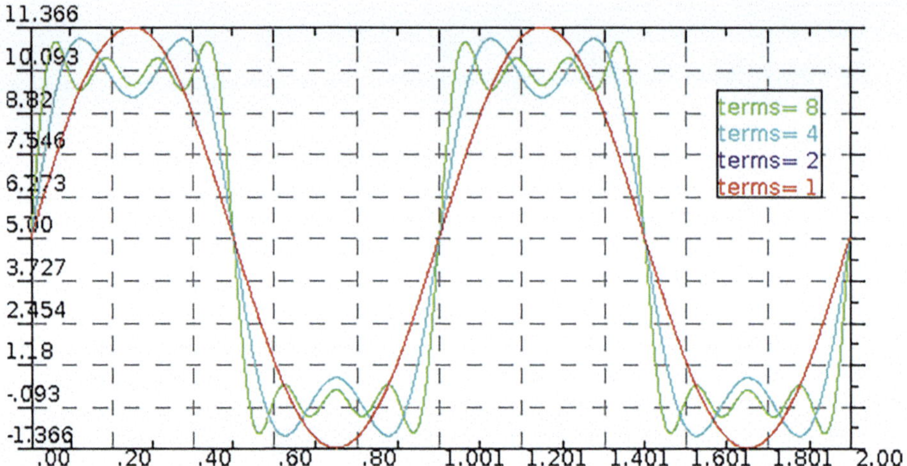

Fig. 2.30 Fourier series approximation of an external periodic forcing function $f(t)$ employing 1, 2, 4 and 8 terms, see Sect. 2.14.5

```
23          f.add(val)
24        }else{
25          f.add(f[k])
26        }
27      }
28    }
29    pf = new plotfunction(dt,f as double[])
30    pf.setName("terms=  "+Nterms)
31    TimeHistoryPlot.addFunction(pf)
32    Nterms=Nterms*2
33  }
34  TimeHistoryPlot.setAutoColor(true)
35  TimeHistoryPlot.makeLegend(true)
36  TimeHistoryPlot.show()
```

Listing 2.8 Reconstruction of a periodic orthogonal function as a series summation of harmonic terms

Consider next an SDOF system with natural period T_0, stiffness k and damping ratio ξ (as percent of critical damping). The time history of the SDOF displacement response, once the Fourier expansion coefficients of the external forcing function have been evaluated, can be computed with reference to Eq. (2.39) as follows:

$$u(t) = \frac{f_0}{2k} + \sum_{n=1,3,5,\ldots}^{\infty} \frac{2f_0}{n\pi k} \frac{(1-\beta_n^2)\sin(n\omega_p t) - 2\xi\beta_n \cos(n\omega_p t)}{(1-\beta_n^2)^2 + (2\xi\beta_n)^2}.$$

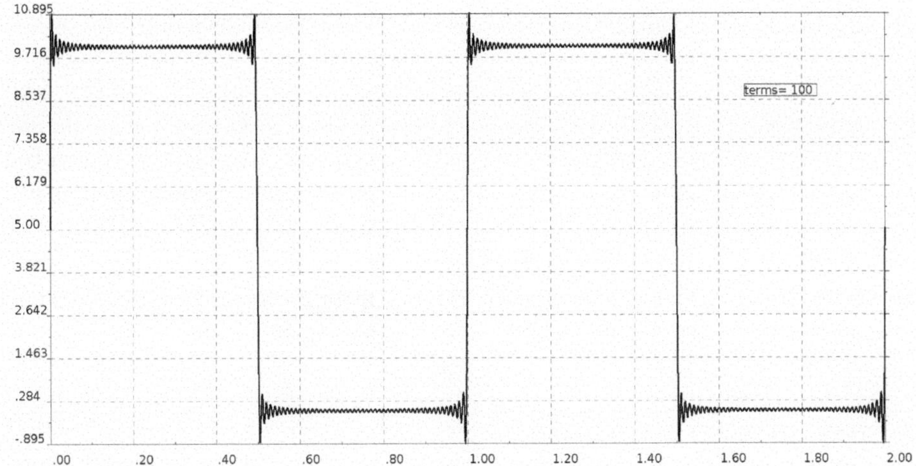

Fig. 2.31 Fourier series representation of an orthogonal periodic forcing function $f(t)$ using 100 terms, see Sect. 2.14.5

2.14.6 SDOF System Response to an Arbitrary External Load Using the Fourier Transform

Consider an SDOF oscillator with material constants k, m, c subjected to an external forcing function $f(t)$ and compute its response using the Fourier transform. This procedure can materialize in the following three steps:

(a) Numerical evaluation of the complex-valued function $\mathcal{F}[f(t)] = F(\omega)$, see Eq. (2.50), using the direct FFT.
(b) For a discrete spectrum of frequency values ω_i, the displacement response is compute by the complex frequency response function of Eq. (2.52) as $U(\omega_i) = F(\omega_i)H(\omega_i)$.
(c) By using the pairs $(\omega_i, U(\omega_i))$ in conjunction with Eq. (2.54) for the inverse FFT, the SDOF transient displacement $u(t)$ is evaluated at a sequence of discrete time points.

The example that follows concerns an SDOF oscillator with stiffness $k = 1000$ (in kN/m), mass $m = 1.0$ (in tn) and damping ratio $\xi = 0.1$. The external load (in kN) is a superposition of four harmonics as

$$f(t) = \sum_{n=1}^{4} a_n \sin(\omega_n t)$$

The pairs (a_i, ω_i) for $i = 1, 2, 3, 4$, are assigned the following numerical values: (10.0, $1.10\omega_0$), (100, $8.0\omega_0$), (15.0, $12.1\omega_0$) and (22.0, $17.0\omega_0$). The solution routine followed is a three-step sequence starting from Listing 2.9 and moving to Listing 2.10 followed by

Listing 2.11 that will be explained in what follows. Note that the solution can be executed as either one entity at a time or as a continuous sequence. Starting with Listing 2.9, the SDOF system is described as an object `sdof` of class `courses.structuraldynamics`. Next, the external load $f(t)$ is specified as a functional object (closure) in Groovy. Finally, the time step dt, the total time interval of interest Tot and the number of time steps Nt are prescribed. It is highly desirable that Nt is a power of two, since the efficiency of computations in FFT requires that the total set of digitized points in a signal must have the structure (2^n). If the incoming signal does not fulfill this requirement, then the addition of trailing zeros will restore this structure. Note that the frequency of sampling of a time signal (sampling rate) is the inverse of the time step $Fs = 1/dt$.

```groovy
import courses.structuraldynamics.*

k=100.0; m=1.0; xsi=0.1
theSDOF = new sdof(k,m)
theSDOF.setCritDampRatio(xsi)

om_0=theSDOF.getNaturalFrequency(); println ("om_0= "
    +om_0)
om_ext_1=1.1*om_0; a_1=10.0; println ("om_ext_1= "+
    om_ext_1)
om_ext_2=8.0*om_0; a_2=100.0; println ("om_ext_2= "+
    om_ext_2)
om_ext_3=12.1*om_0; a_3=15.0; println ("om_ext_1= "+
    om_ext_1)
om_ext_4=17.0*om_0; a_4=22.0; println ("om_ext_2= "+
    om_ext_2)

f={
   (a_1*sin(om_ext_1*it)+a_2*sin(om_ext_2*it)
    +a_3*sin(om_ext_3*it)+a_4*sin(om_ext_4*it))
}

Nt=2**14-1
dt=0.001d
Tot=Nt*dt
```

Listing 2.9 Specification of the SDOF system properties, of the forcing function and of the time discretization

Subsequently, Listing 2.10 performs the direct Fourier $F(\omega)$ transformation of $f(t)$ using module FFT, thus initiating the second step mentioned above (see Fig. 2.32). Following that, we activate object `TrasferFunction` in class `sdof` that provides the transfer function $H(\omega)$ at any given frequency value ω (see Fig. 2.33). Thus, the SDOF response is computed in the frequency domain as the product $F(\omega)H(\omega)$ (see Fig. 2.34), which completes the second step.

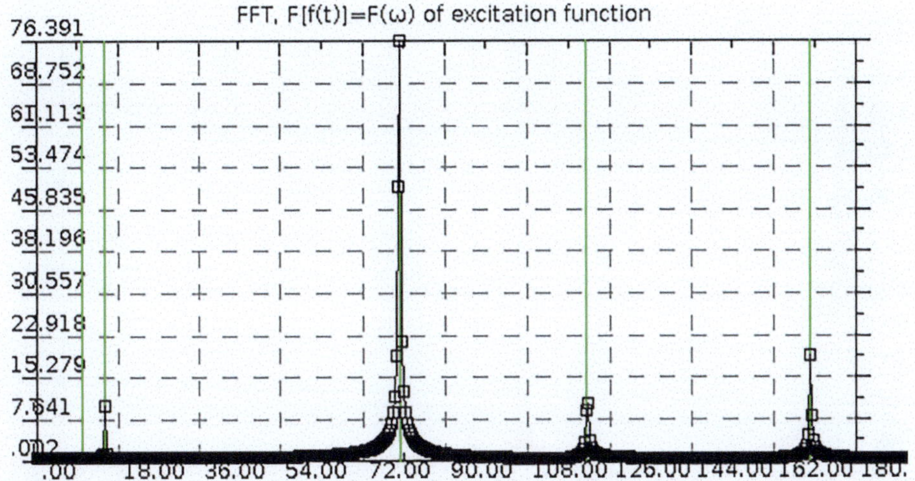

Fig. 2.32 Fourier transformation of time function $f(t)$ showing the absolute value of the complex-valued transform $F(\omega)$. *Note* The vertical lines identify the natural frequency ω_0 of the SDOF oscillator plus the forcing function frequencies ω_1 through ω_4 (in green color)

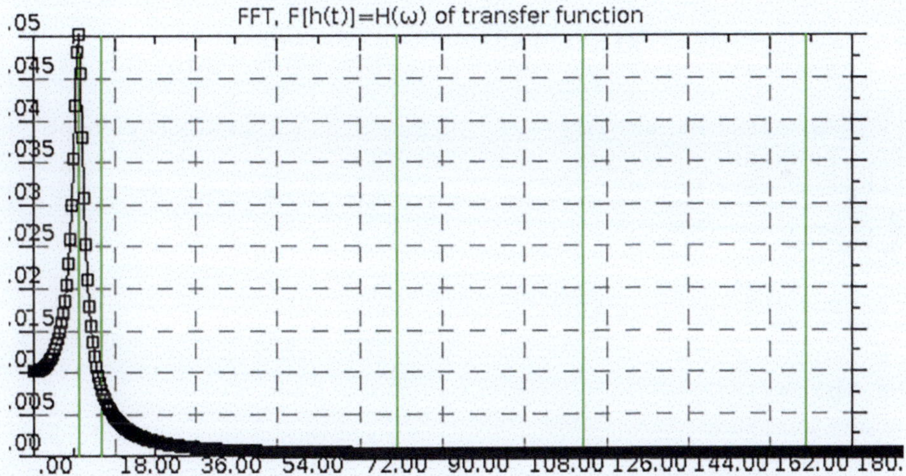

Fig. 2.33 The complex-valued transfer function $H(\omega)$ of the SDOF oscillator in the frequency domain

```
22  FFT_f_Plot= new  PlotFrame()
23  FFT_h_Plot= new  PlotFrame()
24  FFT_u_Plot= new  PlotFrame()
25  fext=[]
```

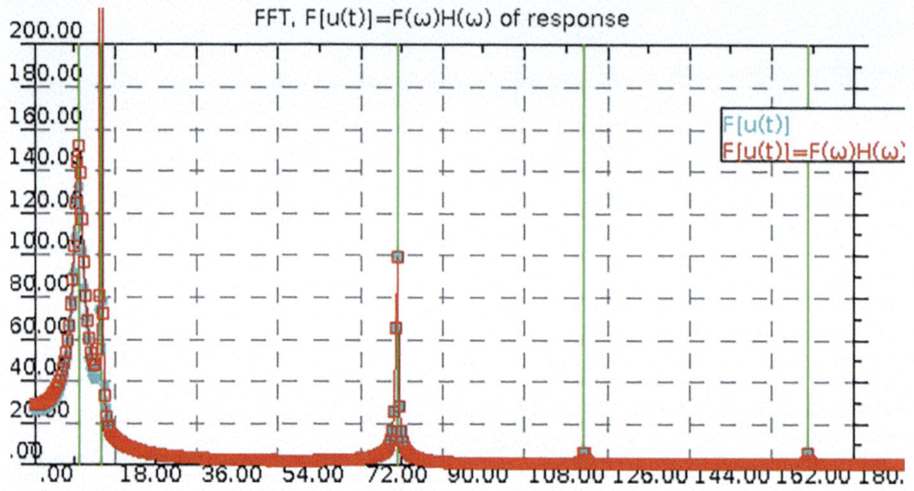

Fig. 2.34 The frequency response for an SDOF system $U(\omega)$ computed as the product $F(\omega)H(\omega)$ in the frequency domain. *Note* The vertical lines identify the natural frequency ω_0 of the SDOF oscillator plus the forcing function frequencies ω_1 through ω_4 (in green color)

```
26  (0..Nt).each{fext.add(f(it*dt))}
27
28  fft= new FastFourierTransformer(DftNormalization.
       STANDARD)
29
30  F_f = fft.transform(fext as double[], TransformType.
       FORWARD)
31  Lp=F_f.size()
32
33  freq=[]
34  freqplt=[]
35  F_fplt=[]
36
37  F_h=[]
38  F_hplt=[]
39  F_u=[]
40  F_uplt=[]
41
42  Fs=1.0/dt
43  (0..(Lp)/2).each{
44    freqplt.add(2.0*PI*it*Fs/Lp)
45    F_fplt.add(F_f[it].abs()/(Lp/2.0))
46
47    freq.add(freqplt[it])
48
49    F_h.add(theSDOF.TransferFunction(freqplt[it]))
```

```
50    F_hplt.add(F_h[it].abs())
51
52    F_u.add(F_h[it]*F_f[it])
53    F_uplt.add(F_u[it].abs())
54 }
55
56 (2..(Lp)/2).each{
57    freq.add(-freqplt[-it])
58    F_h.add(theSDOF.TransferFunction(-freqplt[-it]))
59    F_u.add(theSDOF.TransferFunction(-freqplt[-it])*F_f
       [-it])
60 }
61
62 FFT_f_Plot.addFunction(new plotfunction(freqplt as
       double[], F_fplt))
63 FFT_f_Plot.setMarker(true)
64 FFT_f_Plot.Title("FFT, F[f(t)]=F(omega) of excitation
       function")
65 FFT_f_Plot.show()
66
67 FFT_h_Plot.addFunction(new plotfunction(freqplt as
       double[], F_hplt))
68 FFT_h_Plot.setMarker(true)
69 FFT_h_Plot.Title("FFT, F[h(t)]=H(omega) of transfer
       function")
70 FFT_h_Plot.show()
71
72 pf=new plotfunction(freqplt as double[], F_uplt)
73 pf.setName("F[u(t)]=F(omega)H(omega)")
74 FFT_u_Plot.addFunction(pf)
75 FFT_u_Plot.Title("FFT, F[u(t)]=F(omega)H(omega) of
       response")
76 FFT_u_Plot.show()
```

Listing 2.10 Discrete transformation of function $f(t)$ through FFT followed by computation of the response as $U(\omega) = F(\omega)H(\omega)$ and plotting

Following the recovery of the SDOF solution $U(\omega)$ in the frequency domain as the product $F(\omega)H(\omega)$, we can now complete the third step and reconstitute the response in the time domain using the inverse transformation Fourier. This last step materializes in Code [2.11].

```
78 u = fft.transform(F_u as Complex[], TransformType.
       INVERSE)
79 ResponsePlot= new PlotFrame()
80 uplt=[]
81 (0..<u.length).each{
82    uplt.add(2.0*u[it].re())
83 }
84 pf=new plotfunction(dt,uplt as double[])
```

```
85  pf.setName("IFFT[U(omega)]=u(t)")
86  ResponsePlot.addFunction(pf)
87
88  theSDOF.setRHS(f as DoubleFunction)
89  theSDOF.duhamel(Tot,dt)
90
91  pf=new plotfunction(dt,theSDOF.Disp())
92  pf.setName("Duhamel u(t)")
93  ResponsePlot.addFunction(pf)
94  ResponsePlot.setMarker(true)
95  ResponsePlot.Title("IFFT, F^(-1)[U(omega)]=u(t) of
        response")
96  ResponsePlot.setAutoColor(true)
97  ResponsePlot.makeLegend(true)
98  ResponsePlot.hline(0.0)
99  ResponsePlot.show()
```

Listing 2.11 Inverse Fourier transformation of the response $U(\omega)$ that recovers $u(t)$ and subsequent plotting

At the same time, for comparison reasons and for result validation, the response is computed from a numerical evaluation of the SDOF closed form solution given as a time convolution integral using object Duhamel, see Listing 2.12. These results appear in Fig. 2.35.

```
101  F_un = fft.transform(theSDOF.Disp(), TransformType.
         FORWARD)
102  Lp=F_un.size()
103
104  F_unplt=[]
105
106  Fs=1.0/dt
107  (0..(Lp)/2).each{
108     F_unplt.add(F_un[it].abs())
109  }
110
111  pf=new plotfunction(freqplt as double[], F_unplt)
112  pf.setMarkerStyle(1)
113  pf.setMarkerFill(true)
114  pf.setName("F[u(t)]")
115  FFT_u_Plot.addFunction(pf)
116  FFT_u_Plot.setMarker(true)
117  FFT_u_Plot.setAutoColor(true)
118  FFT_u_Plot.makeLegend(true)
119  FFT_u_Plot.show()
```

Listing 2.12 Direct Fourier transformation on solution $u(t)$ that is numerically evaluated from the Duhamel integral, thus yielding the response $U(\omega)$. Note: Concurrently plotted is the frequency domain solution $F(\omega)H(\omega)$ for comparison purposes

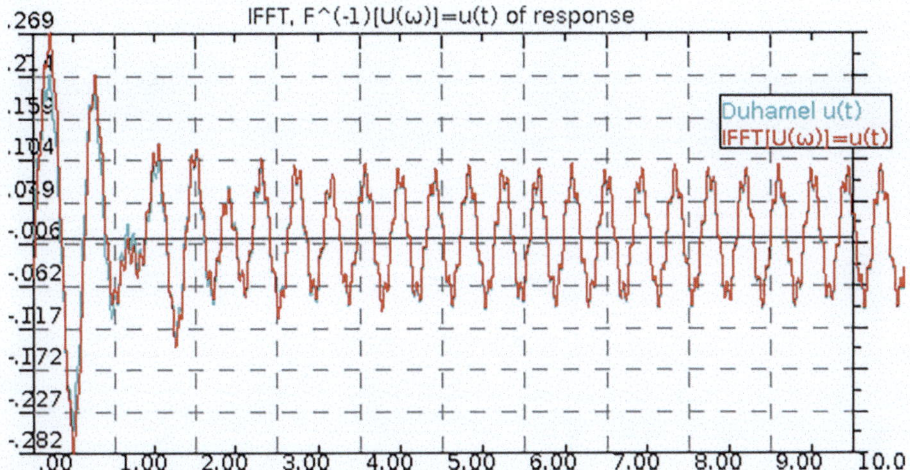

Fig. 2.35 SDOF response in the time domain using the inverse Fourier transformation of the solution $U(\omega) = F(\omega)H(\omega)$ in the frequency domain. Concurrently plotted is the solution derived from a numerical implementation of the Duhamel integral

2.14.7 The β-Newmark Method

In this Section, the numerical implementation of the method β-Newmark will be presented, based on the steps previously outlined in Table 2.5. The application example is for an SDOF system with material properties (namely mass, stiffness and damping) defined in Listing 2.13. The method is essentially a linear acceleration algorithm, implemented here for a gradually increasing time step that is defined with respect to the critical time step given by Eq. (2.59), and ranging from $0.1\Delta t_{cr}$ to $1.01\Delta t_{cr}$. More specifically, in Fig. 2.36 the left column plots the displacement $u(t)$ versus time t diagram, the middle column plots the velocity $v(t)$ versus time t diagram, while the right column is a phase diagram plotting displacement versus velocity, i.e., $u(t)$ versus $v(t)$. The time step values Δt used in the numerical integration are listed from top to bottom as ratios 0.1, 0.2, 0.4, 0.8, 0.99, 1.0 and 1.01 of the critical time step Δt_{cr}, respectively. It is observed that up until the time step value of $\Delta t \leq \Delta t_{cr}$ is reached, the solution is stable, meaning no spurious energy is added to the SDOF system. Of course, accuracy in the results drops as the time step Δt value increases past its critical value.

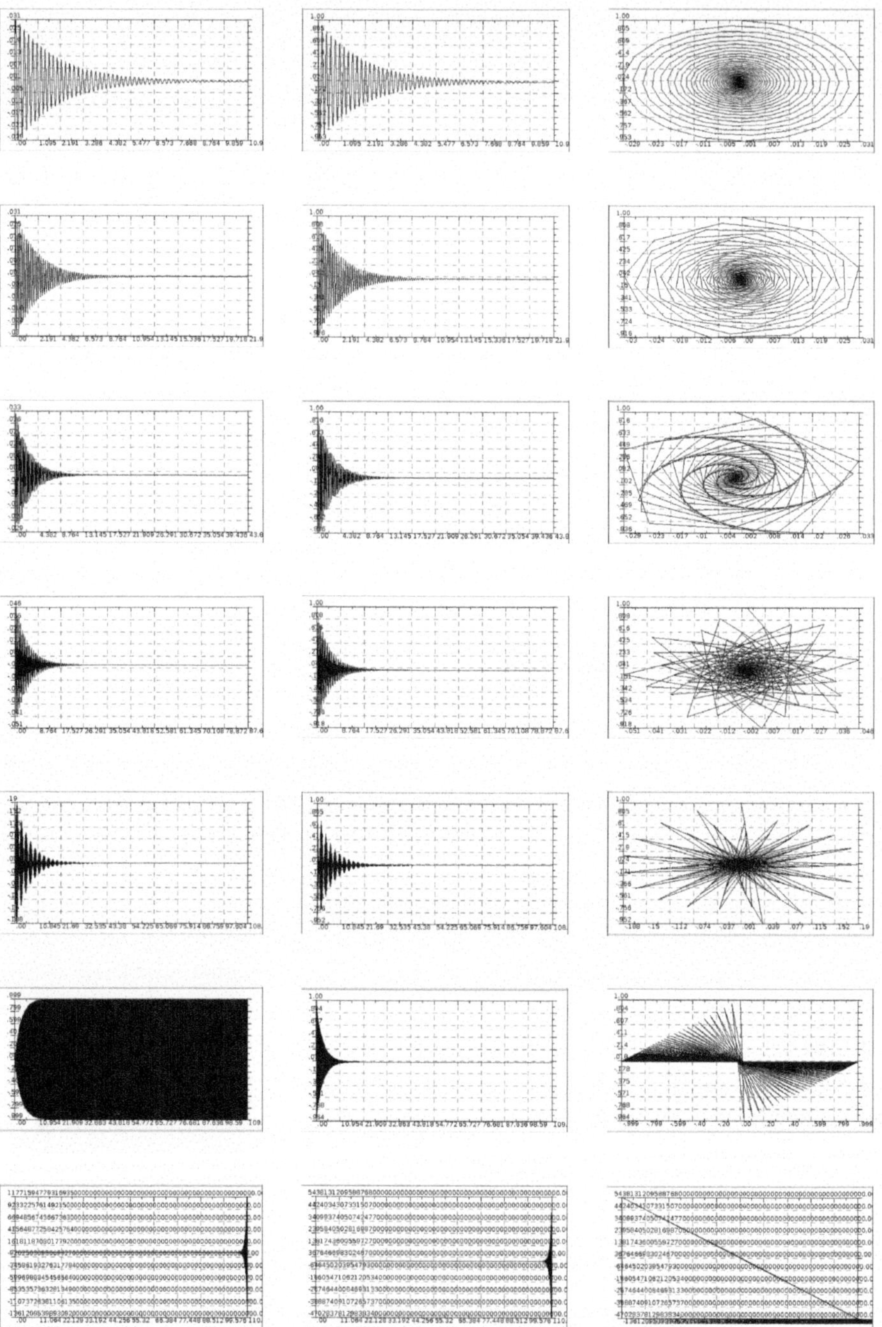

Fig. 2.36 Left column: $u(t) - t$; middle column $v(t) - t$; right column $u(t) - v(t)$. Moving from top to bottom are time step values of $\Delta t = (0.1, 0.2, 0.4, 0.8, 0.99, 1.0, 1.01)$ as a percentage of Δt_{cr}

```
k=1000.0; m=1.0; c=1.0
u0=0.0; v0=1.0
// external loading could be any function of time
// e.g.: f={sin(2.0*sqrt(k/m)*it)}
// however here we assume free vibration, therefore:
f={0.0}

g=1.0/2.0 // gamma
b=1.0/6.0 // beta

om=sqrt(k/m); xsi=c/(2*m*om)
dtcr=1/om*( xsi*(g-0.5)+sqrt(0.5*g-b+(xsi*xsi*(g-0.5)*(g
    -0.5))) )/(0.5*g-b)

dt=1.0*dtcr
N=1000

u=new double[N+1]; u[0]=u0
v=new double[N+1]; v[0]=v0
a=new double[N+1]; a[0]=(f(0)-c*v[0]-k*u[0])/m

//beta-Newmark coefficients
w0=1/(b*dt*dt); w1=g/(b*dt);        w2=1/(b*dt);   w3=1.0/(2*b
    )-1.0
w4=g/b-1.0;    w5=(g/b-2.0)*dt/2.0;  w6=dt*(1.0-g);  w7=dt*g

// K_effective
kef=k+w0*m+w1*c

// iterative computations
(1..N).each{
  // effective load
  fef=f(it*dt)+m*(w0*u[it-1]+w2*v[it-1]+w3*a[it-1])+c*(w1*u
    [it-1]+w4*v[it-1]+w5*a[it-1])

  // solve for current displacement
  u[it]=fef/kef

  // current acceleration and velocity
  a[it]=w0*(u[it]-u[it-1])-w2*v[it-1]-w3*a[it-1]
  v[it]=v[it-1]+w6*a[it-1]+w7*a[it]
}

x=u; y=v // x,y could take the values of dt, u, v
respPlot= new PlotFrame()
respPlot.addFunction(new plotfunction(x,y))
respPlot.setMarker(false)
respPlot.show()
```

Listing 2.13 Numerical implementation of the β-Newmark method

2.14.8 SDOF Oscillator with a Non-linear Spring

A more general form of the the elastic-perfectly plastic spring model discussed previously in reference to nonlinear SDOF systems is the spring exhibiting a post-elastic behavior capable of absorbing further forces, a property known as work hardening. The force-displacement diagram of such non-linear springs is shown in Fig. 2.37.

In this example we will use the beta-Newmark written in incremental form. Specifically, the active stiffness at any time step n is given as

$$\hat{k}_n = k_n + \frac{\gamma}{\beta \Delta t} c + \frac{1}{\beta \Delta t^2} m \tag{2.83}$$

At the same time, the equivalent forcing function acting on the SDOF oscillator is

$$\Delta \hat{f}_n = f_{n+1} - f_n + \left(\Delta t \left(\frac{\gamma}{2\beta} - 1 \right) c + \frac{1}{2\beta} m \right) \ddot{u}_n$$

$$+ \left(\frac{\gamma}{\beta} c + \frac{1}{\beta \Delta t} m \right) \dot{u}_n \tag{2.84}$$

The resulting equation in incremental form is

$$\hat{k}_n \Delta u_n = \Delta \hat{f}_n \tag{2.85}$$

Fig. 2.37 Material law for a nonlinear spring plotting the restoring force versus displacement. Note: The model is bilinear with initial slope k in the elastic regime and slope k_h after first yield is exceeded

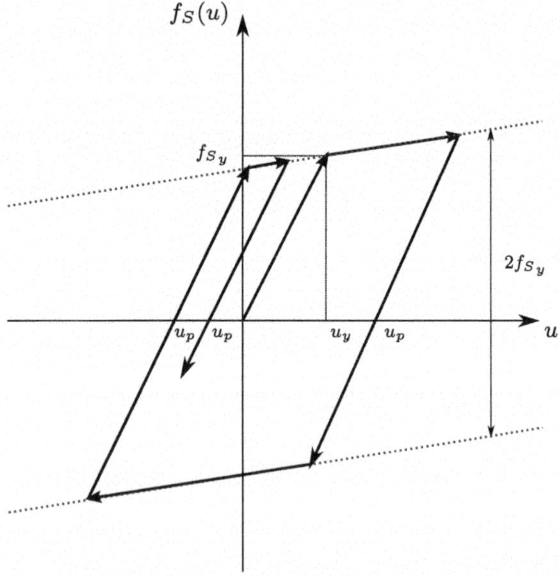

from which the increment in the displacement can be computed as Δu_n. The corresponding velocity increment is

$$\Delta \dot{u}_n = \frac{\gamma}{\beta \Delta t} \Delta u_n - \frac{\gamma}{\beta} \dot{u}_n + \Delta t \left(1 - \frac{\gamma}{2\beta}\right) \ddot{u}_n \tag{2.86}$$

Finally, Eq. (2.74) defined for time step n can be used to compute the acceleration increment $\Delta \ddot{u}$ as follows:

$$\Delta \ddot{u}_n = \frac{\Delta f_n}{m} - \frac{c}{m} \Delta \dot{u}_n - \frac{k_n}{m} \Delta u_n. \tag{2.87}$$

Following solution during a time increment Δt_n, it becomes necessary to check possible yielding in the spring and whether or not the SDOF system has entered the unloading path. These checks are necessary for evaluating the current value of the spring stiffness k_{ep}, which falls between the initial stiffness k and the work hardened stiffness k_h. This necessitates computation of the level of yielding from equation

$$f^{yield}(u_{n+1}, \dot{u}_{n+1}) = k_h u_{n+1} + (k - k_h)u_y \text{sgn}(\dot{u}_{n+1}) \tag{2.88}$$

followed by a check of three possibilities.
We first start with the plastic flow check:

(a) If $k_{ep} = k$ and $\dot{u}_{n+1} > 0$ plus $k_{ep}(u_{n+1} - u_p) > f^{yield}$, the SDOF oscillator has crossed for the first time the plastic flow level moving in the positive direction. We now set $k_{ep} = k_h$ and $u_p = (1 - k/k_h)u_y$.
(b) If $k_{ep} = k$ and $\dot{u}_{n+1} < 0$ plus $k_{ep}(u_{n+1} - u_p) < f^{yield}$, the SDOF oscillator has crossed for the first time the plastic flow level moving in the negative direction. We now set $k_{ep} = k_h$ and $u_p = (k/k_h - 1)u_y$.

We next check for the unloading path:

(c) If $k = k_h$ and $\dot{u}_{n+1}\dot{u}_n < 0$, we set $k_{ep} = k$ and $u_p = u_{n+1} - (k_h/k - 1)(u_{n+1} - u_p)$.

The restoring force in the spring is now equal to $f_{S_{n+1}} = k_{ep}(u_{n+1} - u_p)$.

In what follows, we present results for a nonlinear SDOF oscillator with a work-hardening spring ($k = 1000, k_{ep} = 250$ (kN/m)), a mass of $m = 10(tn)$ and a damper $c = 1$ (kN s/m) that is subjected to a harmonic force of the type $f(t) = 100sin(10t)$ (kN). Details of the solution steps can be found in Listing 2.14 given below. Also, the numerical results produced for this example are shown in Figs. 2.38, 2.39 and 2.40.

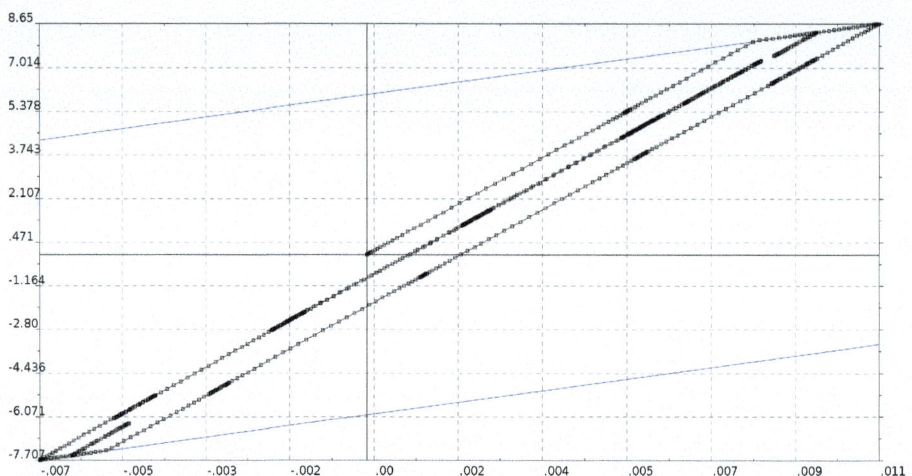

Fig. 2.38 Restoring force in the SDOF oscillator with work hardening as a function of the displacement for a harmonic input $f = 100sin(10t)$. *Note* The two parallel lines (in blue color) are the upper and lower plastic flow lines

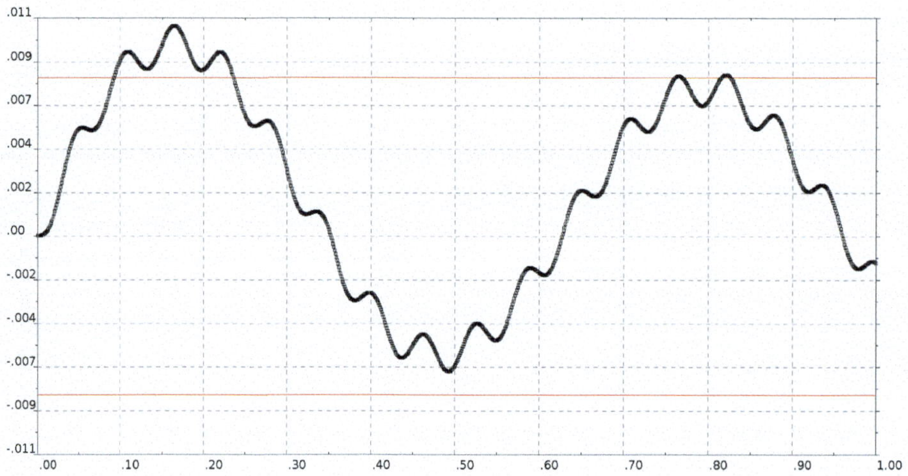

Fig. 2.39 Displacement versus time for the elastoplastic SDOF oscillator with kinematic hardening for a harmonic input $f = 100sin(10t)$. *Note* The two parallel lines (in red color) cross the yield displacement u_y

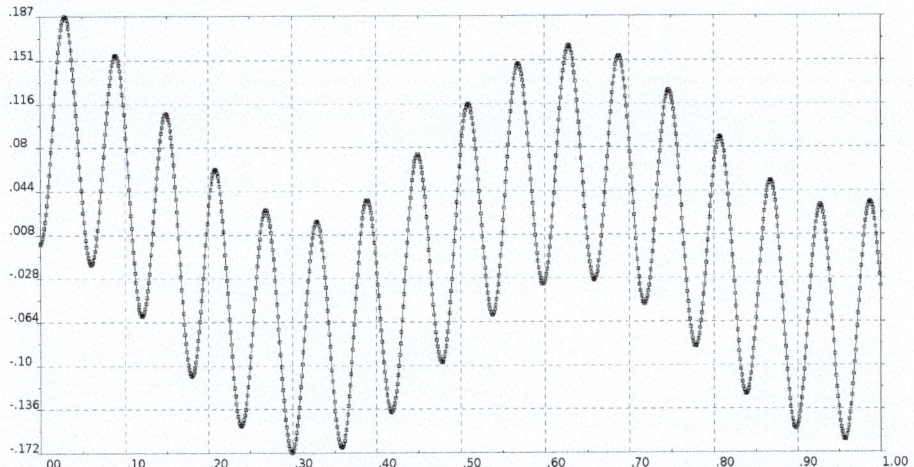

Fig. 2.40 Velocity versus time for the elastoplastic SDOF oscillator with kinematic hardening for a harmonic input $f = 100 sin(10t)$

```
k=1000.0;  kh=0.25*k;  m=10.0;  c=1.0
fsy=8.0;  uy=fsy/k
u0=0.0  ; v0=0.0// initial displacement u0 should be
    lower than uy
upl=0.0
f={100.0*sin(10.5*sqrt(k/m)*it)}

g=1.0/2.0 // gamma
b=1.0/4.0 // beta

dt=0.001
N=1000

u=new double[N+1];  u[0]=u0
v=new double[N+1];  v[0]=v0
a=new double[N+1];  a[0]=(f(0)-c*v[0]-k*u[0])/m
fs=new double[N+1];  fs[0]=k*u0

plastic=false
kep=k
// iterative computations
(1..N).each{
    // K_effective
    kef=kep+m/(b*dt*dt)+c*g/(b*dt)

    // effective load
```

```
26    fef=f(it*dt)-f((it-1)*dt)+m*(v[it-1]/(b*dt)+a[it
         -1]/(2.0*b))+c*(v[it-1]*g/b+a[it-1]*dt*(g/(2.0*b)
         -1.0))
27
28    // solve for current displacement step
29    du=fef/kef
30    u[it]=u[it-1]+du
31
32    // current acceleration and velocity
33    v[it]=v[it-1]+du*g/(b*dt)-v[it-1]*g/b+a[it-1]*g
         *(1.0-g/(2.0*b))
34    a[it]=a[it-1]+(f(it*dt)-f((it-1)*dt) - c*(v[it]-v[
         it-1]) - kep*(u[it]-u[it-1]))/m
35
36    // check yielding
37    yield=kh*u[it]+(k-kh)*uy*signum(v[it])
38    if(!plastic){
39      if(v[it]>0.0 && kep*(u[it]-upl)>yield){
40        kep=kh
41        upl=(1.0-k/kh)*uy
42        plastic=true
43        alter=true
44      }
45      if(v[it]<0.0 && kep*(u[it]-upl)<yield){
46        kep=kh
47        upl=(k/kh-1.0)*uy
48        plastic=true
49        alter=true
50      }
51    }else{
52      if(v[it]*v[it-1]<0){
53        kep=k
54        upl=u[it]-(u[it]-upl)*kh/k
55        plastic=false
56        alter=true
57      }
58    }
59    fs[it]=kep*(u[it]-upl)
60 }
61 ufPlot = new PlotFrame()
62 ufPlot.addFunction(new plotfunction(u,fs))
63 ufPlot.vline(0.0)
64 ufPlot.hline(0.0)
65 ufPlot.incline(kh,uy,fsy,Color.blue)
66 ufPlot.incline(kh,-uy,-fsy,Color.blue)
67 ufPlot.setMarker(true)
68 ufPlot.show()
69
70 tuPlot = new PlotFrame()
```

```
71  tuPlot.addFunction(new plotfunction(dt,u))
72  tuPlot.hline(uy,Color.red)
73  tuPlot.hline(-uy,Color.red)
74  tuPlot.setMarker(true)
75  tuPlot.show()
76
77  tvPlot = new PlotFrame()
78  tvPlot.addFunction(new plotfunction(dt,v))
79  tvPlot.setMarker(true)
80  tvPlot.show()
```

Listing 2.14 Elastoplastic SDOF system solution using the method β-Newmark in incremental form

2.14.9 Comparison Between Non-linear and Linearized Solutions for the Simple Pendulum

In here, we will compare the response computed for the simple pendulum under the assumption of non-linear behavior versus the reference linearized behavior, see Fig. 2.41. The numerical solution in both cases is recovered by using object `pendulum` and instance `solve` which implements the algorithm `Runge–Kutta` of fourth-order. The pendulum has a unit length, zero initial angular velocity and is assigned a series of values of the initial angular displacement.

As can be seen in the diagrams plotted in Fig. 2.42, linearization of the equation of motion of the simple pendulum yields acceptable results only for very small values of the magnitude

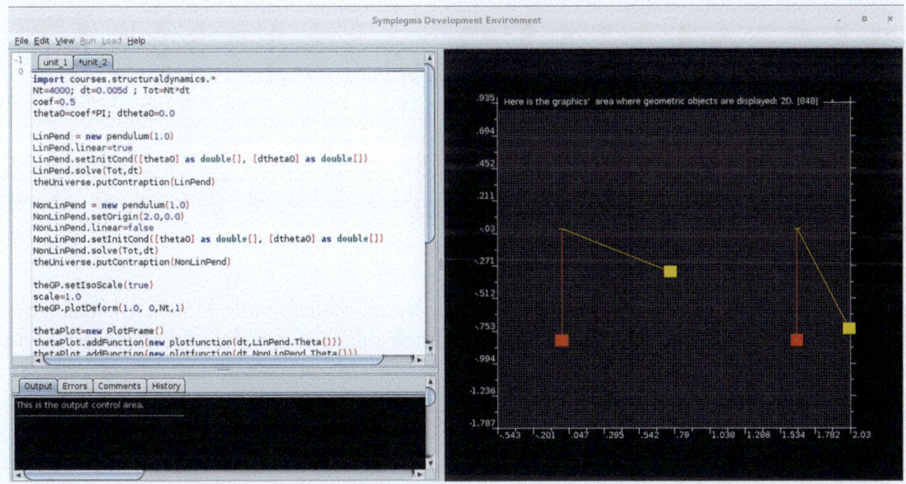

Fig. 2.41 The simple pendulum nonlinear versus linear solutions as implemented in environment SDE

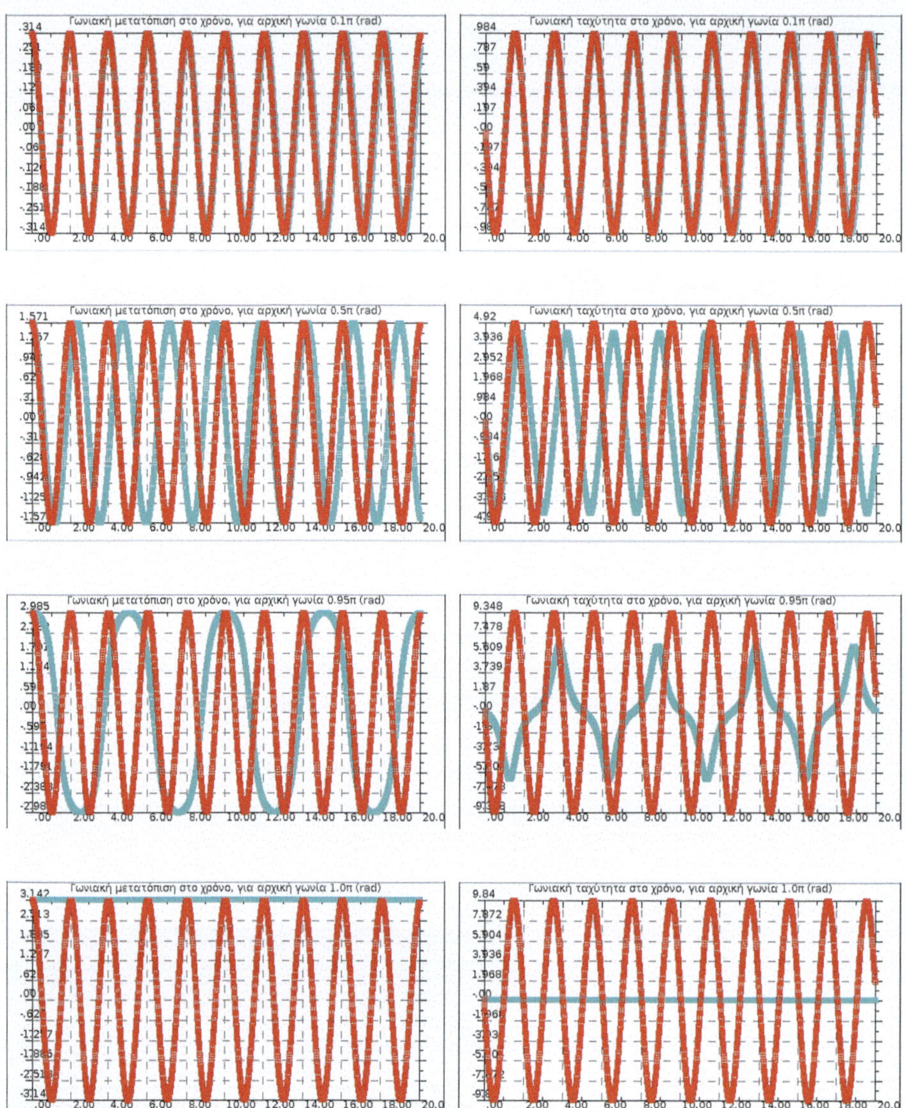

Fig. 2.42 Plots of $\theta - t$ diagrams (left column) and $\dot{\theta} - t$ diagrams (right column), where concurrently shown are both linearized (red) and non-linear (green) model results. Motion is initiated from an initial angular displacement corresponding to an angle of $(0.1, 0.5, 0.95, 1.0)\pi$, moving from top to bottom

of the initial angular displacement. A limit case is that of the an initial angular displacement corresponding to $180°$, i.e., the vertical position, which is however beyond the range of validity of the linearized theory. Nonlinear theory correctly predicts the stationary position of the pendulum, which is unstable, while linearized theory predicts circular motion only.

2.14.10 Numerical Solution of the Duffing Oscillator

In certain cases, nonlinear SDOF systems under periodic excitation respond in a non-periodic way. This type of response is labelled as *chaotic*. A typical example is the Ueda chaotic response manifested by the Duffing oscillator when excited by a harmonic force. The basic non-linear differential equation given by Eq. (2.81) now assumes the form

$$m\ddot{u} + c\dot{u} + ku + \mu u^3 = f_0 \cos(\bar{\omega}t) \tag{2.89}$$

A numerical solution of this equation derives by recasting it as a first-order differential equation system, as outlined in Eq. (2.82). To this end, we use the library Apache Commons Mathematics as listed algorithmically in Listing 2.15.

```
import org.apache.commons.math3.ode.
    FirstOrderDifferentialEquations
import org.apache.commons.math3.ode.nonstiff.
    ClassicalRungeKuttaIntegrator
import org.apache.commons.math3.ode.
    ContinuousOutputModel

class duffing implements
    FirstOrderDifferentialEquations{

  Closure f;
  double m,c,k,mu;

  duffing(double m, double k, double mu, double c,
    Closure f) {
      this.f = f;
      this.m = m;
      this.k = k;
      this.mu = mu;
      this.c = c;
  }

  int getDimension() {
    return 2;
  }

  void computeDerivatives(double t, double[] y,
    double[] yDot) {
      yDot[0] = y[1]
```

```
24      yDot[1] = -c*y[1]/m-k*y[0]/m-mu*y[0]*y[0]*y[0]/m+
        f(t)/m
25    }
26
27 }
28
29 dt=0.01
30 CRK = new ClassicalRungeKuttaIntegrator(dt)
31 gamma=7.5
32 omega=1.0
33 f={gamma*cos(omega*it)}
34
35 ode = new duffing(1.0,0.0,1.0,0.05,f);
36 double[] y = [3.1, 4.1]; // initial state
37
38 tracker = new ContinuousOutputModel()
39
40 CRK.addStepHandler(tracker);
41
42 tot=250.0*PI
43 CRK.integrate(ode, 0.0, y, tot, y);
44 u=[]
45 v=[]
46 (0..tot/dt).each{
47    tracker.setInterpolatedTime(it*dt)
48    res=tracker.getInterpolatedState()
49    u.add(res[0])
50    v.add(res[1])
51 }
52
53 plot(); plot(u,v)
```

Listing 2.15 Numerical solution of the Duffing oscillator using the fourth-order Runge–Kutta method by recourse to library Apache Commons Mathematics directly accessed from the development environment SDE

We now present results for the SDOF oscillator with (dimensionless) stiffness $k = 0.05$, mass $m = 1$, damping $c = 0$ and nonlinear parameter $\mu = 1.0$. The external excitation has a harmonic time variation of $f(t) = 7.5 \cos t$. Two sets of initial conditions are examined, namely (a) $(u_0, v_0) = (3.0, 4.0)$ and (b) $(u_0, v_0) = (3.01, 4.01)$, that are quite similar but serve to show that in the presence of nonlinearities, even small perturbations in the input data may lead to a markedly different response. The results are plotted in Figs. 2.43, 2.44 and 2.45 given below.

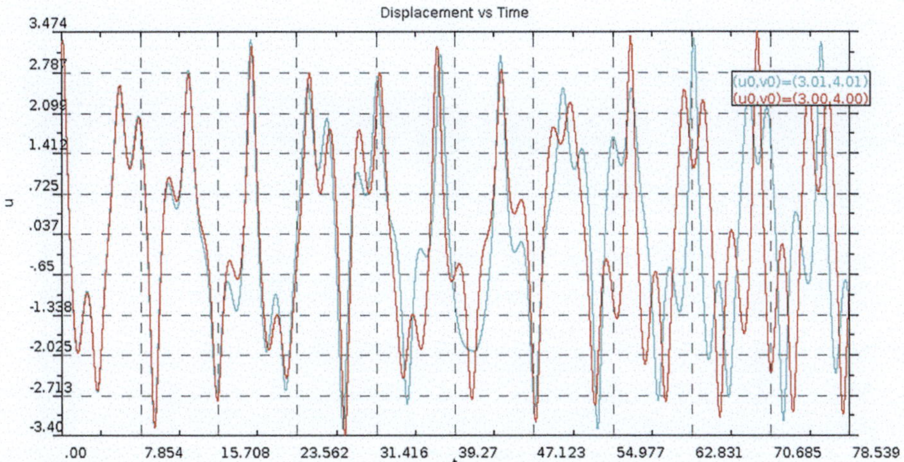

Fig. 2.43 Displacement response of the nonlinear Duffing oscillator with $k = 0.05$, $m = 1$, $c = 0$ and $\mu = 1.0$: Parametric study for minor changes in the initial conditions

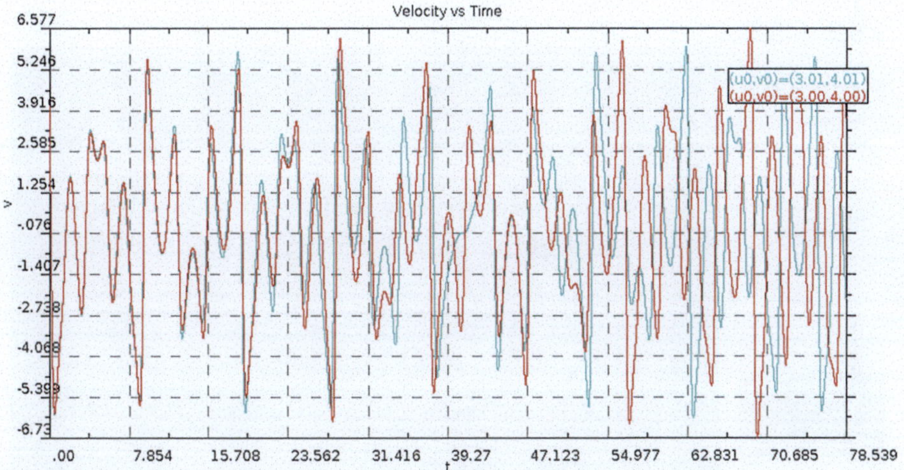

Fig. 2.44 Velocity response of the nonlinear Duffing oscillator with $k = 0.05$, $m = 1$, $c = 0$ and $\mu = 1.0$: Parametric study for minor changes in the initial conditions

In sum, a characteristic property of a nonlinear SDOF system (e.g., a Duffing oscillator) exhibiting chaotic behavior is its sensitivity to the initial conditions. It is observed that minor perturbations in the initial conditions, which in practice may be unavoidable, lead to a major divergence in the kinematic response of the SDOF oscillator. This can be observed in Figs. 2.43, 2.44 and 2.45, where at the begining of time the response to a perturbation of the initial conditions leads to an almost identical response, but diverges as time passes.

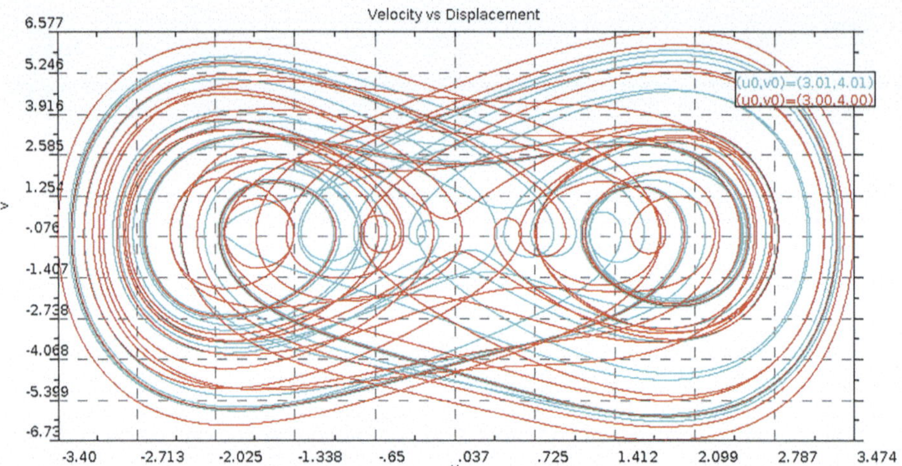

Fig. 2.45 Phase diagrams of the response of the nonlinear Duffing oscillator with $k = 0.05, m = 1$, $c = 0$ and $\mu = 1.0$: Parametric study for minor changes in the initial conditions

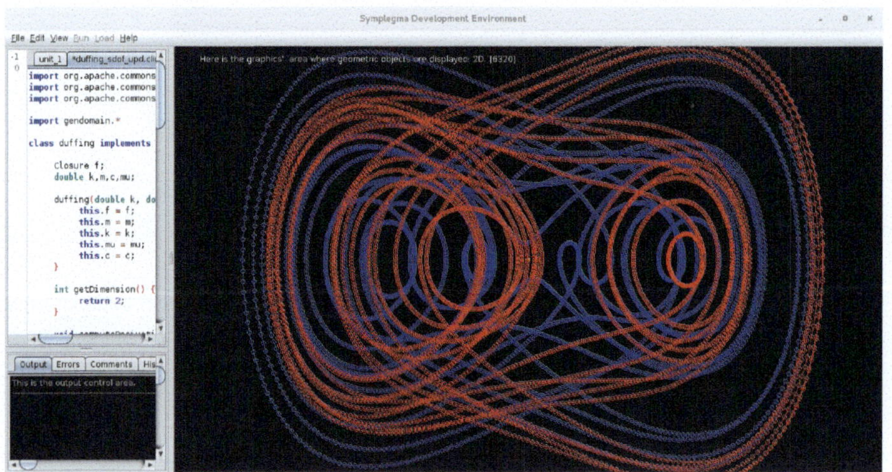

Fig. 2.46 Phase diagrams for the Duffing oscillator under harmonic input solved in the development environment SDE. Video snapshot from produced animation

We note in parallel that the code used for the solution of the Duffing oscillator can, with minor changes, be used to create a video, which further advances the use of the development environment SDE. A snapshot of this video can be seen of Fig. 2.46.

Multiple Degree-of-Freedom Systems

3

3.1 MDOF System Oscillations

We begin with the analysis of a two DOF system containing masses, springs and dashpots that are linked together, as shown in Fig. 3.1. Both system DOF pertain to the horizontal displacements of each of the two masses, in which case Newton's law of dynamic equilibrium can be applied in reference to the free-body diagram. More specifically, we have for the first mass

$$
m_1\ddot{u}_1(t) = f_1(t) - k_1 u_1(t) - c_1\dot{u}_1(t) - k_2\left(u_1(t) - u_2(t)\right) - c_2\left(\dot{u}_1(t) - \dot{u}_2(t)\right) \rightarrow
$$

$$
m_1\ddot{u}_1(t) + (c_1 + c_2)\dot{u}_1(t) - c_2\dot{u}_2(t) + (k_1 + k_2)u_1(t) - k_2 u_2(t) = f_1(t) \tag{3.1}
$$

and similarly for the second mass

$$
m_2\ddot{u}_2(t) + (c_2 + c_3)\dot{u}_2(t) - c_2\dot{u}_1(t) + (k_2 + k_3)u_2(t) - k_2 u_1(t) = f_2(t) \tag{3.2}
$$

The above two equilibrium equations, (3.1) and (3.2), can be written in matrix form as follows:

$$
\begin{bmatrix} m_1 & 0 \\ 0 & m_2 \end{bmatrix}
\begin{Bmatrix} \ddot{u}_1(t) \\ \ddot{u}_2(t) \end{Bmatrix}
+
\begin{bmatrix} c_1 + c_2 & -c_2 \\ -c_2 & c_2 + c_3 \end{bmatrix}
\begin{Bmatrix} \dot{u}_1(t) \\ \dot{u}_2(t) \end{Bmatrix}
+
$$

$$
\begin{bmatrix} k_1 + k_2 & -k_2 \\ -k_2 & k_2 + k_3 \end{bmatrix}
\begin{Bmatrix} u_1(t) \\ u_2(t) \end{Bmatrix}
=
\begin{Bmatrix} f_1(t) \\ f_2(t) \end{Bmatrix} \tag{3.3}
$$

Using compact notation, the matrix equilibrium equation is now

$$
M\ddot{u}(t) + C\dot{u}(t) + K u(t) = f(t) \tag{3.4}
$$

© The Author(s), under exclusive license to Springer Nature Switzerland AG 2026
G. Manolis and C. Panagiotopoulos, *Vibrations of Structural Systems*, Synthesis Lectures on Mechanical Engineering, https://doi.org/10.1007/978-3-032-12279-7_3

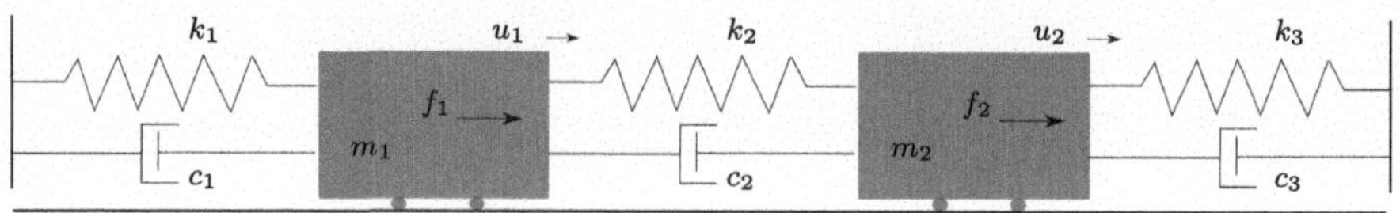

Fig. 3.1 A two DOF system

where K, M, C are the **matrices** pertaining to the system's stiffness, mass and damping, respectively. Furthermore, we have the **vectors** for the displacements u, velocities $\dot{u}$ and accelerations $\ddot{u}$, with f the **vector** containing the external force components. In sum, the equations of motion comprise a system of matrix differential equations of second order (in time) that must be accompanied by initial conditions at time t_0, which marks the instant the vibrations commence. These initial conditions are defined as

$$u(t_0) = \left\{ \begin{array}{c} u_1(t_0) \\ u_2(t_0) \end{array} \right\} = \left\{ \begin{array}{c} u_{1,0} \\ u_{2,0} \end{array} \right\} \tag{3.5}$$

for the displacements and

$$\dot{u}(t_0) = \left\{ \begin{array}{c} \dot{u}_1(t_0) \\ \dot{u}_2(t_0) \end{array} \right\} = \left\{ \begin{array}{c} \dot{u}_{1,0} \\ \dot{u}_{2,0} \end{array} \right\} \tag{3.6}$$

for the velocities. An alternative notation for the velocities is

$$\dot{u}(t_0) = v(t_0) = \left\{ \begin{array}{c} v_1(t_0) \\ v_2(t_0) \end{array} \right\} = \left\{ \begin{array}{c} v_{1,0} \\ v_{2,0} \end{array} \right\}. \tag{3.7}$$

The two DOF presented above can easily be extended to encompass an arbitrary number N of masses, springs and dashpots that comprise a structural system. The number of DOF coincides with the number of discrete masses N for the simple model, where the lumped mass executes one component of motion from a total of six possible components in the three-dimensional space, namely three translational and three rotational ones. The equations of motion for a MDOF system defined in the 2D plane can therefore be derived by considering the equilibrium equations of the free body diagram centered around each mass. This procedure is outline in module `courses.structuraldynamics` for class `sdofSeries` that will be explained below.

Class `sdofSeries`

This is a first brief presentation of module `courses.structuraldynamics` that addresses the case of a dynamic system formed by the coupling of a series of masses, springs and dahspots. This MDOF system is realized through class `sdofSeries` and all references made here are to Listing 3.1. In order to use class `sdofSeries` we must first activate module `courses.structuraldynamics`, see line 1. Next, the numbering of the springs must be introduced in line 3, and similarly for the masses in line 4. This linear dynamic system comprising coupled masses is now defined in line 6 by setting up a numbered directory for the masses and stiffnesses. This way, the snapshot mSeries of object `sdofSeries` will be created and will correspond to zero damping. In line 8, the thickness of the masses is increased for plotting higher quality graphs, while in line 9 we introduce the `Universe` of SDE to activate the graphics. Finally, for an optimal graphical representation, the image framework theGP of SDE at state IsoScale is defined in line 11, which specifies the same scaling factor for both vertical directions x and y. Upon execution of this code, the graphical environment of SDE will produce a figure similar to Fig. 3.2 given below.

```
1   import courses.structuraldynamics.*
2
3   double[] springs = [1000.0,500.0,1000.0]
4   double[] masses = [10.0,5.0]
5
6   mSeries = new sdofSeries(masses, springs)
7
8   mSeries.dx=2.0*mSeries.dx
9   theUniverse.putContraption(mSeries)
10
11  theGP.setIsoScale(true)
```

Listing 3.1 Brief presentation of an MDOF comprising two coupled masses, see object `sdofSeries`

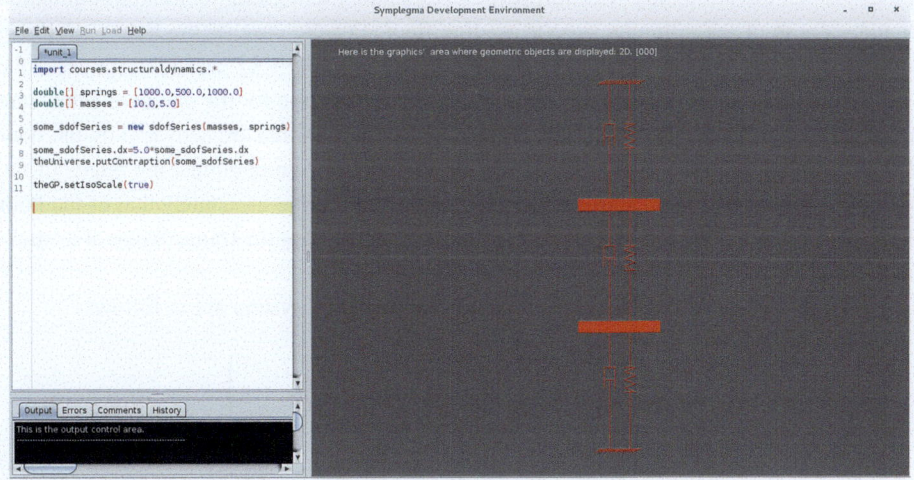

Fig. 3.2 Representation of a dynamic system comprising two coupled masses from `sdofSeries` of SDE

The system of equations (3.4) can be solved numerically by the methods previously described in Sect. 2.11.1. However, MDOF systems can be treated by other methods as well, primarily through the solution of the eigenvalue problem that is based on separation of variables. At this point, we observe from Eq. 3.3 that the mass, stiffness and damping matrices are all symmetric. This symmetry appears in a large category of mechanical systems to which we will restrict the analysis that follows. Furthermore, matrices K and M are semi-positive definite, meaning that the following relations hold true

$$x^T K x \geq 0 \quad \text{and} \quad x^T M x > 0$$

for any arbitrary vector x.

3.2 Vibrations Planar Shear-type Frames

The planar shear-type frame is a simple structural system well suited for studying its dynamic response as a MDOF system. The assumptions behind modelling a frame as being of the shear-type are as follows:

- All mass is concentrated at the floor levels.
- The horizontal elements, i.e., the beams supporting the floors, are assumed to be rigid.
- Any deformations induced in either the columns or the beams and slabs due to axial loads are neglected.

A typical shear frame is shown in Fig. 3.3. Every floor is represented as a mass with motion in the horizontal direction only. The stiffness corresponding to a given floor is the summation of the stifness of the columns supporting that floor. For the case of the frame shown in Fig. 3.3, we start with the ground floor stiffness as $k_1 = k_1^\delta + k_1^\alpha$ and then move to the first floor stiffness $k_2 = k_2^\delta + k_2^\alpha$. A similar situation holds true for the dampers, which are activated by the floor velocities, in contrast to the stiffnesses that are activated by the floor displacements. Since we neglect any axial deformation in the columns, the edges of the columns abutting the floors all have the same lateral displacement, namely that of the floor itself. This allows a simplification of the mathematical model of the frame to that of a 'stick' model, see Fig. 3.3.

The equations of motion in matrix form for the two DOF system are as follows:

$$\begin{bmatrix} m_1 & 0 \\ 0 & m_2 \end{bmatrix} \begin{Bmatrix} \ddot{u}_1(t) \\ \ddot{u}_2(t) \end{Bmatrix} + \begin{bmatrix} c_1 + c_2 & -c_2 \\ -c_2 & c_2 \end{bmatrix} \begin{Bmatrix} \dot{u}_1(t) \\ \dot{u}_2(t) \end{Bmatrix} +$$

$$\begin{bmatrix} k_1 + k_2 & -k_2 \\ -k_2 & k_2 \end{bmatrix} \begin{Bmatrix} u_1(t) \\ u_2(t) \end{Bmatrix} = \begin{Bmatrix} f_1(t) \\ f_2(t) \end{Bmatrix} \tag{3.8}$$

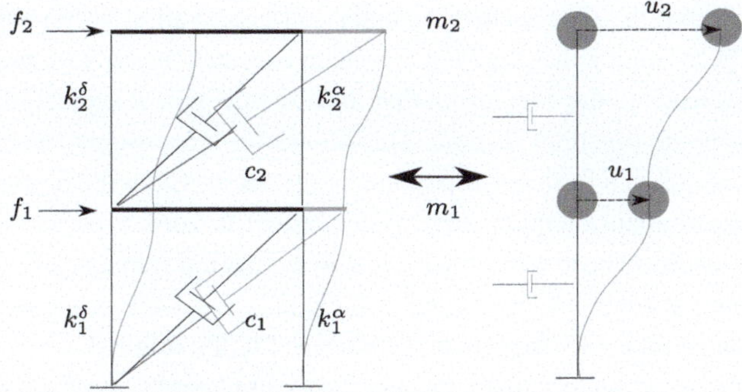

Fig. 3.3 Vibrating shear frame with two DOF

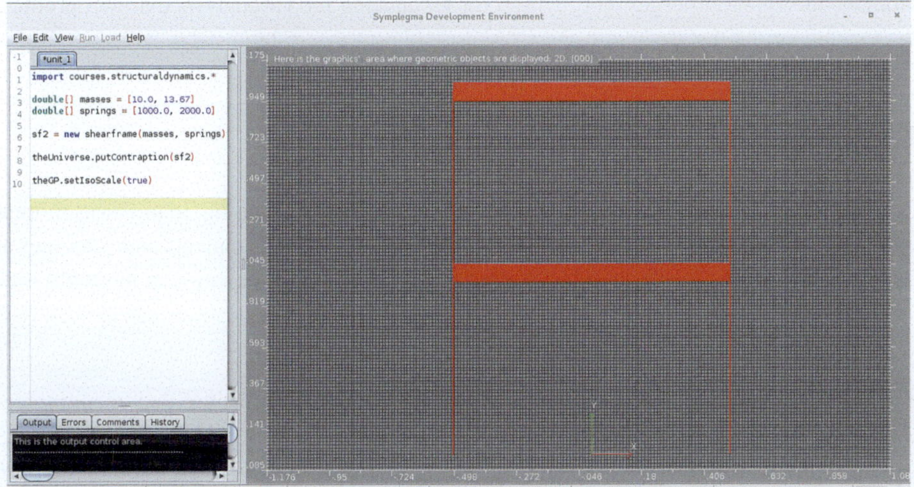

Fig. 3.4 Representation of the two DOF dynamic system `massesshearframe` in SDE

These are accompanied by the appropriate initial conditions for the displacement vector $u(0)$ and the velocity vector $\dot{u}(t)$. Obviously, this approach can be generalized to cover a shear frame with N floors.

3.3   The Eigenproblem

The eigenproblem, or to be precise the extraction of eigenvalues (natural frequencies) followed by the recovery of the corresponding eigenvectors (modal shapes) refers to the solution of the homogeneous (i.e., absence of external forces) system of differential equations representing the dynamic response of the MDOF system. A first reading of this problem leads to the conclusion that the displacemet, velocity and acceleration vectors are all equal to zero. However, by assuming that the MDOF system executes free vibrations in the absence of external forces leads to the conclusion that the aforementioned vectors of the kinematic variables are harmonic functions of time. This allows for a recasting of the matrix system of equations of motion in a form corresponding to the Sturm-Liouville problem. Therefore, by setting the determinant of this matrix system equal to zero, a polynomial in the frequency variable results, whose roots correspond to the natural frequencies of vibration. the number of the natural frequencies is equal to the number of DOF of the dynamic system. Back-substitution of each and every natural frequency in the original matrix system yields the corresponding modal shapes that represent the motion that the system would execute if the external frequency of vibration coincided with a natural frequency. Besides the importance of the eigenvalue problem in understanding the fundamental way a dynamical system vibrates, the eigenproperties can be used for an ensuing modal analysis that will produce the time history of the kinematic and stress variables of the system.

3.3.1   MDOF System Without Damping

Consider Eq. 3.4 in the absence of external forces and without damping. In this case, motion will be possible if the initial conditions are non-zero, that is displacements and velocities at time t_0, which marks the onset of the dynamic phenomenon. The equation of motion is now

$$M\ddot{u}(t) + Ku(t) = 0 \tag{3.9}$$

and we seek solutions in the form $u(t) = \phi q(t)$, where ϕ is a **vector** that encompasses the spatial variation of the displacements, while the time variation is represented by the generalized coordinate $q(t)$ (**a scalar** quantity). In other words, we seek a solution whose basic spatial variation ϕ remains constant, but changes in overall magnitude with the passage of time because of measure $q(t)$. This solution is essentially a separation of variables that is commonly used as a method of solution for partial differential equations. Upon substitution in the equation of motion, we have

$$M\phi\ddot{q}(t) + K\phi q(t) = 0 \tag{3.10}$$

Pre-multiplication with the transpose of the modal shape ϕ^T yields

$$\phi^T M\phi\ddot{q}(t) + \phi^T K\phi q(t) = 0 \tag{3.11}$$

followed by separation of the temporal and spatial parts as follows:

$$-\frac{\ddot{q}(t)}{q(t)} = \frac{\phi^T K\phi}{\phi^T M\phi} \tag{3.12}$$

In order for Eq. 3.12 to be hold true, both its left and right hand sides must be equal to the same constant λ, i.e.,

$$-\frac{\ddot{q}(t)}{q(t)} = \frac{\phi^T K \phi}{\phi^T M \phi} = \lambda \tag{3.13}$$

From the time-dependent term we have

$$-\frac{\ddot{q}(t)}{q(t)} = \lambda \rightarrow \ddot{q}(t) = -\lambda q(t) \tag{3.14}$$

By replacing Eqs. (3.14) in (3.10), we have that

$$M\phi\lambda q(t) - K\phi q(t) = 0 \rightarrow$$
$$(K - \lambda M)\phi q(t) = 0 \tag{3.15}$$

For this to hold true for $\forall t \geq 0$, then

$$(K - \lambda M)\phi = 0 \tag{3.16}$$

Equation (3.16) defines the **generalized linear eigenvalue problem**. For a solution to exist, other than the trivial one, the determinant of the above matrix must be set equal to zero, i.e.,

$$\det(K - \lambda M) = 0, \tag{3.17}$$

which results in the *characteristic equation* for the MDOF system. Upon sorting terms in the characteristic equation, a polynomial $\Pi(\lambda) = 0$ emerges which is of order N in terms of originally defined constant λ. The roots of the polynomial are the eigenvalues λ_i, while their square roots $\omega_i = \sqrt{\lambda_i}$ are the eigenfrequencies (or natural frequencies) of the MDOF system, with the smallest in value eigenfrequency ω_1 being the dominant (or fist) eigenfrequency of the MDOF system. We note in passing that for MDOF systems with no damping, the eigenvalue problem is often defined directly in terms of eigenfrequencies instead of eigenvalues as follows:

$$(K - \omega^2 M)\phi = 0$$

We will not delve any further in the solution of the eigenproblem, because much information can be found in books on structural dynamics and also on books in linear algebra. Instead, we will focus on solving the eigenproblem by recourse to our software package SDE with its accompanying library containing various useful modules. More specifically, following the solution of the eigenproblem as shown in figure (3.16), a set of N eigenvalue pairs λ_i along with their corresponding eigenvectors ϕ_i are recovered, where N is the number of DOF of the structural system in question. These eigenvalue-eigenvector pairs are called the eigensolution of the problem. Depending on the particular structural system, the eigenfrequencies may be discrete or may present a multiplicity p. Even then, however, to an eigenfrequency of multiplicity p there correspond p linearly independent eigenvectors. It should be noted that the eigenvalue problem has a degree of indeterminacy since it is a solution to the free vibration problem of Eq. (3.16). This implies that it is not possible to determine an exact form of the eigenvector ϕ_i, because if it is multiplied by a constant α, the new vector $\alpha\phi_i$ is also an eigenvector since it too satisfies Eq. (3.16). What is possible however, is to compute the $N - 1$ components of the eigenvector relative to one component, whose selection is arbitrary. This procedure is known as **normalization**. Finally, the term eigenshape is sometimes used interchangeably with the term eigenvector.

The most common eigenvector normalization is one where the maximum value of a component is unity. This implies dividing the nth eigenvector ϕ_n recovered from the eigenvalue analysis by its component with the maximum absolute value as follows:

$$\phi_n = \frac{\phi_n}{\max(\phi_n)} \tag{3.18}$$

This new set of eigenvectors are known as normalized eigenvectors. From all N eigenvectors it is possible to construct the modal matrix Φ containing these eigenvectors as columns:

$$\Phi = \begin{bmatrix} \phi_1 & \phi_2 & \cdots & \phi_N \end{bmatrix} = \begin{bmatrix} \phi_{11} & \phi_{11} & \cdots & \phi_{1N} \\ \phi_{21} & \phi_{22} & \cdots & \phi_{2N} \\ \cdots & \cdots & \cdots & \cdots \\ \phi_{N1} & \phi_{N2} & \cdots & \phi_{NN} \end{bmatrix}$$

An alternative normalization procedure is for a given eigenvector to have a unit measure with the mass matrix as the weighing function, i.e.,

$$\phi_i^T M \phi_i = 1.$$

To accomplish this we divide each component of the n-th eigenvector ϕ_n with the weighted measure

$$\phi_n := \frac{\phi_n}{\sqrt{\phi_n^T M \phi_n}} \tag{3.19}$$

Another important property associated with the eigenvalue problem is the **normality** of the masses, according to which

$$\phi_i^T M \phi_j = \begin{cases} m_i^*, & \text{if } i = j \\ 0, & \text{if } i \neq j \end{cases} \tag{3.20}$$

and the normality of the stiffness matrix

$$\phi_i^T K \phi_j = \begin{cases} k_i^* = \omega_i^2 m_i^*, & \text{if } i = j \\ 0, & \text{if } i \neq j \end{cases} \tag{3.21}$$

In the above, m_i^* and k_i^* respectively are the modal mass and the modal stiffness corresponding to the i-the eigenvalue. They are also known as the generalized mass and the generalized stiffness associated with the ith eigenvalue. A final remark is that the eigenvectors ϕ_i comprise an orthogonal base in an N-dimensional vector space, which implies that any N dimensional vector associated with the particular MDOF system can be expressed as a linear combination of these base vectors. In other words, an N-dimensional vector x (e.g., the displacement vector) can be synthesized by superposing the N eigenvectors ϕ_i as follows:

$$x = \sum_{i=1}^{N} \alpha_i \phi_i \tag{3.22}$$

In the above, components α_i are the projections of the vector x on each eigenvector ϕ_i of the vector base.

3.3.2   MDOF System with Damping

In the presence of material damping, matrix C of the MDOF system is non-zero, and the equation of motion reads as follows:

$$M\ddot{u}(t) + C\dot{u}(t) + Ku(t) = 0 \tag{3.23}$$

The damping matrix is symmetric and semi-positive definite, which implies the following relations

$$C^T = C \quad \text{and} \quad x^T C x \geq 0$$

for any arbitrary non-zero real vector x. The eigenvalue problem is now different as compared to the undamped case given in Eq. (3.16) and reads as

$$(K + \lambda C + \lambda^2 M)\psi = 0 \tag{3.24}$$

As before, λ is an eignevalue, while ψ is the corresponding eignevector. The eigenvalues λ are the roots of the characteristic equation

$$\det (K + \lambda C + \lambda^2 M) = 0, \tag{3.25}$$

which upon expansion leads to a polynomial of degree $2N$ in terms of λ.

MDOF System Classification Based on the Presence of Damping

- Undamped system: $C = 0$
- System with proportional damping:

  - Rayleigh-type damping, with $C = \alpha_m M + \alpha_k K$
  - Modal damping defined as a percent of critical damping in each and every mode
  - Generalized proportional damping: $CM^{-1}K = KM^{-1}C$

- Non-proportional damping: $CM^{-1}K \neq KM^{-1}C$.

Note that the eigenvectors of the undamped MDOF system and those of the proportionally damped one are real quantities and coincide. In the case of non-proportional damping, the eigenvectors are complex quantities. It can be shown that in the presence of proportional damping, the normality rule holds and we have that

$$\phi_i^T C \phi_j = \begin{cases} c_i^*, & \text{if } i = j \\ 0, & \text{if } i \neq j \end{cases} \tag{3.26}$$

The two Rayleigh factors a_m and a_k can be evaluated by matching terms with the modal damping factor ξ and eigenfrequency ω assumed for the any two modes of the MDOF system, which usually are the first (dominant) and second modes, i.e., those with the lowest values for the eigenfrequency. This choice is dictated by the fact that most external loads energize the lowest modes, the exception being very high frequency loads. Thus, the aforementioned constants are related to the eigenproperties as follows:

$$\xi = \frac{a_m}{2\omega} + \frac{a_k \omega}{2} \tag{3.27}$$

Figure 3.5 plots the relation between the stiffness-related Rayleigh factor ($\alpha_m = 0, \alpha_k \neq 0$), the mass-related Rayleigh factor ($\alpha_m \neq 0, \alpha_k = 0$) and their combination ($\alpha_m \neq 0, \alpha_k \neq 0$) as functions of frequency.

Returning to the modal damping hypothesis, we assume that each eigenmode, say the ith for generality, has a specific amount of damping quantified by the damping ratio ξ_i, so that we have

$$\Phi^T C \Phi = C^* \tag{3.28}$$

where C^* is a diagonal matrix with the elements given as $C_{ii}^* = c_i^* = 2\xi_i \omega_i m_i^*$. The damping matrix C in the physical coordinate space that corresponds to the modal damping matrix C^* is computed from the realtion

$$C = M\Phi C^* \Phi^T M \tag{3.29}$$

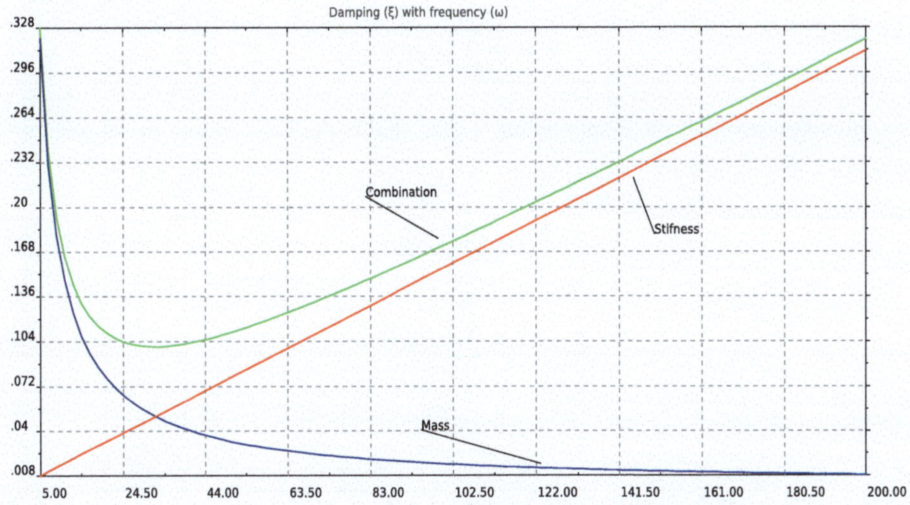

Fig. 3.5 Green line: Damping proportional to the stiffness ($\xi = a_k\omega/2$); blue line: damping proportional to the mass ($\xi = \frac{a_m}{2\omega}$); red line: combination of both cases ($\xi = \frac{1}{2}\left(\frac{a_m}{\omega} + a_k\omega\right)$)

The result given by Eq. (3.29) holds true for eigenvectors Φ that are normalized with respect to the mass matrix. If this is not the case, then we have that

$$C = M\Phi M^{*-1}C^*M^{*-1}\Phi^T M \qquad (3.30)$$

Note that the inverse of the mass matrix M^* is trivial to compute since it is diagonal. The above relations can be explicitly written as follows:

$$C = M\left(\sum_{i=1}^{N}\frac{2\xi_i\omega_i}{m_i^*}\phi_i\phi_i^T\right)M \qquad (3.31)$$

3.4   Modal Analysis

Modal analysis is perhaps the most common way for solving for the response of MDOF systems to external dynamic loads. Mathematically speaking, modal analysis is basically the separation of variables method, as applied to partial differential equations. Going back to the equation of motion, namely Eq. (3.4) plus the initial conditions on the displacemnts, Eq. (3.5) and on the velocities Eq. (3.6) we recall that any response vector of the particular MDOF system under study can be decomposed in terms of its eigenvectors ϕ_i as indicated by Eq. (3.22). Starting with the displacement vector, we have that

$$u(t) = \sum_{i=1}^{N} \phi_i q_i(t), \tag{3.32}$$

The time derivatives of the displacement vector, that is the velocity and the acceleration vectors are therefore written as

$$\dot{u}(t) = \sum_{i=1}^{N} \phi_i \dot{q}_i(t), \tag{3.33}$$

$$\ddot{u}(t) = \sum_{i=1}^{N} \phi_i \ddot{q}_i(t). \tag{3.34}$$

Substitution of Eqs. (3.32) and (3.33) in the equation of motion, Eq. (3.4) yields

$$M \sum_{i=1}^{N} \phi_i \ddot{q}_i(t) + C \sum_{i=1}^{N} \phi_i \dot{q}_i(t) + K \sum_{i=1}^{N} \phi_i q_i(t) = f(t) \tag{3.35}$$

A left-side multiplication of the above equation by the n-th eigenvector ϕ_n^T and invocation of the normality conditions given by Eqs.(3.20)–(3.21) and (3.26) results in

$$m_n^* \ddot{q}_n(t) + c_n^* \dot{q}_n(t) + k_n^* q_n(t) = \phi_n^T f(t) \tag{3.36}$$

Note that at this point we assume that ϕ is the correct eigenvector that encompasses the most general case of non-proportional damping. In what follows, we work primarily with the eigenvectors ϕ that belong to the category of proportionally damped MDOF systems. This way, we have the same eigenvectors as for the undamped case. If we now divide the above equation with the modal mass m_n^*, we have

$$\ddot{q}_n(t) + 2\xi_n \omega_n \dot{q}_n(t) + \omega_n^2 q_n(t) = f_n^*(t) \tag{3.37}$$

where $f_n^*(t) = \frac{\phi_n^T f(t)}{m_n^*}$ is the forcing function in modal coordinates. The initial conditions, as they now apply to the generalized coordinates q_n, are

$$q_n(0) = \frac{\phi_n^T M u(0)}{m_n^*} \tag{3.38}$$

for the displacements and

$$\dot{q}_n(0) = \frac{\phi_n^T M \dot{u}(0)}{m_n^*}, \tag{3.39}$$

for the velocities. It is reminded at this point that $u(0)$ and $\dot{u}(0)$ are the vectors of the initial conditions as they apply to the physical coordinates and M is the mass matrix of the MDOF system.

Concluding, we observe that the original problem involving a MDOF system that yielded a second-order, matrix differential equation has now degenerated to a solution of N differential equations, each addressing an SDOF-like oscillator. These SDOF-like oscillators obey Eq. (3.37) along with the relevant initial conditions (3.38) and (3.39). The solution to each such equation representing a given generalized (or modal) coordinate can be achieved either analytically or numerically, as was discussed in the previous chapter on SDOF systems. Once the solutions for all generalized coordinates $q_i(t)$ have been derived, the complete vector solution in the physical coordinate space is reconstituted by superposition, see Eq. (3.32).

3.5   Harmonic Vibrations

We now focus on harmonic vibrations of MDOF systems, which obey the equation of motion in the following form:

$$M\ddot{u}(t) + C\dot{u}(t) + Ku(t) = B_f e^{i\omega t} \tag{3.40}$$

In the above, B_f is the vector of the amplitudes of the time-harmonic, external forcing function. We assume that the MDOF system response to harmonic excitation is also harmonic. Thus, the displacement vector for steady-state (SS) vibrations is given as

$$u_{ss}(t) = U(\omega)e^{i\omega t} \tag{3.41}$$

where U is the frequency-dependent displacement amplitude. This assumption implies that the first time derivative of the displacement vector is simply $i\omega$ times U and similarly for the higher-order derivatives. Returning to the equation of motion, the solution is given as

$$U(\omega) = H(\omega)B_f \tag{3.42}$$

where $H(\omega)$ is the matrix transfer function of the MDOF system in the frequency domain. This is given as

$$H(\omega) = \left(-\omega^2 M + i\omega C + K\right)^{-1}, \tag{3.43}$$

3.6   The One-Storey 3D Building

The single-storey is a structural system which due to its simplicity it is very instructive for understanding basic concepts of the dynamics of structures [1]. Static and dynamica analysis of the three-dimensional single-storey structure, is the base for antiseismic design of structures, especially in earthquake prone countries such as Greece [2].

The single-storey building considered here comprises an arbitrary number of columns that remain axially undeformed and respond only in bending. These columns are either hinged or fixed and support a roof slab that acts as a diaphragm (rigid surface). The motion of the roof slab is fully described by the two horizontal translations and a rotation about the vertical z-axis. The reference system is at the origin the geometric center of the slab. It is obvious that this system has three degrees of freedom, the two translational displacements $u_x = u_1$ and $u_y = u_2$ of the center and the angle of rotation $\theta_z = u_3$ about the vertical axis normal to the slab's plane. The *stiffness matrix* K is a 3×3 symmetric with elements defined as follows:

$$k_{xx} = \sum_i k_{xi}, \quad k_{yy} = \sum_i k_{yi}, \quad k_{xy} = k_{yx} = 0$$

$$k_{xz} = k_{zx} = -\sum_i y_i k_{xi}, \quad k_{yz} = k_{zy} = \sum_i x_i k_{yi} \tag{3.44}$$

$$k_{zz} = \sum_i \left(k_{zi} + y_i^2 k_{xi} + x_i^2 k_{yi} \right)$$

where k_{xi}, k_{yi} and k_{zi} stiffness coefficients of the ith column. The stiffness matrix could be diagonalized with respect to the principal reference system (e.g., elastic center) whose coordinates are given by

$$(x_k, y_k) = \left(\frac{k_{zy}}{k_{yy}}, -\frac{k_{zx}}{k_{xx}} \right). \tag{3.45}$$

The three diagonal terms of the stiffness matrix are now

$$k_I = k_{II} = \frac{k_{xx} + k_{yy}}{2} \pm \sqrt{\left(\frac{k_{xx} - k_{yy}}{2} \right)^2}$$

$$k_{III} = k_{zz} - x_k^2 k_{yy} - y_k^2 k_{xx}. \tag{3.46}$$

For the simplified case in entity storey, the center of mass coincides with the geometric center of the slab. The *mass matrix* is 3×3 and diagonal with elements

$$m_x = m_y = m, \quad m_z = J_m = mr^2 \tag{3.47}$$

where m the total mass of the slab, J_m is its mass moment of inertia and r the radius of inertia (or radius of gyration). The mass moment of inertia for a rectangle $L_x \times L_y$ slab is given by $J_m = m(L_x^2 + L_y^2)/12$. The equation of motion is the standard multiple degree of freedom equation of motion, namely Eq. (3.23).

3.7   Package `courses.structuraldynamics`

3.7.1   Class `sdofSeries`

Description:
This object pertains to the linear MDOF dynamic system comprising a series of masses interconnected through a spring and dashpot mechanism, as can be seen in Fig. 3.2. The present object includes methods of analysis of the MDOF system under any combination of initial conditions and external loads, Furthermore, it solves teh eigenproblem and yields the solution in the form of eigenfrequencies and eigenvectors.

Instance setup:
The constructor of a snapshot of a given object appears in two versions: The first defines an input of three real-values, double precision variables corresponding to the masses, the spring constants and the dashpot constants comprising the MDOF system. The second version is simpler in that it ignores damping, i.e., zero dashpot values.

- `sdofSeries`(double[] `masses`, double[] `springs`, double[] `dashpots`)
- `sdofSeries`(double[] `masses`, double[] `springs`)

Methods of solution:
The naming of the solution methods is such that it best describes these methods.

- void setLengths(double[ ] lengths)
- void setInitCond(double[ ] u0, double[ ] v0)
- void setInitCond(double[ ] u0)
- void setRHS(DoubleFunction df)
- void setRHS(DoubleFunction df, int dof)
- void setRHS(DoubleFunction[] df)

continued ...

Class `sdofSeries`

Methods of solution:

- RealMatrix getMass()
- RealMatrix getStiffness()
- RealMatrix getDamping()
- void setRayleigh(double am, double ak)
- void solve(double tot, double dt)
- double[ ] Disp(int w)
- void eigenAnalysis()
- double getEigenFrequency(int i)
- double[ ] getEigenFrequencies()
- double[ ] getEigenVecArray(int i)
- RealVector getEigenVector(int i)

3.7.2 Class `shearframe`

Description:
This object addresses the solution of the linear, MDOF system comprising a vertical array of interconnected masses, namely a 'stick' model. Each mass is connected to the top and bottom masses by a spring-dashpot link, as can be seen in Fig. 3.4. This object contains methods for the solution of equations of motion for any combination of external loads and initial conditions, as well as for the solution of the eigenvalue problem for determining the eigenvalue-eigenvector pairs.

Instance setup:
The constructor of a snapshot of a given object appears in two versions: The first defines an input of three real-values, double precision variables corresponding to the masses, the spring constants and the dashpot constants comprising the MDOF system. The second version is simpler in that it ignores damping, i.e., for zero dashpot values.

- shearframe(double[] masses, double[] stiff, double[] dashpots)
- shearframe(double[] masses, double[] stiff)

Methods:
The methods contained in class `shearframe` are the same as those appearing in class `sdofSeries` described in Sect. 3.7.1.

3.7.3 Class `mdof`

Description:
The present object is one of the most basic ones: It addresses the linear MDOF system comprising a mass matrix, a stiffness matrix and a damping matrix. The object is equipped with all the basic methods for solving the corresponding eigenproblem.

Instance setup:
The constructor of a snapshot of an object appears in three variants: The first defines an input of three real-values, double precision variables corresponding to the mass, stiffness and damping matrices comprising the MDOF system. The second variant of the constructor omits the damping matrix, since an undamped MDOF system is assumed. In the third variant, only the number of DOF of the system is declared as an integer value.

- `mdof(double[][] K, double[][] M, double[][] C)`
- `mdof(double[][] K, double[][] M)`
- `mdof(int NDOFS)`

Methods:
In as much as possible, the naming of the methods used here approximates their function so as to make them readily understandable.

- void setC(double[ ][ ] c)
- void setC(double am, double ak)
- void setC(double[ ] xis)
- void setC(double[ ] xis, boolean computeC)
- void setDiagonalMass(boolean b)

continued …

Class mdof

Methods:

- void eigenAnalysis()
- double getEigenMass(int i)
- double getEigenValue(int i)
- double[ ] getEigenValues()
- double[ ] getEigenVecArray(int i)
- RealVector getEigenVector(int i)
- Complex[ ][ ] TransferFunction(double omega)
- void setRHS(DoubleFunction df)
- void setRHS(DoubleFunction df, int dof)
- void setRHS(DoubleFunction[] df)
- void solve(double tot, double dt)
- double[ ] Disp(int w)

3.8   User Defined Class as Structural Dynamics SDE Entity

We take advantage of the opportunity to show how one can programme new entities. To that purpose we intetionally keep the one-storey 3D building not as a core part of `structuraldynamics` package. Therefore, the one-storey 3D building is introduced as a user defined entity. While we could keep it simpler, the mechanism for introducing such a new entity the `contraption` interface which mainly serves for a graphical representation and uses the SDE graphics panel.

Although not necesary we choose for the `oneStorey` to be a subclass of the `mdof` one presented in Sect. 3.7.3. The code for `oneStorey` class is given as Appendix A.12. Therefore, `oneStorey` inherits all the variable and methods of its parent's class `mdof`, together with some new variables that are metnioned in the following description. For further assistance one could refer to examples given in Sect. 3.9.2.

3.8.1   Class oneStorey

Description:
The present object is a subclass of mdof presented in Sect. 3.8.1 and is a digital simplification analogue of a three dimensional one storey bulding. As an mdof object it can
be solved for specific time depended loading with specific initil conditions.

Instance setup:
We do not define some specific constructor while we rather use Groovy's map constructors with annotation [3]. The additional to mdof parameters of the oneStorey class
are:

- int id Is the id of the instance f the object.
- double Elast, Poiss=0.25 Stand for the elasticity modulus, the Poisson's ratio
  (given also some default value)
- double Lx, Ly, mass The x- and y-dirrection lengths as well as the total mass.
- double x0=0.0, y0=0.0, h=1.0 Reference point's coordinates (zero by
  default) to specificy location of rest and the height of the storey (with default value
  of one).
- boolean stif=false When false stiffness coefficients are computed using the
  fixed column value, that is $\frac{12EI}{h^3}$ with the E the elasticity modulus and the moments
  of inertia as given below. In case of being true, then it is assumed the values passed
  by the user are refered directy to stiffness coefficients.
- ArrayList columns In that array whose elements are other double array of up to
  five elements, user can pass the data for the columns of the single storey bulding.
  Each of these double arrays contain in their first two elements the cordinates x and
  y of the column, the third element is the I_x moment of inertia, the fourth if exist is
  the I_y moment of inertia while in absence it is assumed to be $I_y = I_x$ and finally the
  fifth element when given is the value of the polar moment of interia that in absence
  assume to be zero valued.

Methods:
Having initialize some instance of the oneStorey where we have given the necesary parameters then we should use the setmdof method to appropriately initialize
oneStorey as an mdof that will be capable of solving the eigenproblem and/or some
transient repsonse.

- void setmdof()

3.9   Application Examples

3.9.1   Coupled Shear Frames Modelled as a 2-DOF Dynamic System

The two shear frames shown in Fig. 3.6 are coupled through a spring-dashpot system. The stiffness of the supporting columns are $k_1 = 13377.78$ kN/m, $k_2 = 12902.51$ kN/m, and $k_0 = 0.5(k_1 + k_2)$. The floor masses are $m_1 = 7.95$ tn and $m_2 = 9.25$ tn. Regarding the damping in this structural system, we assume that the modal damping ratios that correspond to the two modes of the combined system are $\xi_1 = \xi_2 = \xi, \xi = 2\%$. The vector of the external forces is $f_2(t) = f_1(t) = 100$ up until $t \leq T_1/2$, past which the forces are removed from the system. Therefore, the time variation of the forces is the rectangular pulse that appeared in Fig. 2.19 of Sect. 2.14.1. For this example, we will compute its dynamic response using modal superposition.

Solution of the Eigenproblem

At first, we solve the eigenproblem for the undamped 2-DOF system, Eq. (3.16). Following the formulation of the equations of motion describing the 2-DOF system, see Sect. 4.6.3, we substitute numerical values for the mass and stiffness coefficients in the system matrix:

$$\begin{bmatrix} k_1 + k_0 - \lambda_i m_1 & -k_0 \\ -k_0 & k_2 + k_0 - \lambda_i m_2 \end{bmatrix} \begin{Bmatrix} \phi_{1i} \\ \phi_{2i} \end{Bmatrix} = \begin{Bmatrix} 0 \\ 0 \end{Bmatrix}$$

The eigenvalues λ of the 2-DOF system are computed by setting the determinant of the system matrix, Eq. (3.17), equal to zero. This yields

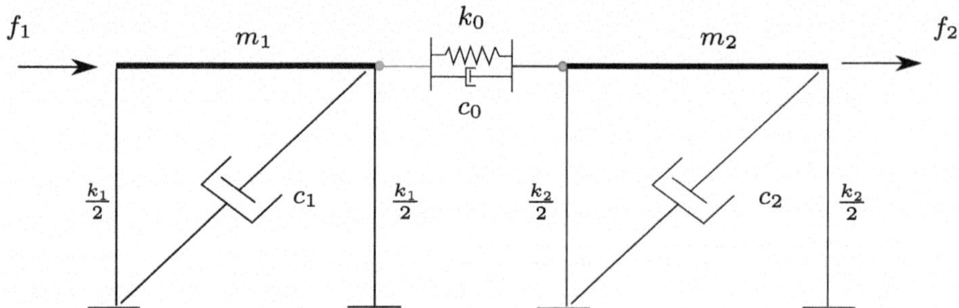

Fig. 3.6 Coupled shear frames modelled as a 2-DOF sytems

$$\begin{vmatrix} k_1 + k_0 - \lambda m_1 & -k_0 \\ -k_0 & k_2 + k_0 - \lambda m_2 \end{vmatrix} = 0 \rightarrow$$

$$\begin{vmatrix} 26517.925 - 7.95\lambda & -13140.145 \\ -13140.145 & 26042.655 - 9.25\lambda \end{vmatrix} = 0 \rightarrow$$

$$73.538\lambda^2 - 452329.914\lambda + 517933761.470 = 0$$

Solving this second-order polynomial in terms of λ, we find the following roots:

$$\lambda_1 = 1521.288, \quad \omega_1 = \sqrt{\lambda_1} = 39.004 \text{ rad/s}, \quad T_1 = \frac{2\pi}{\omega_1} = 0.161 \text{ s}$$

$$\lambda_2 = 4629.681, \quad \omega_2 = \sqrt{\lambda_2} = 68.042 \text{ rad/s}, \quad T_2 = \frac{2\pi}{\omega_2} = 0.092 \text{ s}$$

In order to compute the first eigenvector, we substitute the corresponding eigenvalue

$$\begin{bmatrix} k_1 + k_0 - \lambda_1 m_1 & -k_0 \\ -k_0 & k_2 + k_0 - \lambda_1 m_2 \end{bmatrix} \begin{Bmatrix} \phi_{11} \\ \phi_{21} \end{Bmatrix} = \begin{Bmatrix} 0 \\ 0 \end{Bmatrix}$$

and assume a fixed value (unity) for the first entry of the eigenvector $\phi_{11} = 1$,

$$\begin{bmatrix} 14423.685 & -13140.145 \\ -13140.145 & 11970.741 \end{bmatrix} \begin{Bmatrix} \phi_{11} = 1 \\ \phi_{21} \end{Bmatrix} = \begin{Bmatrix} 0 \\ 0 \end{Bmatrix}$$

The above two equations are linearly dependent, so either one can be solved to yield the value of $\phi_{21} \approx 1.098$. The remaining equation can simply be used for verification, although there might be some small divergence in the recovered value because of round-off. Similarly, we can compute the second eigenvector by substituting the second eigenvalue λ_2 in the system matrix and again assuming a fixed value (unity) for one of the vector components, i.e., for $\phi_{12} = 1$,

$$\begin{bmatrix} -10288.039 & -13140.145 \\ -13140.145 & -16781.894 \end{bmatrix} \begin{Bmatrix} \phi_{12} = 1 \\ \phi_{22} \end{Bmatrix} = \begin{Bmatrix} 0 \\ 0 \end{Bmatrix}$$

From the above, $\phi_{22} \approx -0.783$. Summarizing, the eigenvectors of the 2-DOF system computed with a normalization of the component corresponding to the first DOF of the dynamic system:

$$\phi_1 = \begin{Bmatrix} \phi_{11} \\ \phi_{21} \end{Bmatrix} = \begin{Bmatrix} 1 \\ 1.098 \end{Bmatrix} \text{ and } \phi_2 = \begin{Bmatrix} \phi_{12} \\ \phi_{22} \end{Bmatrix} = \begin{Bmatrix} 1 \\ -0.783 \end{Bmatrix}$$

Both eigenvectors are now sketched in Figs. 3.7 and 3.8.

Finally, the generalized (or modal) masses and their corresponding stiffnesses (for the particular eigenvector normalization used here) are given below as

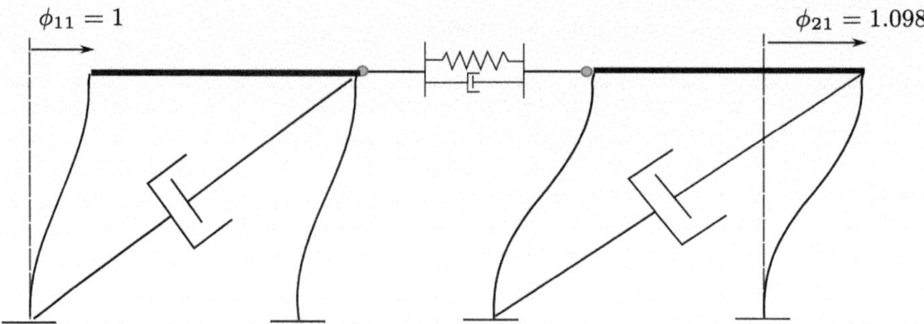

Fig. 3.7 First eigenvector normalized by using a unit value for the vector component corresponding to the first DOF of the dynamic system

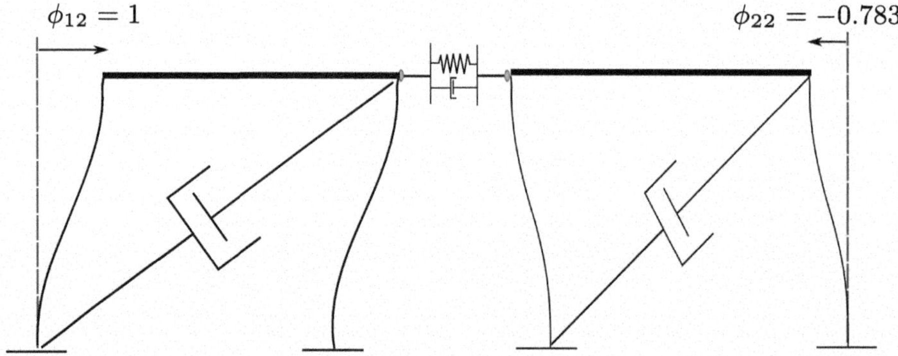

Fig. 3.8 Second eigenvector normalized by using a unit value for the vector component corresponding to the first DOF of the dynamic system

$$m_1^* = \phi_1^T M \phi_1 = 19.102, \quad k_1^* = \phi_1^T K \phi_1 = \lambda_1 m_1^* = \omega_1^2 m_1^* = 29059.854$$
$$m_2^* = \phi_2^T M \phi_2 = 13.621, \quad k_2^* = \phi_2^T K \phi_2 = \lambda_2 m_2^* = \omega_2^2 m_2^* = 63061.670$$

where M and K are the original mass and stiffness matrices.

Eigenvector Normalization With Respect ot the Mass Matrix

In order to normalize the MDOF system eigenvectors with respect to the mass matrix, an appropriate scaling factor must be determined. Specifically, for a given eigenvector ϕ_i the relation $\phi_i^T M \phi_1 == 1$ must hold, implying that the generalized (modal) matrix is now equal to unity. To this end, we commence with the first eigenvector of our 2-DOF system and divide its components by the square root of the corresponding (first) generalized mass. This way, dividing ϕ_1 by the square root $\sqrt{m_1^*}$, the new eigenvector will be normalized:

$$\phi_1 = \left\{ \begin{matrix} \phi_{11}/\sqrt{m_1^*} \\ \phi_{21}/\sqrt{m_1^*} \end{matrix} \right\} = \left\{ \begin{matrix} 1/\sqrt{19.102} \\ 1.098/\sqrt{19.102} \end{matrix} \right\} = \left\{ \begin{matrix} 0.229 \\ 0.251 \end{matrix} \right\}$$

Similarly for the second eigenvector, normalization is achieved as follows

$$\phi_2 = \left\{ \begin{matrix} \phi_{12}/\sqrt{m_2^*} \\ \phi_{22}/\sqrt{m_2^*} \end{matrix} \right\} = \left\{ \begin{matrix} 1/\sqrt{13.621} \\ -0.783/\sqrt{13.621} \end{matrix} \right\} = \left\{ \begin{matrix} 0.271 \\ -0.212 \end{matrix} \right\}$$

Finally, the matrix of the normalized (with respect to the mass) eigenvectors is the following one:

$$\Phi = \begin{bmatrix} \phi_{11} & \phi_{12} \\ \phi_{21} & \phi_{22} \end{bmatrix} = \begin{bmatrix} 0.229 & 0.271 \\ 0.251 & -0.212 \end{bmatrix}.$$

Modal Damping and the Modal Damping Matrix

In the start of the example, a damping ratio of $\xi = 2\%$ was assumed for both modes of the 2-DOF system. This implies that each modal equation will have a corresponding damper equal to $c_i^* = 2m_i^* \omega_i \xi_i$. Keeping in mind that $\xi_i = \xi$ and the eigenvector normalization with respect to the mass, the damper value is now $c_i^* = 2\omega_i \xi_i$. The damping matrix in the generalized (modal) coordinates will be diagonal because of the orthogonality relation holding for the eigenvectors:

$$C^* = \begin{bmatrix} 2\xi\omega_1 & 0 \\ 0 & 2\xi\omega_2 \end{bmatrix} = \begin{bmatrix} 1.560 & 0 \\ 0 & 2.722 \end{bmatrix}.$$

From Eq. (3.29) we can evaluate the damping matrix of the 2-DOF system corresponding to the previously specified damping ratios:

$$C = M\Phi C^* \Phi^T M = \begin{bmatrix} 17.804 & -4.904 \\ -4.904 & 18.876 \end{bmatrix} \tag{3.48}$$

It is noted here that in order for the damping matrix to correspond to the Lagrange equations see Eq. (4.28), see Sect. 4.6.3, the following relations must hold:

$$c_0 = -4.904$$

$$c_1 + c_0 = 17.804 \rightarrow c_1 = 22.708$$

$$c_2 + c_0 = 18.876 \rightarrow c_2 = 23.780.$$

Solution by Modal Superposition

From Eq. (3.32), modal superposition is based on expressing the physical coordinates (the displacement vector in particular) in terms of the generalized coordinates as

$$u(t) = \phi_1 q_1(t) + \phi_2 q_2(t) \rightarrow \begin{Bmatrix} u_1(t) \\ u_2(t) \end{Bmatrix} = \begin{Bmatrix} \phi_{11} \\ \phi_{21} \end{Bmatrix} q_1(t) + \begin{Bmatrix} \phi_{12} \\ \phi_{22} \end{Bmatrix} q_2(t)$$

For this case, the vector base used is has the eigenvectors normalized with respect to the mass. Taking into account that the force vector must be expressed in modal coordinates as well, namely $f_1^*(t) = \phi_1^T f(t)$ and $f_2^*(t) = \phi_2^T f(t)$, the two uncoupled differential equations of motion are

$$\ddot{q}_1(t) + 2\xi\omega_1\dot{q}_1(t) + \omega_1^2 q_1(t) = \phi_{11} f_1(t) + \phi_{21} f_2(t) \tag{3.49}$$

$$\ddot{q}_2(t) + 2\xi\omega_2\dot{q}_2(t) + \omega_2^2 q_2(t) = \phi_{12} f_1(t) + \phi_{22} f_2(t) \tag{3.50}$$

Substituting numerical values for the 2-DOF system parameters yields

$$\ddot{q}_1(t) + 1.560\dot{q}_1(t) + 1521.312 q_1(t) = 48.0 \tag{3.51}$$

$$\ddot{q}_2(t) + 2.722\dot{q}_2(t) + 4629.714 q_2(t) = 5.9 \tag{3.52}$$

for time $t \leq T_1/2 = 0.161/2$, past which we have a free vibration regime. Also, initial conditions are assumed to be zero. At this point, the above system of equations can be solved either analytically (see Sect. 2.14.1) or numerically, keeping in mind that these are essentially two separate SDOF equations. Here we opt to solve them numerically for a time interval exceeding the time of application of the loads, that is for $T_{tot} = 4T_1$. The results for the 2 DOF of the dynamic system are shown separately in Figs. 3.9 and 3.10. Concurrently plotted in these figures are the contributions of the generalized (modal) coordinates, where we observe that the second mode (in green color) contributes very little to the physical displacements. This fact could lead to the omission of the second mode's contribution, something which is of course related to the accuracy level desired in the analysis. In general one advantage of modal analysis is exactly this option, namely the possibility to ignore the contribution of the higher modes in synthesizing the dynamic system's response.

We further include Listing 3.2 with appropriate code to numerically solve the problem and make the some plot of the response of the system using the `mdof` entity presented in Sect. 3.7.3.

```
import courses.structuraldynamics.*
k= [[26517.925,-13140.145],[-13140.145, 26042.655]]
    as double[][]
m = [[7.95,0.0],[0.0, 9.25]] as double[][]
model = new mdof(k,m)
println model.getEigenValues()
model.setC([0.02,0.02] as double[])
```

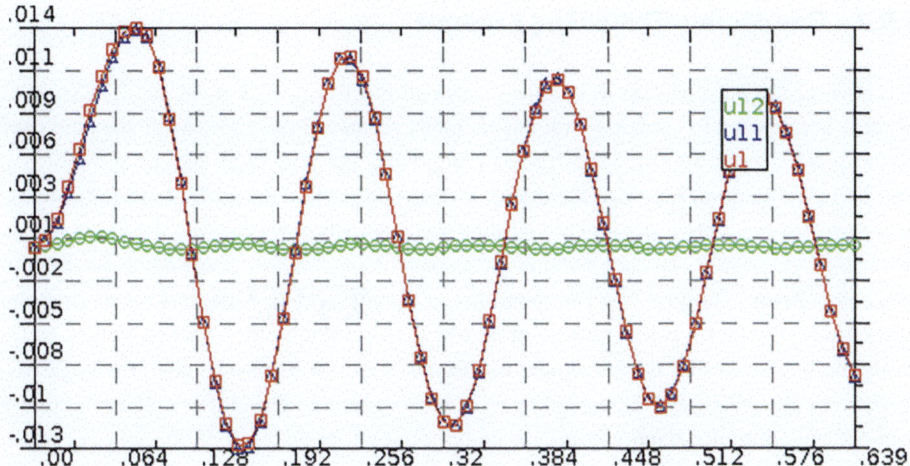

Fig. 3.9 Displacement time history $u_1(t)$ of the left frame. The contribution of the first mode is $u_{11}(t) = \phi_{11}q_1(t)$, while that of the second mode is $u_{12}(t) = \phi_{12}q_2(t)$

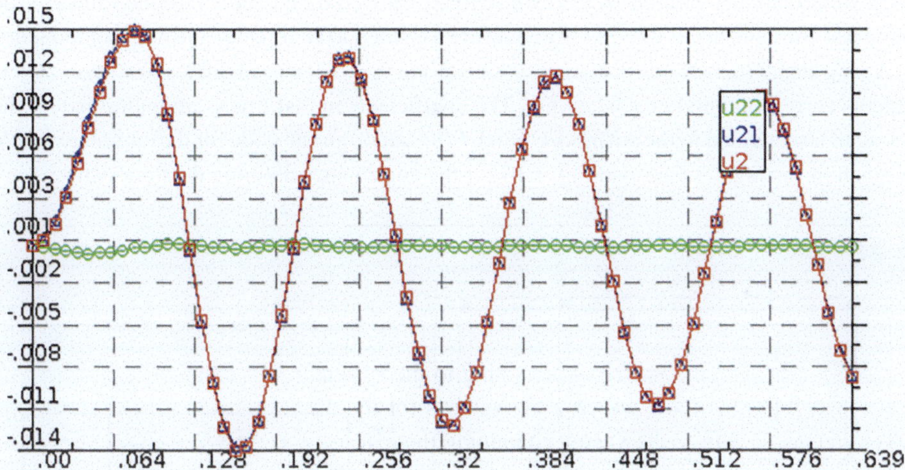

Fig. 3.10 Displacement time history $u_2(t)$ of the right frame. The contribution of the first mode is $u_{21}(t) = \phi_{21}q_1(t)$, while that of the second mode is $u_{22} = \phi_{22}q_2(t)$

```
7   T1=2*pi/sqrt(model.getEigenValues()[0])
8   model.setRHS({if(it<T1/2.0){100.0d}else{0.0d}} as DF)
9   tot=4.0*T1; dt=0.001
10  model.solve(tot, dt)
11  plot(dt,model.Disp(1))
12  plot(dt,model.Disp(2))
```

Listing 3.2 Response calculation of problem in Sect. 3.9.1 using the mdof entity

3.9.2 One-Storey 3D Building Example

As another example, consider the single-storey building of Fig. 3.11 found in [2]. The center of mass is supposed to coincide with the geometric center of the plate, while the total mass of the plate is $m = 590$(tn). The frames comprising the structural system are shown in Fig. 3.11. We mention here that there are two ways to declare the stiffness of the columns of a `oneStorey` entity, the first by defining the moments of inertia into the columns data in variable `columns`. To each column an array of double values corresponds havin up to five components. The first two corrspond to coordinates x and y, the thrid is the moment of inertia I_x, the fourth is the moment of inertia I_y (for a symmetric cross-section $I_y = I_x$), while the fifth is for the polar moment of inertia I_p, which is zero if torsion is neglected. In such a case the `oneStorey` entity in order to compute the stiffness of each column uses the equation of a fixed element given by expressions $12EI_i/h^3$ for $i = x, y$ and for the torsional stiffness GI_d/h, where h the height of the storey, E the elasticity modulus and G the shear modulus. In this case we should define in `oneStorey` entity the values for variable 'h' of height, 'Elast' of elasticity modulus and 'Pois' of Poisson's ratio. In case, we declare the variable of the `oneStorey` entity `stif` to be `true` the values of the three last components of the `column` data are assumed to be direclty stiffness components.

Using standard methods of statics, as shown in Fig. 3.12, we calculate the stiffness coefficients corresponding to each column. The coefficients in that figure are multiplied by the constant term related to the stiffness $k = 6EI/h^2$. Appropriate code for the current example is given in Listing 3.3 (see also Fig. 3.13 for the coordinates defining the floor plan).

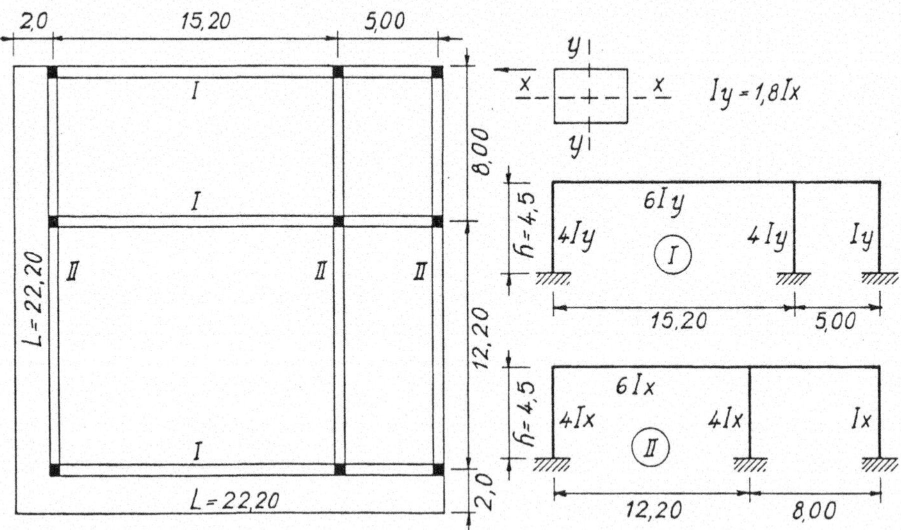

Fig. 3.11 The one storey building appearing in [2]

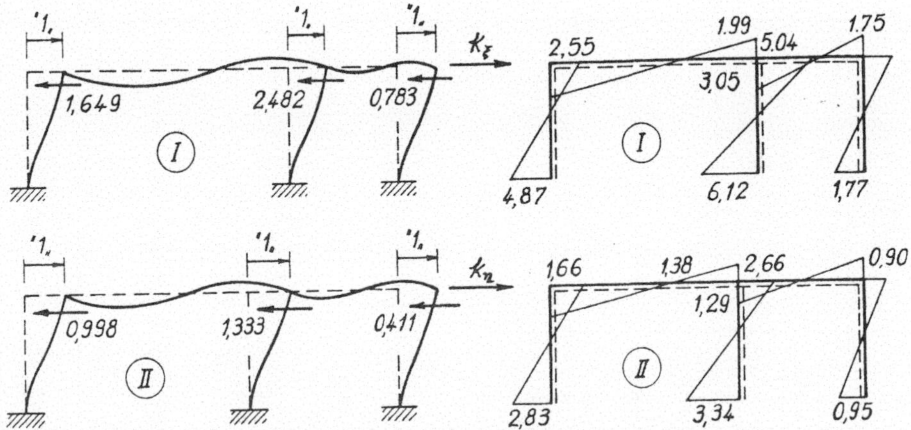

Fig. 3.12 Cross-sections I and II depicting the underlying frames showing the deformation to a unit horizontal displacement (left) and the corresponding shear force diagram (right)

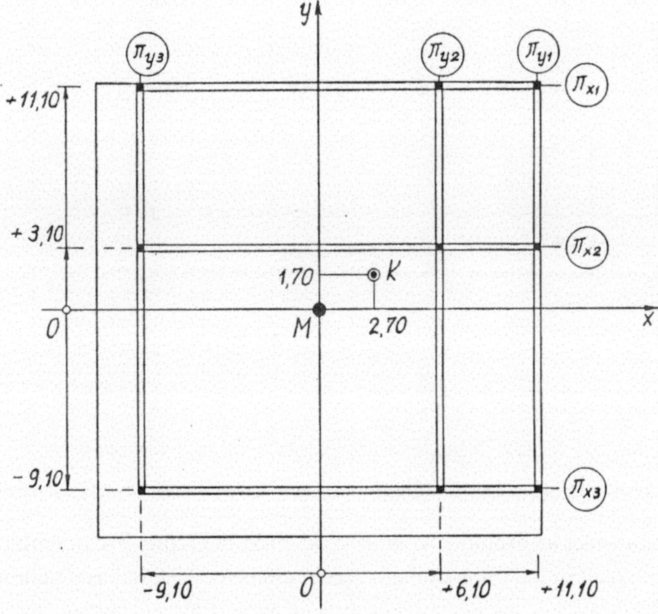

Fig. 3.13 Plan view of the floor with coordinates used to define the `oneStorey` entity. The coordinates of the expected elastic center are also presented

```
theUniverse.cls()
k=15.05e3
Ix147=1.649*k
Iy123=0.998*k
Ix258=2.482*k
Iy456=1.333*k
Ix369=0.783*k
Iy789=0.411*k
L=22.2
Building3D1St = new oneStorey(Lx:L, Ly:L)
Building3D1St.columns=[
    [-9.1,-9.1,Ix147,Iy123],
    [6.1,-9.1,Ix258,Iy123],
    [11.1,-9.1,Ix369,Iy123],
    [-9.1,3.1,Ix147,Iy456],
    [6.1,3.1,Ix258,Iy456],
    [11.1,3.1,Ix369,Iy456],
    [-9.1,11.1,Ix147,Iy789],
    [6.1,11.1,Ix258,Iy789],
    [11.1,11.1,Ix369,Iy789]
]
Building3D1St.mass=560.0
Building3D1St.stif=true
theUniverse.putContraption(Building3D1St)
Building3D1St.setmdof()
println Building3D1St.K
println Building3D1St.M
println "elastic center at:"+Building3D1St.
    getElasticCenter()
println Building3D1St.getEigenValues()

scale=1.0; which=2
theGP.title="eigenshape: ${which+1} for
    eigenfrequency "+Building3D1St.getEigenValues()[
    which]+"(rad/sec)"
theGP.plotDeform(scale,which)
```

Listing 3.3 Introducing one storey building of Fig. 3.13 into `oneStorey` entity

We further consider an example, with its code given in Listing 3.4, of multiple one storey 3D buildings aligned in the same domain that could be considered as a building block, see Fig. 3.14.

```
bx=0.4; by=0.55
Ix=by*bx**3/12;  Iy=bx*by**3/12;

theUniverse.cls()
Lx=5.0; Ly=3.0;
Building3D1St = new oneStorey(Lx:Lx, Ly:Ly)
```

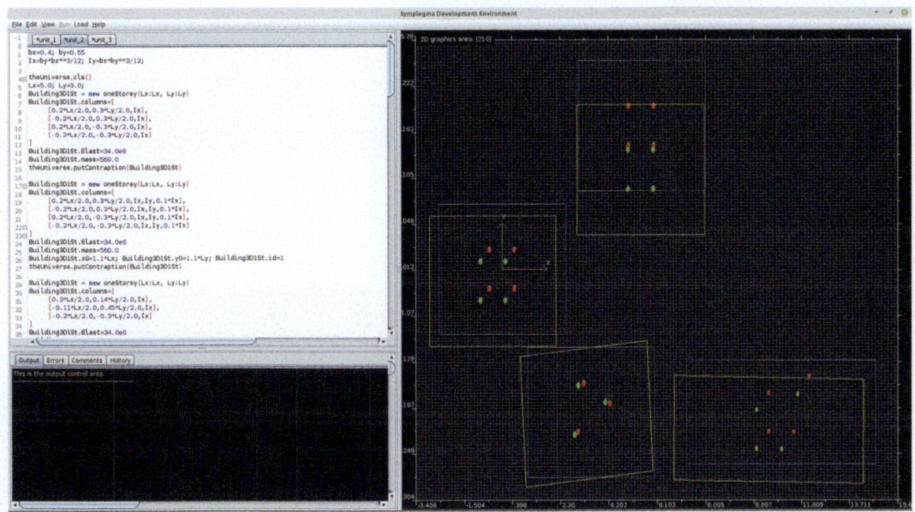

Fig. 3.14 Screenshot of response of the building block of distinct one storey building under random initial conditions

```
 7  Building3D1St.columns=[
 8       [0.2*Lx/2.0,0.3*Ly/2.0,Ix],
 9       [-0.2*Lx/2.0,0.3*Ly/2.0,Ix],
10       [0.2*Lx/2.0,-0.3*Ly/2.0,Ix],
11       [-0.2*Lx/2.0,-0.3*Ly/2.0,Ix]
12  ]
13  Building3D1St.Elast=34.0e6
14  Building3D1St.mass=560.0
15  theUniverse.putContraption(Building3D1St)
16
17  Building3D1St = new oneStorey(Lx:Lx, Ly:Ly)
18  Building3D1St.columns=[
19       [0.2*Lx/2.0,0.3*Ly/2.0,Ix,Iy,0.1*Ix],
20       [-0.2*Lx/2.0,0.3*Ly/2.0,Ix,Iy,0.1*Ix],
21       [0.2*Lx/2.0,-0.3*Ly/2.0,Ix,Iy,0.1*Ix],
22       [-0.2*Lx/2.0,-0.3*Ly/2.0,Ix,Iy,0.1*Ix]
23  ]
24  Building3D1St.Elast=34.0e6
25  Building3D1St.mass=560.0
26  Building3D1St.x0=1.1*Lx;  Building3D1St.y0=1.1*Ly;
        Building3D1St.id=1
27  theUniverse.putContraption(Building3D1St)
28
29  Building3D1St = new oneStorey(Lx:Lx, Ly:Ly)
30  Building3D1St.columns=[
31       [0.3*Lx/2.0,0.14*Ly/2.0,Ix],
```

```
32      [-0.11*Lx/2.0,0.45*Ly/2.0,Ix],
33      [-0.2*Lx/2.0,-0.3*Ly/2.0,Ix]
34  ]
35  Building3D1St.Elast=34.0e6
36  Building3D1St.mass=560.0
37  Building3D1St.x0=0.7*Lx; Building3D1St.y0=-1.1*Ly;
       Building3D1St.id=2
38  theUniverse.putContraption(Building3D1St)
39
40  Building3D1St = new oneStorey(Lx:Lx*1.5, Ly:Ly*0.8)
41  Building3D1St.columns=[
42      [0.45*Lx/2.0,0.55*Ly/2.0,Ix],
43      [-0.2*Lx/2.0,0.3*Ly/2.0,Ix],
44      [0.2*Lx/2.0,-0.3*Ly/2.0,Ix],
45      [-0.2*Lx/2.0,-0.3*Ly/2.0,Ix]
46  ]
47  Building3D1St.Elast=34.0e6
48  Building3D1St.mass=560.0
49  Building3D1St.x0=2.2*Lx; Building3D1St.y0=-1.1*Ly;
       Building3D1St.id=3
50  theUniverse.putContraption(Building3D1St)
51
52  tot=60.0; dt=0.001; Nt=floor(tot/dt) as int
53  theUniverse.getContraptions().each{
54      it.value.h=4.5
55      it.value.setmdof()
56      it.value.setC([0.05]*3 as double[], true)
57      it.value.u0[0]=random()
58      it.value.u0[1]=random()
59      it.value.solve(tot, dt)
60  }
61
62  scale=1.0;
63  theGP.plotDeform(scale,0,Nt,1)
```

Listing 3.4 A building block of one storey buildings using `oneStorey` entity

References

1. Manolis G, Koliopoulos P, Panagiotopoulos C (2015) Dynamics of structures. Hellenic Academic Libraries Link. Kallipos electronic book project, National Technical University of Athens (in Greek)
2. Anastasiadis K (1999) Antiseismic structures. Zitis Publications, Thessaloniki, Greece (in Greek) In greek
3. Hubert A, Ikkink K (2018) Add map constructor with annotation. Cambridge University Press, Groovy Goodness

Formulation of Equations of Motion 4

4.1 D'Alembert's Principle

This principle is the outcome of Newton's second law that states that the rate of change of the momentum of a moving mass m equals the sum of forces acting upon it. In the case of an SDOF system, the principle can be written as follows:

$$m\frac{\partial \dot{u}}{\partial t} = \sum_i f_i(t) = f(t) - f_D(t) - f_S(t) \rightarrow$$

$$m\ddot{u}(t) + c\dot{u} + ku(t) = f(t)$$

This yields a dynamic equilibrium equation where at any time instant t, the sum of the inertia force plus the damping force and the restoring force equal the externally imposed force on the SDOF oscillator.

4.2 The Principle of Virtual Work

According to this principle, if a virtual (i.e., hypothetical) displacement is imposed on an elastic body acted upon by a system of external forces, then the work produced by these forces equals zero. This is true provided that the virtual displacement does not violate the boundary conditions. Specifically, if a virtual displacement δu, consistent with the boundary conditions, is imposed on the SDOF system, the point of application of all forces moves and the total work produced is given as

$$\delta \mathcal{E} = -m\ddot{u}\delta u - c\dot{u}\delta u - ku\delta u + f\delta u = 0$$

© The Author(s), under exclusive license to Springer Nature Switzerland AG 2026
G. Manolis and C. Panagiotopoulos, *Vibrations of Structural Systems*, Synthesis Lectures
on Mechanical Engineering, https://doi.org/10.1007/978-3-032-12279-7_4

Upon cancellation of the common virtual displacement, the equation of motion results, as was previously derived, see Eq. (2.1). Note that in the above virtual work $\delta\mathcal{E}$ statement, the negative sign implies that the force involved has an opposite direction to the virtual displacement δu.

4.3　Hamilton's Principle

According to Hamilton's principle, if the change in the kinetic and potential energy in an elastic body during time interval (t_1, t_2) is added to the work produced by external, non-conservative forces, then this sum is equal to zero. More specifically, we define the kinetic energy as

$$T = \frac{1}{2}m\dot{u}^2$$

and the potential energy as

$$V = \frac{1}{2}ku^2$$

that includes all restoring and conservative force in the elastic body. Furthermore, we define as W the work produced by external non-conservative forces. Introducing the mathematical concept of variation, δ, allows for writing the resulting equation for the SDOF oscillator as follows:

$$\int_{t_1}^{t_2} \delta(T - V)dt + \int_{t_1}^{t_2} \delta W dt = 0 \tag{4.1}$$

It is straightforward to show that the variation in all terms appearing above are $\delta T = m\dot{u}\delta\dot{u}$, $\delta V = ku\delta u$ and $\delta W = f(t)\delta u - c\dot{u}\delta u$. Following substitution in Hamilton's equation, we now have

$$\int_{t_1}^{t_2} \left(-m\ddot{u} - c\dot{u} - ku + f(t)\right)\delta u\, dt = 0$$

Since the above equation holds true for any non-zero, but otherwise arbitrary variation δu, then the terms inside the parenthesis must be zero. This way, we end up with the same Eq. (2.1) as the one dervived by the previous methods.

4.4   Lagrange's Equations

Lagrange's equations are a direct application of Eq. 4.1, where the kinetic energy $\mathcal{T}$, the potential energy $\mathcal{V}$ and the external work term $\delta \mathcal{W}$ are all expressed in terms of the generalized coordinates $q_1, q_2, \ldots, q_N$. By the term generalized coordinates we refer to the necessary number of parameters needed to define the motion of the dynamic system, that is the minimum number of DOF, which may or may not have a physical meaning. For the majority of structural and mechanical systems, their kinetic energy may be expressed in terms of generalized coordinates plus their first time derivatives, while their potential energy in terms of the generalized coordinates only. Finally, the external work can be expressed as a linear combination of these coordinates. We therefore have

$$\mathcal{T} = \mathcal{T}(q_1, q_2, \ldots, q_N, \dot{q}_1, \dot{q}_2, \ldots, \dot{q}_N) \tag{4.2}$$

$$\mathcal{V} = \mathcal{V}(q_1, q_2, \ldots, q_N) \tag{4.3}$$

$$\delta \mathcal{W} = Q_1 \delta q_1 + Q_2 \delta q_2 + \ldots + Q_N \delta q_N \tag{4.4}$$

where the terms $Q_1, Q_2, \ldots, Q_N$ are the generalized external load generalized coordinates which correspond to the original generalized coordinates $q_1, q_2, \ldots, q_N$. Substitution of the above equations in Hamilton's principle, Eq. (4.1), followed by integration bu parts of the velocity terms and setting the variations $\delta q_i(t_1) = \delta q_i(t_2) = 0$, the following equations are derived:

$$\frac{d}{dt}\left(\frac{\partial \mathcal{T}}{\partial \dot{q}_i}\right) - \frac{\partial \mathcal{T}}{\partial q_i} + \frac{\partial \mathcal{V}}{\partial q_i} = Q_i \tag{4.5}$$

These are known as Lagrange's equations of motion, which have a wide range of application in engineering science.

4.5   Package `courses.structuraldynamics`

4.5.1   Class `pendulum`

Description:
This object addresses the problem of a double pendulum comprising two masses inter-connected by rigid and weightless rods, see Fig. 4.1. Each mass has a value of m_i and each rod has a length of of l_i which starts from an initial value of unity. In the absence of external forces, this pendulum executes free vibrations. The present object contains methods for the solution of initial value problems of both linear and quasi-linear differential equations. The current version of this object allows for the solution of a simple pendulum with one mass and also of the double pendulum (two masses).

Creation of a snapshot:
The constructor of the snapshot of an object appears in three versions: In the first version we define sets of two real-number, double precision variables which correspond to the masses and length of the connecting rods. In the second version, these numbers are omitted as they are assumed to correspond to unit values. The third version addresses the simple pendulum and requires only the definition of the length of the connecting rod.

- `pendulum(double[] masses, double[] lengths)`
- `pendulum(double[] masses)`
- `pendulum(double len)`

Methods:
The naming of the methods used here is one in a manner that alludes to their function.

- void setLengths(double[ ] lengths)
- void setInitCond(double[ ] th0, double[ ] dth0)
- void setInitCond(double[ ] th0)
- void solve(double tot, double dt)

continued …

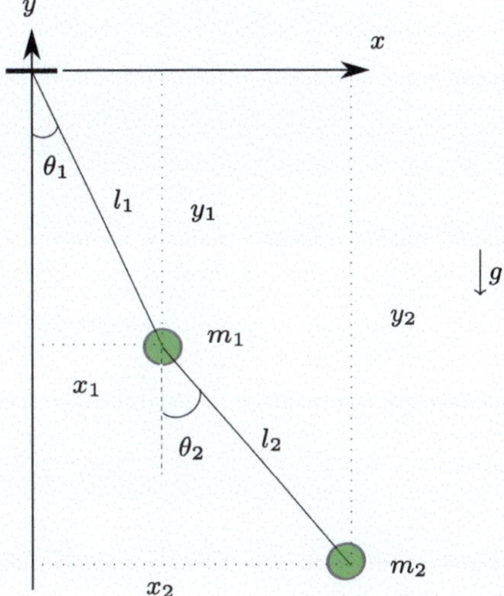

Fig. 4.1 Double pendulum system comprising two masses connected by rigid and weightless rods

Class pendulum

Methods:

- void setGravity(double g)
- void setOrigin(double x0, double y0)
- double[ ] Theta(int dof)
- double[ ] Theta()
- double[ ] DTheta(int dof)
- double[ ] DTheta()
- double[ ] Xdef(int dof)
- double[ ] Ydef(int dof)

4.6   Application Examples

4.6.1   The Double Pendulum

At first, the equations of motion of the double pendulum under free vibrations will be derived from the Lagrange equations, after the constituent energy quantities, Eqs. (4.2)–(4.4) are expressed as functions of the generalized coordinates and their time derivatives. To this end, we will select the angles traced between the two rigid rods and the vertical direction, namely θ_1 and θ_2, as the generalized coordinates. The Cartesian coordinates of masses m_1 and m_2 can then be expressed in terms of the generalized coordinates as follows:

$$x_1 = l_1 \sin \theta_1 \qquad\qquad \dot{x}_1 = l_1 \dot{\theta}_1 \cos \theta_1$$

$$y_1 = -l_1 \cos \theta_1 \qquad\qquad \dot{y}_1 = l_1 \dot{\theta}_1 \sin \theta_1$$

$$x_2 = x_1 + l_2 \sin \theta_2 \qquad\qquad \dot{x}_2 = \dot{x}_1 + l_2 \dot{\theta}_2 \cos \theta_2$$

$$y_2 = y_1 - l_2 \cos \theta_2 \qquad\qquad \dot{y}_2 = \dot{y}_1 + l_2 \dot{\theta}_2 \sin \theta_2 \tag{4.6}$$

Next, the kinetic energy of the dynamic system is

$$T = \frac{1}{2} m_1 (\dot{x}_1^2 + \dot{y}_1^2) + \frac{1}{2} m_1 (\dot{x}_2^2 + \dot{y}_2^2)$$

Following substitution of the relations given in Eq. (4.6), we obtain

$$T = \frac{1}{2}(m_1 + m_2) l_1^2 \dot{\theta}_1^2 + m_2 l_1 l_2 \dot{\theta}_1 \dot{\theta}_2 \cos (\theta_1 - \theta_2) + \frac{1}{2} m_2 l_2^2 \dot{\theta}_2^2.$$

The potential energy of the system is due to the gravity field, so with g the acceleration of gravity we have that

$$V = m_1 g y_1 + m_2 g y_2 + \text{constant}$$

$$= -(m_1 + m_2) g l_1 \cos \theta_1 - m_2 g l_2 \cos \theta_2 + \text{constant}$$

Regarding the work of the external forces δW, it is zero since no such forces and other types of non-conservative forces exist.

The next steep is to write Lagrange's equations, as given in Eq. (4.5), for each of the two generalized coordinates. Starting with $q_i := \theta_1$, we derive

$$\frac{\partial T}{\partial \dot{\theta}_1} = (m_1 + m_2) l_1^2 \dot{\theta}_1 + m_2 l_1 l_2 \dot{\theta}_2 \cos (\theta_1 - \theta_2)$$

The time derivative of the first generalized coordinate is

$$\frac{d}{dt}\left(\frac{\partial T}{\partial \dot{\theta}_1} \right) = (m_1 + m_2) l_1^2 \ddot{\theta}_1 + m_2 l_1 l_2 \ddot{\theta}_2 \cos (\theta_1 - \theta_2)$$

$$- m_2 l_1 l_2 \dot{\theta}_2 (\dot{\theta}_1 - \dot{\theta}_2) \sin (\theta_1 - \theta_2) \tag{4.7}$$

Also, the derivative of the kinetic energy with respect to the angular coordinate θ_1 is

$$\frac{\partial T}{\partial \theta_1} = -m_2 l_1 l_2 \dot{\theta}_1 \dot{\theta}_2 \sin (\theta_1 - \theta_2), \tag{4.8}$$

while the derivative of the potential energy with respect to θ_1 is

$$\frac{\partial V}{\partial \theta_1} = (m_1 + m_2) g l_1 \sin \theta_1. \tag{4.9}$$

Continuing with the second generalized coordiante $q_i := \theta_2$, we compute

$$\frac{\partial \mathcal{T}}{\partial \dot{\theta}_2} = m_2 l_1 l_2 \dot{\theta}_1 \cos(\theta_1 - \theta_2) + m_2 l_2^2 \dot{\theta}_2$$

The time derivative of the second generalized coordinate is

$$\frac{d}{dt}\left(\frac{\partial \mathcal{T}}{\partial \dot{\theta}_2}\right) = m_2 l_1 l_2 \ddot{\theta}_1 \cos(\theta_1 - \theta_2) + m_2 l_2^2 \ddot{\theta}_2$$
$$- m_2 l_1 l_2 \dot{\theta}_1 (\dot{\theta}_1 - \dot{\theta}_2) \sin(\theta_1 - \theta_2) \tag{4.10}$$

Furthermore, the derivative of the kinetic energy with respect to θ_2 is

$$\frac{\partial \mathcal{T}}{\partial \theta_2} = m_2 l_1 l_2 \dot{\theta}_1 \dot{\theta}_2 \sin(\theta_1 - \theta_2), \tag{4.11}$$

while that of the potential energy is with respect to θ_2 is

$$\frac{\partial \mathcal{V}}{\partial \theta_2} = m_2 g l_2 \sin \theta_2. \tag{4.12}$$

According to Eq. (4.5), the two governing equations of motion of the double pendulum are given by adding terms as (4.7) – (4.8) + (4.9) and (4.10) – (4.11) + (4.12),

$$(m_1 + m_2) l_1^2 \ddot{\theta}_1 + m_2 l_1 l_2 \ddot{\theta}_2 \cos(\theta_1 - \theta_2) + m_2 l_1 l_2 \dot{\theta}_2^2 \sin(\theta_1 - \theta_2)$$
$$+ (m_1 + m_2) l_1 g \sin \theta_1 = 0 \tag{4.13}$$

$$m_2 l_1 \ddot{\theta}_1 \cos(\theta_1 - \theta_2) + m_2 l_2 \ddot{\theta}_2 - m_2 l_1 \dot{\theta}_1^2 \sin(\theta_1 - \theta_2) + m_2 g \sin \theta_2 = 0 \tag{4.14}$$

These differential equations are highly non-linear and exhibit chaotic behavior. In solving these equations numerically, e.g., by the Runge-Kutta method, further care must be exercised. As example, we show a snapshot of the solution of the double pendulum in Fig. 4.2, where the trajectories of the two masses are delineated. In order to achieve this solution, the first governing equation is solved for $\ddot{\theta}_1$ and the second one for $\ddot{\theta}_2$. Careful substitution on one expression into the other leads to uncoupling and we now have two equations, one involving $\ddot{\theta}_1$ and the other $\ddot{\theta}_2$. In module `courses.structuraldynamics` and more specifically in object `pendulum`, we use the relations appearing in the internet site.[1] We reproduce here the relations which yield the non-linear governing equation of the double pendulum in a form appropriate for numerical solution using the Runge-Kutta method, that is by converting the second order differential equation system into a first order one. By introducing the time derivatives of the angular displacements $\dot{\theta}_1 = \omega_1$ and $\dot{\theta}_2 = \omega_2$, i.e., the angular velocities, we write the original system of equations Eq. (2.64) as follows:

---

[1] https://myphysicslab.com/pendulum/double-pendulum-en.html.

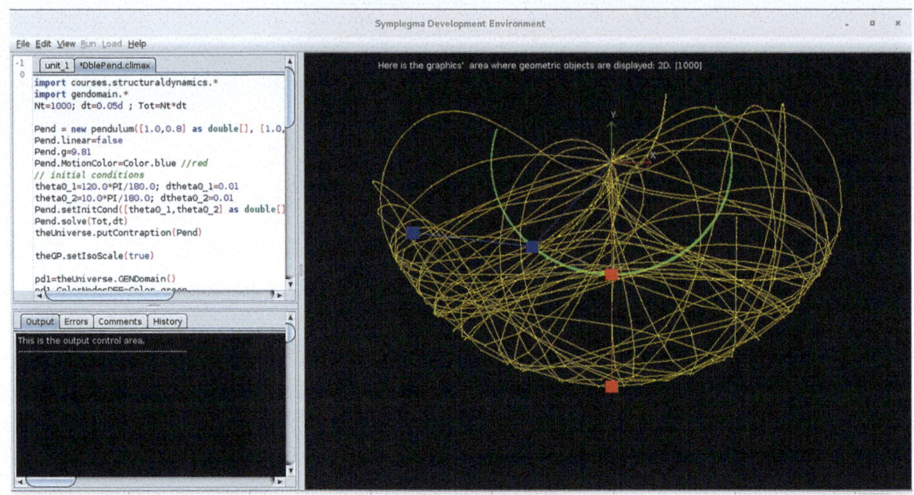

Fig. 4.2 Solution of the double pendulum in environment SDE using object function `pendulum`

$$\dot{\theta}_1 = \omega_1$$
$$\dot{\theta}_2 = \omega_2$$
$$\dot{\omega}_1 = \frac{A_1 + A_2}{l_1 \Gamma}$$
$$\dot{\omega}_2 = \frac{B_1 + B_2}{l_2 \Gamma}$$

Re-arranging terms, we have

$$A_1 = -g(2m_1 + m_2)\sin\theta_1 - m_2 g \sin(\theta_1 - 2\theta_2),$$
$$A_2 = -2\sin(\theta_1 - \theta_2)m_2(\omega_2^2 l_2 + \omega_1^2 l_1 \cos(\theta_1 - \theta_2)),$$

Similarly,

$$B_1 = 2\sin(\theta_1 - \theta_2)(\omega_1^2 l_1(m_1 + m_2),$$
$$B_2 = g(m_1 + m_2)\cos\theta_1 + \omega_2^2 l_2 m_2 \cos(\theta_1 - \theta_2)).$$

and finally

$$\Gamma = (2m_1 + m_2 - m_2 \cos(2\theta_1 - 2\theta_2)).$$

It is noted here that the assumption of small angles θ_1, θ_2 in the pendulums' motions, the governing equations can be linearized if we assume that for a small angle θ (in rad), we have the approximations

$$\sin\theta \approx \theta$$

$$\cos\theta \approx 1 - \frac{\theta^2}{2}.$$

In this case the governing equations assume the form

$$(m_1 + m_2)l_1^2\ddot{\theta}_1 + m_2 l_1 l_2 \ddot{\theta}_2 + (m_1 + m_2)g l_1 \theta_1 = 0$$

$$m_2 l_1 l_2 \ddot{\theta}_1 + m_2 l_2^2 \ddot{\theta}_2 + m_2 g l_2 \theta_2 = 0$$

that can be written in a more compact form using matrix notation:

$$\begin{bmatrix} (m_1 + m_2)l_1^2 & m_2 l_1 l_2 \\ m_2 l_1 l_2 & m_2 l_2^2 \end{bmatrix} \begin{Bmatrix} \ddot{\theta}_1(t) \\ \ddot{\theta}_2(t) \end{Bmatrix} + \begin{bmatrix} (m_1 + m_2)g l_1 & 0 \\ 0 & m_2 g l_2 \end{bmatrix} \begin{Bmatrix} \theta_1(t) \\ \theta_2(t) \end{Bmatrix} = \begin{Bmatrix} 0 \\ 0 \end{Bmatrix}$$

This form is now familiar, i.e., $M\ddot{\theta}(t) + K\theta(t) = 0$, which can be solved by methods previously discussed in reference to SDOF systems, such as the β-Newmark.

4.6.2   The Inverted Pendulum

The inverted pendulum has its mass placed above the equilibrium position. It is supported on a trailer that can move in a horizontal direction through a rigid and massless rod. The difference with the conventional pendulum is that the upper equilibrium position is unstable, in contrast to the lower one that is stable. The result is that the inverted pendulum can be destabilized by a small perturbation about its upper equilibrium position. The inverted pendulum is a classical problem in dynamics used within the context of passive control of structural systems and is often used to calibrate various control algorithms. In what follows, we will model the inverted pendulum as an SDOF system consisting of the trailer's mass m, plus a restoring spring k and a damper c which oppose the motion of the trailer. We also have a hinged connection between the trailer and the rod that supports the mass m_p of the pendulum, with the attachment of a rotational spring k_θ and a corresponding damper c_θ. Motion in the trailer is initiated by an external force f, as can be seen in Fig. 4.3.

In order to derive the equations of motion for the inverted pendulum under free vibrations, we will formulate Lagrange's equations after deriving expressions for the energy quantities (4.2)–(4.4) in terms of the generalized coordinates and their time derivatives. Specifically, we will select the angle traced by the supporting rod with respect to the vertical direction θ and the horizontal displacement of the trailer u as the generalized coordinates. The Cartesian coordinates of the mass of the pendulum m_p and the velocities are given as

$$x_p = u - l\sin\theta \qquad\qquad \dot{x}_p = \dot{u} - l\dot{\theta}\cos\theta$$

$$y_p = l\cos\theta \qquad\qquad \dot{y}_p = -l\dot{\theta}\sin\theta \qquad (4.15)$$

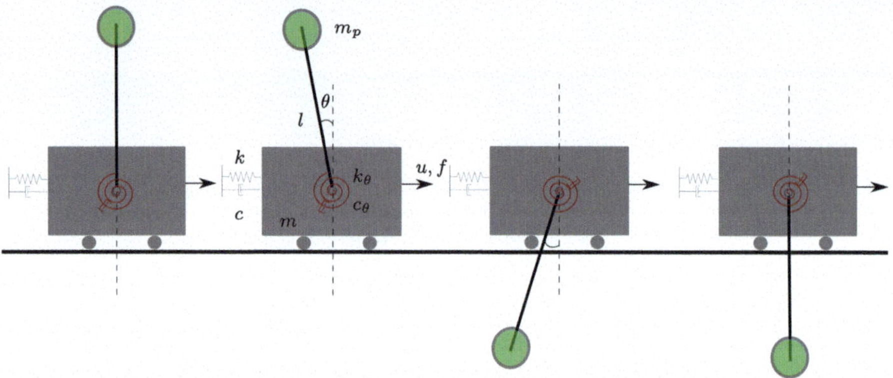

Fig. 4.3 A system of inverted pendulii with elastic restraints and damping mechanisms under an external loading

Next, the kinetic energy of the system is given as ,

$$T = \frac{1}{2}m\dot{u}^2 + \frac{1}{2}m_p(\dot{x}_p^2 + \dot{y}_p^2)$$

Following a substitution of the relations given in Eq. (4.15), we have

$$T = \frac{1}{2}(m + m_p)\dot{u}^2 + \frac{1}{2}m_p(l^2\dot{\theta}^2 - 2\dot{u}l\dot{\theta}\cos\theta)$$

The potential energy of the dynamic system is due to the gravity field with g as the acceleration of gravity and to the elastic energy stored in the springs:

$$V = m_p g y_p + \frac{1}{2}ku^2 + \frac{1}{2}k_\theta\theta^2 + \text{constant}$$

$$= m_p g l \cos\theta + \frac{1}{2}ku^2 + \frac{1}{2}k_\theta\theta^2 + \text{constant}$$

Next, we formulate Lagrange's equations as given in Eq. (4.5) for each of the two generalized coordinates. Starting with the first one, $q_i := u$, we have

$$\frac{\partial T}{\partial \dot{u}} = (m + m_p)\dot{u} - m_p l\dot{\theta}\cos\theta$$

The time derivative of this equation is now

$$\frac{d}{dt}\left(\frac{\partial T}{\partial \dot{u}}\right) = (m + m_p)\ddot{u} - m_p l\ddot{\theta}\cos\theta + m_p l\dot{\theta}^2\sin\theta \tag{4.16}$$

In addition, the derivative of the kinetic energy with respect to u is

$$\frac{\partial T}{\partial u} = 0 \tag{4.17}$$

while the derivative of the potential energy with respect to u is

$$\frac{\partial V}{\partial u} = ku \tag{4.18}$$

We now introduce a virtual displacement δu corresponding to the generalized coordinate u, so that the external forces and the non-conservative forces produce virtual work as

$$\delta W = Q_u \delta u = (f - c\dot{u})\delta u \rightarrow Q_u = f - c\dot{u} \tag{4.19}$$

We now continue with the second generalized coordinate $q_i := \theta$ and compute

$$\frac{\partial T}{\partial \dot{\theta}} = m_p l^2 \dot{\theta} - m_p \dot{u} l \cos \theta$$

The time derivative of the above equation is

$$\frac{d}{dt}\left(\frac{\partial T}{\partial \dot{\theta}}\right) = m_p l^2 \ddot{\theta} + m_p \dot{u} l \dot{\theta} \sin \theta - m_p \ddot{u} l \cos \theta \tag{4.20}$$

Next, the derivative of the kinetic energy with respect to generalized coordinate θ is

$$\frac{\partial T}{\partial \theta} = m_p \dot{u} l \dot{\theta} \sin \theta, \tag{4.21}$$

while the derivative of the potential energy with respect to θ is

$$\frac{\partial V}{\partial \theta} = -m_p g l \sin \theta + k_\theta \theta. \tag{4.22}$$

Introducing a virtual angle $\delta\theta$ corresponding to the second generalized coordinate θ, the virtual work produced by the external forces and by the non-conservative forces is

$$\delta W = Q_\theta \delta\theta = -c_\theta \dot{\theta} \delta u \rightarrow Q_\theta = -c_\theta \dot{\theta} \tag{4.23}$$

According to Eq. (4.5), the two equations of motion of the double inverted pendulii will result from the combination of equations, (4.16) – (4.17) + (4.18) = (4.19) and (4.20) – (4.21) + (4.22) = (4.23),

$$(m + m_p)\ddot{u} - m_p l \ddot{\theta} \cos \theta + m_p l \dot{\theta}^2 \sin \theta + ku + c\dot{u} = f \tag{4.24}$$

$$m_p l^2 \ddot{\theta} - m_p \ddot{u} l \cos \theta - m_p g l \sin \theta + k_\theta \theta + c_\theta \dot{\theta} = 0 \tag{4.25}$$

These are now the two, non-linear equilibrium equations of the double inverted pendulum, which are often used for structural control purposes.

4.6.3 2-DOF System of Two Coupled Shear Frames

In this sub-section, we will use the energy-based method, as stated by Lagrange's equations, to derive the equations of motion of the system shown in Fig. 3.6, which was originally presented in Sect. 3.9.1. The generalized coordinates that describe the motion of the coupled system are the horizontal displacements of the masses of the constituent frames, namely u_1 and u_2. The total kinetic energy of this system is therefore

$$T = \frac{1}{2}m_1\dot{u}_1^2 + \frac{1}{2}m_1\dot{u}_2^2.$$

As potential energy we consider the energy stored in the dynamic system due to deformations of its structural members

$$V = \frac{1}{2}k_1 u_1^2 + \frac{1}{2}k_2 u_2^2 + \frac{1}{2}k_0(u_2 - u_1)^2.$$

The work done by the external forces and the non-conservative forces is

$$\delta W = f_1\delta u_1 + f_2\delta u_2 - c_1\dot{u}_1\delta u_2 - c_2\dot{u}_2\delta u_2 - c_0(\dot{u}_2 - \dot{u}_1)\delta(u_2 - u_1)$$
$$= \underbrace{(f_1 - c_1\dot{u}_1 + c_0\dot{u}_2 - c_0\dot{u}_1)}_{Q_{u_1}}\delta u_1 + \underbrace{(f_2 - c_2\dot{u}_2 - c_0\dot{u}_2 + c_0\dot{u}_1)}_{Q_{u_2}}\delta u_2$$

Differentiation on the first generalized coordinate $q_i := u_1$, as require by the Lagrange equations, gives

$$\frac{\partial T}{\partial \dot{u}_1} = m_1\dot{u}_1$$

The time derivative of the above expression gives

$$\frac{d}{dt}\frac{\partial T}{\partial \dot{u}_1} = m_1\ddot{u}_1$$

Also, the derivative of the kinetic energy with respect to generalized coordinate u_1 is zero, i.e.,

$$\frac{\partial T}{\partial u_1} = 0$$

Next, the derivative of the potential energy with respect to the first generalized coordinate u_1 is

$$\frac{\partial V}{\partial u_1} = k_1 u_1 - k_0(u_2 - u_1) = k_1 u_1 + k_0 u_1 - k_0 u_2$$

The first of the governing equations of motion of the coupled shear frames, according to Eq. (4.5), is

$$\frac{d}{dt}\left(\frac{\partial T}{\partial \dot{u}_1}\right) - \frac{\partial T}{\partial u_1} + \frac{\partial V}{\partial u_1} = Q_{u_1}$$

Following a re-arrangement of terms, the equation reads as

$$m_1\ddot{u}_1 + c_1\dot{u}_1 + c_0\dot{u}_1 - c_0\dot{u}_2 + k_1 u_1 + k_0 u_1 - k_0 u_2 = f_1 \tag{4.26}$$

The second equation of the dynamic system is the Lagrange equation written in terms of the second generalized coordinate $q_i := u_2$. This gives

$$\frac{\partial T}{\partial \dot{u}_2} = m_2\dot{u}_2$$

whose time derivative is

$$\frac{d}{dt}\frac{\partial T}{\partial \dot{u}_2} = m_2\ddot{u}_2$$

The derivative of the kinetic energy with respect to u_2 is zero,

$$\frac{\partial T}{\partial u_2} = 0$$

while the derivative of the potential energy with respect to u_2 is

$$\frac{\partial V}{\partial u_2} = k_2 u_2 + k_0(u_2 - u_1) = k_2 u_2 + k_0 u_2 - k_0 u_1$$

Next, the second equation of motion can be derived according to Eq. (4.5) as

$$\frac{d}{dt}\left(\frac{\partial T}{\partial \dot{u}_2}\right) - \frac{\partial T}{\partial u_2} + \frac{\partial V}{\partial u_2} = Q_{u_2}$$

Following substitutions and a re-arrangement of terms, we have

$$m_2\ddot{u}_2 + c_2\dot{u}_2 + c_0\dot{u}_2 - c_0\dot{u}_1 + k_2 u_2 + k_0 u_2 - k_0 u_1 = f_2 \tag{4.27}$$

Equations (4.26) and (4.27) can be expressed in terms of matrix notation as

$$\underbrace{\begin{bmatrix} m_1 & 0 \\ 0 & m_2 \end{bmatrix}}_{M}\underbrace{\begin{Bmatrix} \ddot{u}_1 \\ \ddot{u}_2 \end{Bmatrix}}_{\ddot{u}(t)} + \underbrace{\begin{bmatrix} c_1 + c_0 & -c_0 \\ -c_0 & c_2 + c_0 \end{bmatrix}}_{C}\underbrace{\begin{Bmatrix} \dot{u}_1 \\ \dot{u}_2 \end{Bmatrix}}_{\dot{u}(t)} + \underbrace{\begin{bmatrix} k_1 + k_0 & -k_0 \\ -k_0 & k_2 + k_0 \end{bmatrix}}_{K}\underbrace{\begin{Bmatrix} u_1 \\ u_2 \end{Bmatrix}}_{u(t)} = \underbrace{\begin{Bmatrix} f_1 \\ f_2 \end{Bmatrix}}_{f(t)}$$

$$\tag{4.28}$$

which agrees with what was previously derived or the coupled shear frame system.

Continuous Dynamic Systems

5.1   Axial Vibrations in a Rod

As shown in Fig. 5.1, the free-body diagram for a rod element indicates that the inertia force which develops because of the distributed mass m (in tn/m) plus the restoring force (in kN) due to tension/compression across the cross-section area A (in m^2), balance the distributed axial load $q_x(x, t)$ (in kN/m). Thus, the equation of motion written in terms of the axial displacement $u(x, t)$ (in m) is

$$EA\frac{\partial^2 u(x, t)}{\partial x^2} + q_x(x, t) = m\frac{\partial^2 u(x, t)}{\partial t^2} \tag{5.1}$$

In the above, E (in kN/m^2) is the elastic modulus of the material. The aforementioned equation must be accompanied by initial conditions for the displacement $u(x, t)$ and the velocity $v(x, t) = du/dt$ at the onset of time ($t = 0$), plus boundary conditions at the two ends of the rod ($x = 0, x = L$) involving either the displacement $u(x, t)$ or the axial force $N(x, t) = EA\epsilon(x, t)$, where $\epsilon(x, t) = du/dx$ is the axial strain. We now define the velocity of axial waves travelling through the rod as $c = \sqrt{EA/m}$ (in m/s). Following a substitution in Eq. (5.1) and re-arrangement, the equation of motion in the absence of any external force assumes the form

$$\frac{\partial^2 u}{\partial x^2} - \frac{1}{c^2}\frac{\partial^2 u}{\partial t^2} = 0. \tag{5.2}$$

The above equation is known as the wave equation and appears in both fields of mechanics and acoustics.

© The Author(s), under exclusive license to Springer Nature Switzerland AG 2026

G. Manolis and C. Panagiotopoulos, *Vibrations of Structural Systems*, Synthesis Lectures on Mechanical Engineering, https://doi.org/10.1007/978-3-032-12279-7_5

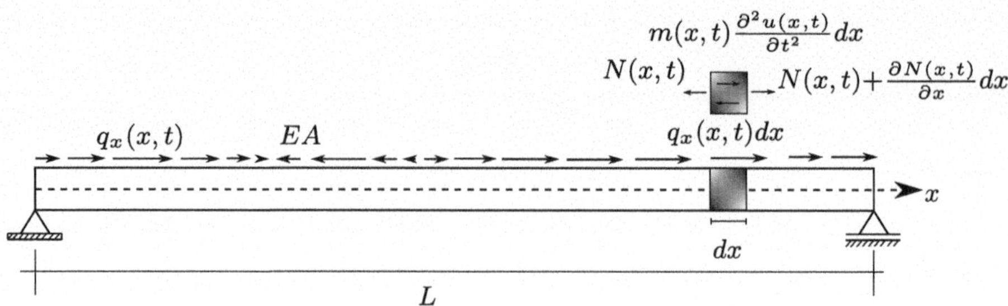

Fig. 5.1 Free body diagram for the rod element

5.2   Flexural Vibrations in a Beam

Regarding the flexural vibrations of the Euler-Bernoulli beam, whose free body diagram is shown in Fig. 5.2, the force balance written in terms of the transverse displacement $w(x, t)$ yields the following fourth-order partial differential equation:

$$EI \frac{\partial^4 w(x, t)}{\partial x^4} + m \frac{\partial^2 w(x, t)}{\partial t^2} = q_y(x, t) \tag{5.3}$$

In the above, I (in m^4) is the moment of inertia of the beam's cross-section about a principal axis, and $q_y(x, t)$ (in kN/m) is the distributed transverse load along the length L. In the absence of external forces, Eq. (5.3) assumes its homogeneous form with parameter $\gamma = \sqrt[4]{EI/m}$ a constant.

$$\gamma^4 \frac{\partial^4 w(x, t)}{\partial x^4} + \frac{\partial^2 w(x, t)}{\partial t^2} = 0 \tag{5.4}$$

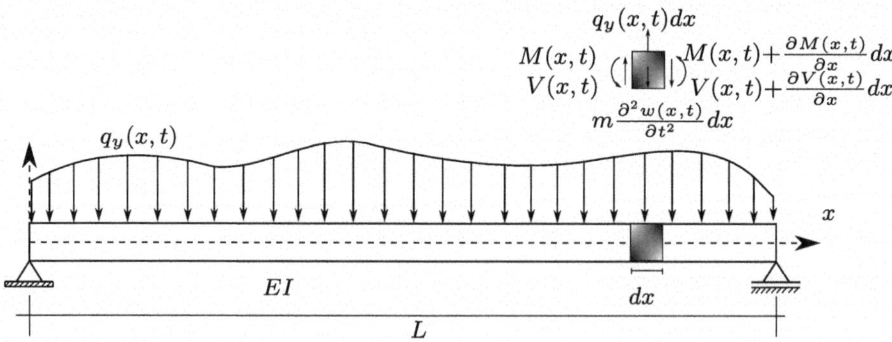

Fig. 5.2 Free body diagram for the beam element

At this point, we define the bending moment that develops across the beam's cross-section in terms of the curvature as $M(x,t) = EI d^2 w/dx^2$ (in kNm) and the corresponding shear force as $V(x,t) = dM/dx = EI d^3 w/dx^3$ (in kN). In terms of boundary conditions, these are defined at the two ends of the beam $x = 0$, $x = L$ and involve spatial derivatives of the displacement function up to the third order, i.e., displacement w, slope w', curvature w'' (referring to the bending moment M) and skewness w''' (referring to the shear force V). The initial conditions involve the displacement and the velocity and are as follows:

$$w(x, t = 0) = w(x, 0) = w_0(x)$$

$$\left. \frac{\partial w(x,t)}{\partial t} \right|_{t=0} = \dot{w}_0(x) \tag{5.5}$$

By adding the contribution of viscous damping in the beam, the partial differential equation (5.3) assumes the form:

$$EI \frac{\partial^4 w(x,t)}{\partial x^4} + m \frac{\partial^2 w(x,t)}{\partial t^2} + c \frac{\partial w(x,t)}{\partial t} = q_y(x,t) \tag{5.6}$$

where c the viscous damping coefficient. Note that this is a linearized form of energy dissipation and the relevant term appearing in the equation of motion is proportional to the velocity.

5.3   Axial Free Vibrations

In a free vibration environment, the rod executes harmonic motions so that separation of variables is possible as $u(x,t) = U(x) exp(i\omega t)$, where ω (in rad/s) is the frequency of vibration. This yields the following equation for the amplitude of vibration $U(x)$:

$$\frac{d^2 U}{dx^2} + k^2 U = 0 \tag{5.7}$$

In the above, $k = \omega/c$ (in 1/m) is the wave number for the longitudinal wave propagating in the rod. Next, the solution of equation (5.7) in the frequency domain for the amplitude of vibration $U(x)$ is

$$u(x,t) = A \sin(kx) + B \cos(kx) \tag{5.8}$$

We discern two basic types of boundary conditions for evaluating the constants A and B, namely (a) fixity when the displacement $U(x) = 0$ and (b) free when the axial force $N(x) = 0$, at either end $x = 0$ and $x = L$ of the rod. Thus, the displacement curve, which is now the eigenfunction of the rod, is either of the sine or the cosine type.

5.4   Flexural Free Vibrations

As before, the free vibrations of the beam result in the absence of external loads, are time harmonic and are distinguished according to the boundary conditions at the two ends. Obviously, in a free vibration environment, initial conditions on the displacement w and the velocity dw/dt are irrelevant, as the beam commenced vibrating at a distant past and has continued ever since. From beam theory, there are two boundary conditions at each end $x = 0$ and $x = L$ coming from either kinematics involving the displacement w and the slope dw/dx, or stresses involving the bending moment M and the shear force V.

Next, separation of variables in reference to Eq. (5.4) is of the form $w(x, t) = \phi(x)y(t)$, i.e., the dependent variable is the product of a spatial function $\phi(x)$ times a temporal function $y(t)$. The relevant space and time derivatives for use in the equation of motion are

$$w''''(x, t) = \phi''''(x)y(t)$$

$$\ddot{w}(x, t) = \phi(x)\ddot{y}(t) \tag{5.9}$$

Following substitution in the equation of motion, Eq. (5.4), we recover by separation two ordinary differential equations in the dependent variables $\phi(x)$ and $y(t)$ as

$$EI\phi''''(x)y(t) + M\phi(x)\ddot{y}(t) = 0 \rightarrow \frac{EI}{m}\frac{\phi''''(x)}{\phi(x)} = -\frac{\ddot{y}(t)}{y(t)} \tag{5.10}$$

These uncouple because each side of the above equation is a function of a different independent variable, and as each side is equal to a constant, here given as teh square of the frequency ω^2. Thus, we recover two uncoupled ordinary differential equations as

$$\frac{EI}{m}\frac{\phi''''(x)}{\phi(x)} = \omega^2$$

$$\frac{\ddot{y}(t)}{y(t)} = \omega^2 \tag{5.11}$$

The homogeneous solutions of these two ordinary differential equations are as follows:

$$y(t) = A \sin(\omega t) + B \cos(\omega t)$$

$$\phi(x) = C_1 \sin(\lambda x) + C_2 \cos(\lambda x) + C_3 \sinh(\lambda x) + C_4 \cosh(\lambda x) \tag{5.12}$$

In the above, sinh and cosh are the hyperbolic sine and cosine functions, while the wave number for transverse vibrations is $\lambda^4 = \omega^2 m/EI = \omega^2\gamma^4$. The solutions to the latter of Eqs. (5.12) are now labelled are eigenfunctions of the flexural beam, while the integration constants C_i, $i = 1, \ldots, 4$ are recovered from the boundary conditions. Regarding the solution of the former of Eqs. (5.12), recovery of constants A, B from the initial conditions merely reconfirms the fact that the time dependent behavior $y(t)$ of the beam is harmonic in the absence of external loads. Next, Table 5.1 lists the beam boundary conditions with

Table 5.1 Boundary conditions for the beam displacement $w(x)$ and slope $w'(x)$ at a boundary point x_Γ

Free displacement at node $x = x_\Gamma$	$w'''(x_\Gamma) = \phi'''(x_\Gamma) = 0$
Free slope at node $x = x_\Gamma$	$w''(x_\Gamma) = \phi''(x_\Gamma) = 0$
Restrained displacement at node $x = x_\Gamma$	$w(x_\Gamma) = \phi(x_\Gamma) = 0$
Restrained slope at node $x = x_\Gamma$	$w'(x_\Gamma) = \phi'(x_\Gamma) = 0$

Table 5.2 Eigenproperties for four boundary conditions cases of the Bernoulli–Euler beam

Restrained beam	Eigenvalues $\lambda_n L$	Eignfunctions $\phi_n(x)$	Norm $\int\limits_0^L \phi_n^2(x)\,dx$
	1.8751 4.6941 7.8547 10.9955 14.1372 for $n > 5 : \frac{(2n-1)\pi}{2}$	$\cosh(\lambda_n x) - \cos(\lambda_n x) -$ $k_n \, (\sinh(\lambda_n x) - \sin(\lambda_n x))$ where $k_n = \frac{\sinh(\lambda_n L) - \sin(\lambda_n L)}{\cosh(\lambda_n L) + \cos(\lambda_n L)}$	L
	3.9266 7.0685 10.2102 13.3518 16.4934 for $n > 5 : \frac{(4n+1)\pi}{4}$	$\cosh(\lambda_n x) - \cos(\lambda_n x) -$ $k_n \, (\sinh(\lambda_n x) - \sin(\lambda_n x))$ where $k_n = \mathrm{ctg}(\lambda_n L)$	L
	4.7300 7.8532 10.9956 14.1372 17.2788 for $n > 5 : \frac{(2n+1)\pi}{2}$	$\cosh(\lambda_n x) - \cos(\lambda_n x) -$ $k_n \, (\sinh(\lambda_n x) - \sin(\lambda_n x))$ where $k_n = \frac{\cosh(\lambda_n L) - \cos(\lambda_n L)}{\sinh(\lambda_n L) - \sin(\lambda_n L)}$	L
	For any $n: n\pi$	$\sin(\lambda_n x)$	$\frac{L}{2}$

Angular eigefrequencies are computed as $\omega_n^2 = \lambda_n^4 \frac{EI}{m}$

reference to the kinematic variables, i.e., the displacement w and the slope $w' = dw/dx$, at a boundary point x_Γ. These can either be free or constrained, thus generating the total of four possible outcomes. It should be noted that the displacement solution for the equation of motion satisfies the same boundary conditions as the eigenfunctions, which are the solutions to the free vibration problem. Of course, more complicated, mixed–type boundary conditions are also possible, but fall outside the range of the present chapter. Next, Table 5.2 lists the first few eigenfrequencies (natural frequencies) and eigenfunctions (modal shapes) for four typical beams, namely the cantilevered, the simply supported, the simply-supported-fixed and the fixed-fixed beam. Of course, more complete tables of this kind can be found in the literature [1].

5.4.1   Orthogonality Property of the Eigenfunctions

It can be shown that the following property holds for two different eigen-pairs $\phi_n(x), \phi_m(x)$ and ω_n, ω_m, where counter $n = 1, 2, 3, \ldots$:

$$(\omega_n^2 - \omega_m^2)\frac{m}{EI}\int_0^L \phi_m\phi_n dx = [\phi_m\phi_n''' - \phi_m'\phi_n'' + \phi_m''\phi_n' - \phi_m'''\phi_n]\Big|_0^L \tag{5.13}$$

For homogeneous boundary conditions, we have that

$$a_1\phi + a_2\phi' + a_3\phi'' + a_4\phi''' = 0 \tag{5.14}$$

yielding a zero right-hand side for Eq. (5.13). This way, we recover the orthogonality property for the eigenfunctions as follows:

$$\int_0^L \phi_m\phi_n dx = 0 \text{ for } m \neq n. \tag{5.15}$$

Note that the boundary condition cases listed in Table 5.1 are time-independent and satisfy Eq. (5.14), while their general form given by Eq. (5.9) does not. This last category requires special treatment and will be shown in the next section. The general solution for Eq. (5.4) is obtained by superposition of an infinite number of eigenfunctions $n = 1, 2, 3, \ldots$ times the corresponding generalized (or modal) coordinates that are functions of time as follows

$$w(x, t) = \sum_{m=1}^{\infty} \phi_m(x)\left(A_m \sin(\omega_m t) + B_m \cos(\omega_m t)\right) \tag{5.16}$$

If Eq. (5.16) is multiplied by an eigenfunction $\phi_n(x)$ and the product is integrated over the length of the beam, the constants A_n, B_n defining a given eigenfunction of order n can be computed by taking Eq. (5.15) into account. The result is the following expressions:

$$A_n = \frac{\int_0^L \phi_n(x)\dot{w}(x, 0)dx}{\omega_n \int_0^L (\phi_n(x))^2 dx}, \quad B_n = \frac{\int_0^L \phi_n(x)w(x, 0)dx}{\int_0^L (\phi_n(x))^2 dx} \tag{5.17}$$

Note that the definite integrals shown above can be evaluated by recourse to Table 5.1. Finally, the standard form of the relations are derived from use of the first of Eq. (5.11) as follows:

$$\int_0^L \phi_m'''' \phi_n dx = 0 \text{ for } m \neq n, \tag{5.18}$$

while for the case of $m = n$,

$$\int_0^L \phi_n'''' \phi_n dx = \frac{\omega_n^2 m}{EI} \int_0^L (\phi_n(x))^2 dx \tag{5.19}$$

5.5   Eigenfunction Method for the Euler-Bernoulli Beam

Starting with Eq. (5.3) and substituting the expansion of the displacement function in terms of the generalized coordinates, we obtain

$$m \sum_{m=1}^{\infty} \phi_i(x) \ddot{y}_i(t) + EI \sum_{m=1}^{\infty} \phi_i''''(x) y_i(t) = q_y(x, t) \tag{5.20}$$

By pre-multiplying Eq. (5.20) with the eigenfunction $\phi_n(x)$ and integrating along the length of the beam, and using Eqs. (5.15) and (5.18)–(5.19), yields the following equation:

$$m \int_0^L (\phi_n(x))^2 dx \, \ddot{y}_n(t) + \omega_n^2 m \int_0^L (\phi_n(x))^2 dx \, y_n(t) = \int_0^L \phi_n(x) q_y(x, t) dx \tag{5.21}$$

The terms appearing above in conjunction with the generalized coordinates are labelled as the generalized mass for the nth eigenfunction,

$$m_n^* = m \int_0^L (\phi_n(x))^2 dx, \tag{5.22}$$

the corresponding generalized stiffness,

$$k_n^* = \omega_n^2 m \int_0^L (\phi_n(x))^2 dx, \tag{5.23}$$

and the generalize forcing term

$$p_n^*(t) = \int_0^L \phi_n(x) q_y(x, t) dx. \tag{5.24}$$

This results in an (theoretically) infinite number of SDOF systems for the generalized coordinates, which assume the standard form

$$m_n^* \ddot{y}_n(t) + k_n^* y_n(t) = p_n(t) \rightarrow \ddot{y}_n(t) + \omega_n^2 y_n(t) = p_n^*(t) \tag{5.25}$$

under the following initial conditions:

$$\dot{y}_n(0) = \frac{\int_0^L \phi_n(x)\dot{w}(x,0)dx}{\omega_n \int_0^L (\phi_n(x))^2 dx}, \quad y_n(0) = \frac{\int_0^L \phi_n(x)w(x,0)dx}{\int_0^L (\phi_n(x))^2 dx} \tag{5.26}$$

Equation (5.25) under the above initial conditions can be solved using either Duhamel's integral, which is an analytical solution discussed in Sect. 2.4, or numerically using the time integration techniques discussed in Sect. 2.11. Reconstitution of the time–dependent displacement of the continuous beam is finally given as follows:

$$w(x,t) = \sum_{n=1}^{\infty} \phi_n(x)y_n(t) \tag{5.27}$$

In practice however, and for low frequency dynamic problems, only a few terms N_{modes} are necessary for a satisfactory approximation of the displacement solution, i.e.,

$$w(x,t) = \sum_{n=1}^{N_{modes}} \phi_n(x)y_n(t). \tag{5.28}$$

5.6 Class `prismaticBeam`

Description: The `prismaticBeam` appearing in the software package `courses.structuraldynamics` is an implementation of the analytical expressions for the eigenfunctions and eigenvalues of a continuous prismatic beam based on Euler–Bernoulli beam theory. Moreover, class `prismaticBeam` takes advantage of the eigenfunction method in order to approximately compute any kind of imposed continuously distributed transverse loading, as well as initial conditions for the displacement and velocity fields. Response under imposition of time dependent boundary conditions can be also considered as a special case of forced vibrations using appropriate transformation as shown in Sect. 5.7.

Attributes

Class `prismaticBeam` attributes include the length of the beam (double `len`) and the id (integer `id`), the elasticity modulus (double `Elast`), the moment of inertia (double `Im`) as well the mass per unit length (double `mass`) together with the definition of left (Integer `bc1`) and right (Integer `bc2`) boundary conditions. The later two attributes signal boundary condition types, namely the value of 0 implying a free end, the value of 1 implying a fixed end and finally the value of 2 for a pinned end.

Instance

No specific constructors have been considered, so a Groovy's map constructor with annotation might be used [2].

Methods

- double eigenfrequency(int n)
- DoubleFunction eigenfunction(int n)
- double eigenfunction(int n, double x)
- double eigenvalues(int n)
- double eigenvalues2len(int n)
- double norm(int n)
- void setCriticalModalDamping(double xsi)
- void setViscousDamping(double dv)
- double[ ] response(DoubleFunction fs, DoubleFunction ft, DoubleFunction dsp0, DoubleFunction vlc0, double TotalTime, double x)
- double[ ] dresponse(DoubleFunction fs, DoubleFunction ft, DoubleFunction dsp0, DoubleFunction vlc0, double TotalTime, double x)
- double[ ] ddresponse(DoubleFunction fs, DoubleFunction ft, DoubleFunction dsp0, DoubleFunction vlc0, double TotalTime, double x)
- double[ ] snapshot(DoubleFunction fs, DoubleFunction ft, DoubleFunction dsp0, DoubleFunction vlc0, double time)
- double[ ][ ] compute(DoubleFunction fs, DoubleFunction ft, DoubleFunction dsp0, DoubleFunction vlc0, double Totaltime)

Note that the methods `response`, `dresponse`, `ddresponse`, `snapshot` and `compute`, can be called with the argument `DoubleFunction fst` instead of the distinct `DoubleFunction fs` and `DoubleFunction ft`, meaning that we have to face a non-separable excitation function of space and time (see Sect. 5.8 for an example). This version of the above methods is more computational intensive its respective separated in space and time load version.

5.6.1 Computation of the Modal Properties of a Beam

As shown in Sect. 5.6, we may use the `prismaticBeam` class to define a prismatic beam and determine its dynamic properties, such is the eigenfrequencies and eigenfunctions. In this example we consider a fixed-end (e.g., simply-supported) beam with modulus of elasticity $E = 10.3$ GPa, moment of inertia $I = 3.125e - 3\mathrm{m}^4$ and mass per unit of length $m = 112.5$ kg/m. A simple script defining such a beam entity and producing a graph of the first three eigenfunctions is given below.

```
import courses.structuraldynamics.prismaticBeam
hsec=0.5; bsec=0.3
thePlot.setAutoColor(true)
thePlot.makeLegend()
thePlot.setTextfontsize(24)
beam=new prismaticBeam(len:5.0, id:1, Elast
    :10300000.0, Im:bsec*hsec**3/12.0, mass:750.0*
    hsec*bsec, bcl:1, bcr:1)
(1..3).each{
  pf = new plotfunction(linspace(0.0, beam.len,1001),
    beam.eigenfunction(it))
  pf.setName(" _"+it+"="+String.format("%.3f", beam.
    eigenfrequency(it))+"rad/s")
  plot(pf)
}
```

Listing 5.1 Define a continuous beam with both ends fixed.

The plot produced by the above given script is shown in Fig. 5.3, where the three first eigenfunctions are plotted in red, blue and green, respectively. Actually, the class of `prismaticBeam` implements the information given in Table 5.2 and sets this information available for further usage.

5.6.2 Imposition of an Initial Disturbance

In this section we consider the simply-supported beam of the previous Sect. 5.6.1 and solve the case of an initial disturbance imposed along its length of the form

$$w(x, 0) = f_0 \exp\left(\frac{-x^2}{4b^2}\right), \qquad \dot{w}(x, 0) = 0.$$

The initial displacement has a spatial variation which corresponds to a Gaussian distribution. A solution for the case of an infinite beam is given in Graff [3] as follows:

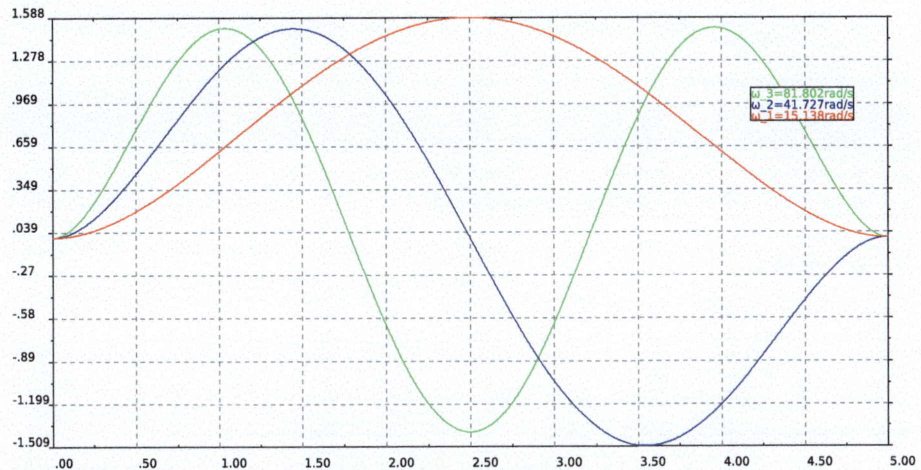

Fig. 5.3 The three first eigenfunctions for a prismatic beam with both ends fixed

$$w(x,t) = \frac{f_0 \exp\left(\frac{-x^2 b^2}{4(b^4 + a^2 t^2)}\right)}{(1 + a^2 t^2/b^4)^{\frac{1}{4}}} \cos\left(\frac{a t x^2}{4(b^4 + a^2 t^2)} - \frac{1}{2}\arctan\left(\frac{at}{b^2}\right)\right)$$

where parameter $a^2 = \frac{EI}{m}$. This solution will be used here as a reference for a qualitative comparison during the time interval preceding any wave reflections from the fixed boundaries.

```
import courses.structuraldynamics.prismaticBeam
hsec=0.5; bsec=0.3
Elast=10300000.0
Im=bsec*hsec**3/12.0
mass=750.0*hsec*bsec
beam=new prismaticBeam(len:5.0, id:1, Elast
    :10300000.0, Im:bsec*hsec**3/12.0, mass:750.0*
    hsec*bsec, bcl:1, bcr:1)
plot()
b=0.1/sqrt(2.0)
a=sqrt(beam.Elast*beam.Im/beam.mass)
w0=1.0

w={x,t->
    w0/sqrt(sqrt(1.0+a*a*t*t/(b*b*b*b)))*exp(-x*x*b*b
    /(4*(b*b*b*b+a*a*t*t)))*cos(a*t*x*x/(4*(b*b*b*b+a
    *a*t*t))-0.5*atan(a*t/(b*b)))
}

axis=linspace(-5.0,5.0,5001)
thePlot.hline(0.0); thePlot.vline(0.0)
```

```
18  thePlot.setTextfontsize(24)
19  thePlot.setAutoColor(true)
20  thePlot.makeLegend(true)
21  [0,0.0005,0.001,0.002,0.003,0.004,0.005].each{
22    shot=it
23    disp={w(it,shot)} as DoubleFunction
24    pf=new plotfunction(axis,disp);  pf.setName("time=  "
       +shot+"s")
25    plot(pf)
26  }
```

Listing 5.2 Scirpt for producing the results shown in Fig. 5.4 (left plot)

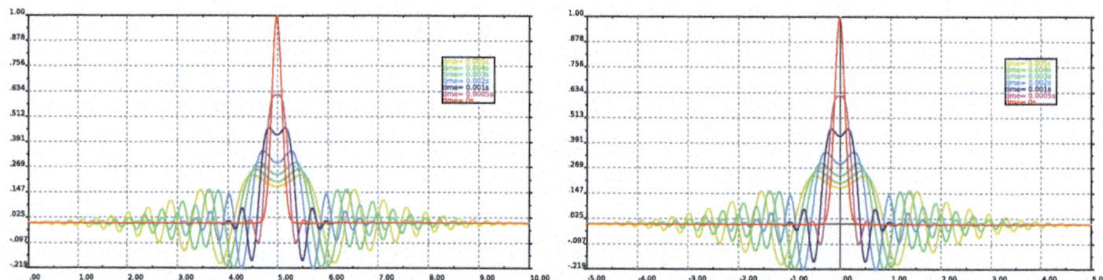

Fig. 5.4 Displacement distribution snapshots along the simply-supported beam under an initial disturbance: (left) The exact solution for an infinite beam; (right) Use of the eigenfunction method for 128 modes of the finite-length beam

Solving an identical problem, yet for a finite length beam using the eigenfunction method as it has been implement in the software package `courses.structuraldynamics` we get a very satisfactory agreement as shown in plots of Fig. 5.4. The scriptr nw can use for the eigenfunction method solution of that problem is given in code Listing 5.3. Here we have used the `snapshot` method, of the `prismaticBeam` class, and more specific the separated force in space and time version since we face a free vibration problem under initial conditions. The `snapshot` method provides us with the displacement solution on a specific instant of time using the eigenfunction method and numerical integration for integrals appeared in the right hand side of equation (5.21) as well as in Eq. (5.26) of initial conditions.

```
1  import  courses.structuraldynamics.prismaticBeam
2  theUniverse.cls()
3  hsec=0.5;  bsec=0.3
4  beam=new prismaticBeam(len:10.0,  id:1,  Elast
       :10300000.0,  Im:bsec*hsec**3/12.0,  mass:750.0*
       hsec*bsec,  bcl:2,  bcr:2)
5  theUniverse.putContraption(beam)
6
7  mesh=32
```

```
 8
 9  Tot = 0.005 //total time
10  Nt = mesh * 200   //discrete time steps (is defined by
        user)
11  beam.dt = Tot/Nt as double
12  beam.Nmodes = 4 * mesh
13  beam.Nx = 200 * mesh
14  fs = {0.0d} as DoubleFunction
15  ft = {0.0d} as DoubleFunction
16  d0 = new mathman.functions.GaussianFunction(1.0,beam.
        len/2.0,0.1)
17  v0 = {0.0d} as DoubleFunction
18
19  if(true){
20     snapshotPlot = new PlotFrame()
21     snapshotPlot.setTextfontsize(24)
22     [0,0.0005,0.001,0.002,0.003,0.004,0.005].each{
23        time = it
24        defshhot = beam.snapshot(fs, ft, d0, v0, time)
25        pf = new plotfunction(beam.len/beam.Nx,defshhot)
26        pf.setName("time= "+time+"s")
27        snapshotPlot.addFunction(pf)
28     }
29     snapshotPlot.makeLegend()
30     snapshotPlot.setAutoColor(true)
31     snapshotPlot.show()
32  }
```

Listing 5.3 Scirpt for an initially disturbed beam

5.6.3 Forced Motion of a Beam

In this section, we present a forced vibration example of a beam that is loaded by a Gaussian
distribution function in both space and time. What is of interest in this example is that by
appropriately controlling the loading function configuration, the limit case of a concentrated
impact load in the form of a Dirac delta function can be recovered. It is obvious that having
such generalized functions describing the forcing term, it is possible to take advantage of
their properties when they are placed as kernels in integral equation formulations. Starting
with the simply-supported beam, the limiting case is a concentrated, short-duration impulse
load applied at $x = L/2$, with L the beam's length. Then, the force can be represented
as $q_y(x, t) = P_0 \delta(x - L/2)\delta(t)$ and following Graff [3], the exact solution, considering a
point impulse load, for the transverse displacement is

$$w(x, t) = \frac{2P_0}{mL} \sum_{n=1}^{\infty} \frac{\sin \frac{n\pi}{2} \sin \lambda_n x \sin \omega_n t}{\omega_n}$$

The wave number λ_n and the eigenfrequency ω_n are given in Table 5.2 for the case of simple supported beam.

```
L=5.0
hsec=0.5; bsec=0.3
Elast=10300000.0
Im=bsec*hsec**3/12.0
mass=750.0*hsec*bsec
P0=10.0

omega={
   sqrt((it*pi)**4*Elast*Im/(mass*L**4))
}

Nmodes=28

xsi=L/2.0; x=L/2.0
y={t->
   val=0.0
   Nmodes.times{
      n=it+1
      val+=sin(n*pi*xsi/L)*sin(n*pi*x/L)*sin(omega(n)*t
      )/omega(n)
   }
   return (val*P0/(mass*L))
} as DoubleFunction

axis=linspace(0.0,2.0,1001)
plot(axis,y)
```

Listing 5.4 Scirpt to produce analytical results for a localised impulse force

The results shown in Fig. 5.5 below are recovered by using the module `prismaticBeam` from package `courses.structuraldynamics`. We observe that in the solution for the problem of a beam under a Gaussian distribution of the loading term in both space and time

$$q_y(x,t) = \left(\frac{1}{\sigma\sqrt{2\pi}} e^{-\frac{x-L/2}{2\sigma^2}} \right) \left(\frac{P_0}{\sigma\sqrt{2\pi}} e^{-\frac{t}{2\sigma^2}} \right) \tag{5.29}$$

approaches that of an impulsive concentrated load as the standard deviation σ decreases. The integral of the given force distribution in space equals unit so as to be equivalent to the total power. Yet, when using the `prismaticBeam.response` method for the spatial load's distribution we can still use a Dirac's delta function as shown in line 17 of the source code Listing 5.5, which leads to similar resutls. However, we can not solve numerically for the limiting case where σ approaches zero, since we use standard trapezoidal rule integration while analytical integration or special numerical treatment would be necessary for such a case.

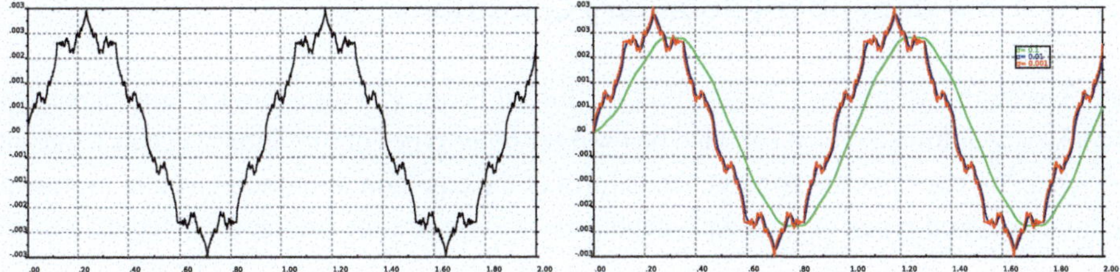

Fig. 5.5 Displacement of the central point in time: (left) Analytical solution due to concentrated impulse load; (right) Solution for a Gaussian force distribution in both space and time with decreasing values of σ

```
import courses.structuraldynamics.prismaticBeam
hsec=0.5; bsec=0.3
beam=new prismaticBeam(len:5.0, id:1, Elast
    :10300000.0, Im:bsec*hsec**3/12.0, mass:750.0*
    hsec*bsec, bcl:2, bcr:2)

mesh=20
println "mesh parameter: "+mesh
P0=10.0; init=0.00
Tot = 2.0 //total time
Nt=mesh*200   //discrete time steps (is defined by
    user)
beam.dt=Tot/Nt as double
beam.Nmodes=4*mesh
beam.Nx=200*mesh
plot()
MDMplot= new PlotFrame()
[0.1,0.01,0.001].each{
  fs=new mathman.functions.GaussianFunction(1.0/(sqrt
    (2.0*pi)*it),beam.len/2.0,it)
  // fs=new mathman.functions.Dirac(beam.len/2.0)
  ft=new mathman.functions.GaussianFunction(P0/(sqrt
    (2.0*pi)*it),init,it)
  d0= {0.0d} as DoubleFunction
  v0={0.0d} as DoubleFunction
  timehistory=beam.response(fs, ft, d0, v0, Tot, beam
    .len/2.0)
  pf=new plotfunction(beam.dt,timehistory)
  pf.setName("sigma= "+it)
  MDMplot.addFunction(pf)
}
MDMplot.makeLegend()
MDMplot.setAutoColor(true)
MDMplot.show()
```

Listing 5.5 Scirpt for a Gauss distibution load in space and time applied to a simply-supported beam

5.7   Time-dependent Boundary Conditions

In this section, we present the analysis for forced beam vibrations under time-dependent boundary conditions, which can also be used for other types of vibrations such as axial and torsional. We consider the homogeneous form of the equation of motion given in (5.4) with the more general types of boundary conditions. Concentrating on the fixed-fixed case, we have

$$w(0, t) = f_a(t) \text{ and } w'(0, t) = q_a(t) \text{ at } x = 0$$
$$w(L, t) = f_b(t) \text{ and } w'(L, t) = q_b(t) \text{ at } x = L \tag{5.30}$$

In reference to previous sections, for modal analysis to be possible the boundary conditions must be homogeneous, see Eq. (5.14), so that the orthogonality property of the eigenfunctions, Eq. (5.15), holds true. In the above case, this condition does not hold, so it is necessary to define an auxiliary function that will be used for the following transformation of the dependent variable:

$$w(x, t) = \tilde{w}(x, t) + \xi(x, t) \rightarrow \tilde{w}(x, t) = w(x, t) - \xi(x, t) \tag{5.31}$$

Function $\tilde{w}(x, t)$ and its spatial derivative $\tilde{w}'(x, t)$ now satisfy homogeneous conditions at boundaries $x = 0$ and $x = L$. Thus, we may write

$$
\begin{aligned}
\tilde{w}(0, t) = 0 \rightarrow & \qquad w(0, t) - \xi(0, t) = 0 \rightarrow & \qquad \xi(0, t) = f_a(t) \\
\tilde{w}'(0, t) = 0 \rightarrow & \qquad w'(0, t) - \xi'(0, t) = 0 \rightarrow & \qquad \xi'(0, t) = q_a(t) \\
\tilde{w}(L, t) = 0 \rightarrow & \qquad w(L, t) - \xi(L, t) = 0 \rightarrow & \qquad \xi(L, t) = f_b(t) \\
\tilde{w}'(L, t) = 0 \rightarrow & \qquad w'(L, t) - \xi'(L, t) = 0 \rightarrow & \qquad \xi'(L, t) = q_b(t)
\end{aligned}
\tag{5.32}
$$

Substituting Eq. (5.31) in the homogeneous form of the equation of motion Eq. (5.3) yields

$$EI\frac{\partial^4 \tilde{w}(x, t)}{\partial x^4} + m\frac{\partial^2 \tilde{w}(x, t)}{\partial t^2} = -EI\frac{\partial^4 \xi(x, t)}{\partial x^4} - m\frac{\partial^2 \xi(x, t)}{\partial t^2} \tag{5.33}$$

obeying homogeneous boundary conditions on $\tilde{w}$ and $\tilde{w}'$ and any other initial conditions. Once the above equation is solved for function $\tilde{w}$, we recover the original solution using Eq. (5.31). All that is left now is to define the auxiliary function $\xi(x)$, which must satisfy the four conditions given in the right-hand side of Eq. (5.32). One possible choice is to assume that $\xi(x, t)$ is a polynomial of the third degree, i.e.,

$$\xi(x, t) = c_0(t) + c_1(t)x + c_2(t)x^2 + c_3(t)x^3 \tag{5.34}$$

whose first spatial derivative is

$$\xi'(x,t) = c_1(t) + 2c_2(t)x + 3c_3(t)x^2. \tag{5.35}$$

Substituting in the relevant boundary conditions results in 4×4 system of equations for determining the unknown coefficients c_i. The result is as follows:

$$c_0(t) = f_a(t)$$

$$c_1(t) = q_a(t)$$

$$c_2(t) = -\frac{3}{L^2} f_a(t) - \frac{2}{L} q_a(t) + \frac{3}{L^2} f_b(t) - \frac{1}{L} q_b(t)$$

$$c_3(t) = \frac{2}{L^3} f_a(t) + \frac{1}{L^2} q_a(t) - \frac{2}{L^3} f_b(t) + \frac{1}{L^2} q_b(t) \tag{5.36}$$

5.7.1 Base Motion in Beams

As an example we pick the case of base motion of a cantilever beam for which an analytical solution for the steady state response may be found in the literature [4]. More specifically, for a pinned-pinned beam with the right hand side boundary excited by a harmonic vertical dispalcement $\delta_2 = \bar{\delta}_2 \sin \bar{\omega} t$, the steady state response is given as,

$$w_{an} = \frac{2mL^4\bar{\omega}^2\bar{\delta}_2}{\pi^5 EI} \sum_{n=1}^{\infty} \pm \frac{1}{n^5} \left[\frac{1}{(1 - \beta_n^2)^2 + (2\xi_n\beta_n)^2} \right]$$

$$\times \left[(1 - \beta_n^2) \sin \bar{\omega} t - 2\xi_n b_n \cos \bar{\omega} t \right] \sin \frac{n\pi x}{L}. \tag{5.37}$$

The operator is positive operator when n is odd and negative when n is even.

```
import courses.structuraldynamics.prismaticBeam
hsec=0.5; bsec=0.3
beam=new prismaticBeam(len:5.0, id:1, Elast
    :10300000.0, Im:bsec*hsec**3/12.0, mass:750.0*
    hsec*bsec, bcl:2, bcr:2)

omexc=beam.eigenfrequency(2)*1.01
dexc=0.1; xin=5.0/100.0; nmodes=6
uanal={x,t->
    coef=2.0*beam.mass*beam.len**4*omexc**2*dexc/(pi
    **5*beam.Elast*beam.Im)
    val=0.0
    (1..nmodes).each{
      n=it; c=1.0
      bn=omexc/beam.eigenfrequency(n)
      if(n%2==0)c=-1.0
      val+=c*1.0/((n as double)**5*((1.0-bn*bn)
    **2+(2.0*xin*bn)**2))*((1.0-bn*bn)*sin(omexc*t)
    -2.0*xin*cos(omexc*t))*sin(n*pi*x/beam.len)
    }
```

```
16    return  val*coef
17
18 }
19 axis=linspace(0.0,15.054722418529577,1000)
20 xloc=beam.len/2.0
21 plot(axis, {uanal(xloc,it)} as DoubleFunction)
```

Listing 5.6 Steady state solution for the simply supported beam

We still consider the simply-supported beam which is excited by a harmonic vertical displacement at its right-hand side support given as $\bar{w} \sin \bar{\omega} t$, where $\bar{w}$ is the monochromatic amplitude of the support motion under the harmonically imposed displacement with excitation frequency $\bar{\omega}$. It is interesting to note the possibility that the excitation frequency $\bar{\omega}$ lies close to one of the natural frequencies of the beam, as this gives rise to the resonance phenomenon. Similarly to what was presented in Sect. 5.7, we assume the transformation given by Eq. (5.31) and prescribe that function w must satisfy the following conditions:

$$w(0, t) = 0 \text{ and } w''(0, t) = 0 \text{ at } x = 0$$

$$w(L, t) = \bar{w} \sin \bar{\omega} t \text{ and } w''(L, t) = 0 \text{ at } x = L$$

Following Eq. (5.33), this leads to a transformation function,

$$\xi(x, t) = -\frac{x}{L} \bar{w} \sin \bar{\omega} t$$

with the right-hand side forcing term of equation (5.33) now becoming

$$f(x, t) = \frac{x}{L} m \bar{w} \bar{\omega}^2 \sin \bar{\omega} t$$

Next, we consider critical damping for each mode n as $\xi_n = 5\%$ and create the script that follows in order to compute the time history of the response at center span $x = L/2$. To that purpose we use the `prismaticBeam`'s method `response`, while for creating the relevant animation we use the `compute` method that will calculates the response for any discrete point of the beam under consideration. A comparison of that solution and analytical expression of Eq. (5.37) is given in the plot of Fig. 5.6.

```
1 import courses.structuraldynamics.prismaticBeam
2 hsec=0.5; bsec=0.3
3 beam=new prismaticBeam(len:5.0, id:1, Elast:10300000.0, Im:
      bsec*hsec**3/12.0, mass:750.0*hsec*bsec, bcl:2, bcr:2)
4 beam.solver=1 // [0,1,2] for [Duhamel, b-Newmrk, Runge-Kutta
      ]
5 mesh=10
6 ombar=beam.eigenfrequency(2)*1.01
7 wbar=0.1
8 Tot = 16.0
9 Nt=mesh*200
10 beam.dt=Tot/Nt as double
```

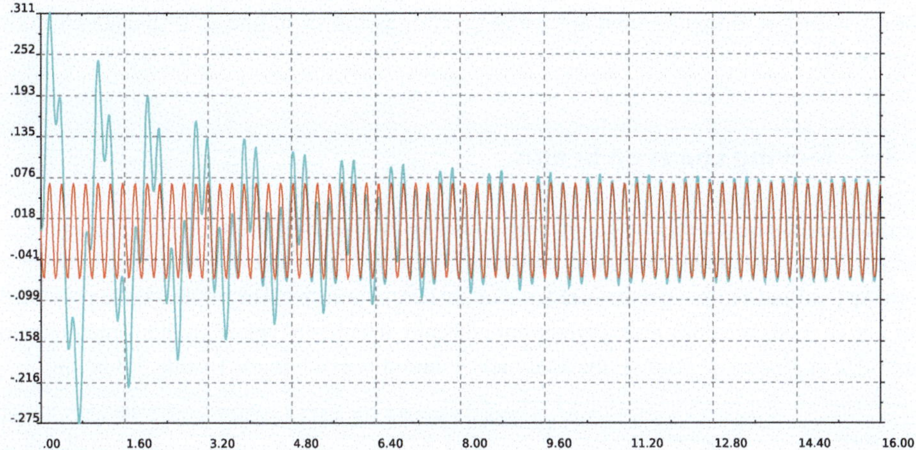

Fig. 5.6 Comparison of solution based on prismaticBeam eigenfunction method implementation with the analytical steady state solution. It is shown that after a first transient regime convergence to the analytical steady state solution takes place

```
11  beam.Nmodes=4*mesh
12  beam.Nx=200*mesh
13
14  MDMplot= new PlotFrame()
15  fs={x->x*wbar*beam.mass/beam.len} as DoubleFunction
16  ft={t->ombar*ombar*sin(ombar*t)} as DoubleFunction
17  d0= {0.0d} as DoubleFunction
18  v0={0.0d} as DoubleFunction
19  beam.setCriticalModalDamping(5.0/100.0)
20  timehistory=beam.response(fs, ft, d0, v0, Tot, beam.len/2.0)
21
22  MDMplot.addFunction(new plotfunction(beam.dt,timehistory))
23  MDMplot.show()
24
25  beam.reference={x,t-> x*wbar*sin(ombar*t)/beam.len} as
        DoubleFunction //
26  if(true){
27     theUniverse.cls()
28     theUniverse.putContraption(beam)
29     results=beam.compute(fs, ft, d0, v0, Tot)
30     theGP.stop()
31     scale=1.0
32     theGP.plotDeform(scale, 0,Nt,1)
33  }
```

Listing 5.7 Solution for support motion of a simply-supported beam

If we wish to animate the total motion of the beam $w(x, t)$ instead of the relative $\bar{w}(x, t)$ one, we should add to the calculated results the function $\tilde{w}(x, t)$. To that purpose we use the `prismaticBeam`'s parameter `reference`, as shown in line 25 of the source code Listing 5.7.

5.8 Moving Loads on Beams

Much information on this topics may be found in the book of Graff [3] (e.g., Example 2, Sect. 3.2.5, p. 169) and the the book by Biggs [5] (Sect. 8.2), where the difference between a moving load and a moving mass is elucidated. In here, a numerical example will be presented for a point load traversing a simply-supported beam, which is a typical problem encountered in bridge engineering. At first, the analytical solution is given in the Listing 5.8 below.

```
import courses.structuraldynamics.prismaticBeam
hsec=0.5; bsec=0.3
beam=new prismaticBeam(len:5.0, id:1, Elast
    :10300000.0, Im:bsec*hsec**3/12.0, mass:750.0*
    hsec*bsec, bcl:2, bcr:2)
Nmodes=8
F0=-10000.0
loadvlc=1.0
OMEGA={it*PI*loadvlc/beam.len}
uanal={x,t->
  coef=2.0*F0/(beam.mass*beam.len)
  val=0.0
  (1..Nmodes).each{
    n=it;
    omg=beam.eigenfrequency(n)
    OMG=OMEGA(n)
    val+=(sin(OMG*t)-OMG*sin(omg*t)/omg)*sin(n*pi*x/
    beam.len)/(omg*omg-OMG*OMG)
  }
  return val*coef

}
TimePlot=new PlotFrame()
axis=linspace(0.0,beam.len/loadvlc,1000)
xloc=beam.len*0.5
TimePlot.addFunction(new plotfunction(axis, {uanal(
    xloc,it)} as DoubleFunction))
TimePlot.setAutoColor(true)
TimePlot.hline(0.0)
TimePlot.show()
```

Listing 5.8 Simply-supported beam analytical solution for a constant velocity moving load (Biggs 8.2)

In what follows, the numerical solution is recovered using the `prismaticBeam` class listed in the following script.

```
1  import courses.structuraldynamics.prismaticBeam
2  theUniverse.cls()
3  hsec=0.5; bsec=0.3
4  beam=new prismaticBeam(len:5.0, id:1, Elast
       :10300000.0, Im:bsec*hsec**3/12.0, mass:750.0*
       hsec*bsec, bcl:2, bcr:2)
5  //theUniverse.putContraption(beam)
6
7  F0=-10000.0d
8  mesh=4
9  beam.Nmodes=1*mesh
10 beam.Nx=50*mesh
11 beam.setCriticalModalDamping(0.0)
12 beam.dt=0.01/mesh
13 loadvlc=1.0
14 tot=1.3*beam.len/loadvlc
15
16 fst={x,t->
17    loadloc=loadvlc*t
18    sz=0.1*beam.len/8.0
19    val=0.0d
20    if((x>=loadloc-sz) && (x<=loadloc+sz))val=F0/(2.0*
        sz)
21    return val
22 }
23
24 timehistory=beam.response(fst as DoubleFunction, {0.0
       d} as DoubleFunction, {0.0d} as DoubleFunction,
       tot, beam.len/2.0)
25 TimePlot=new PlotFrame()
26 TimePlot.addFunction(new plotfunction(beam.dt,
       timehistory))
27 TimePlot.setTextfontsize(24)
28 TimePlot.hline(0.0); TimePlot.vline(beam.len/loadvlc)
       ;
29 TimePlot.show()
```

Listing 5.9 Simple beam's numerical solution for the same configuration

It should be noted that in this example we used the `response` method of the `prismaticBeam` class that accommodates as argument a single loading function in both time and space, as can be seen in lines 16 and 24 of Listing 5.9 source code. The result of the analysis in the form of the time evolution of the transverse displacement of the beam is shown in Fig. 5.7.

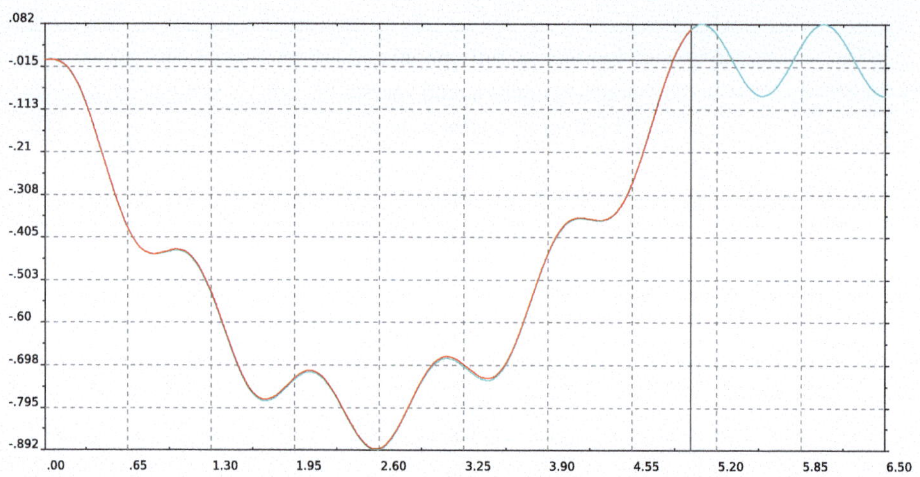

Fig. 5.7 Transverse displacement of a simply-supported beam as it is traversed by a moving point force. Red curve: Forced vibration regime; blue curve: Free vibration regime

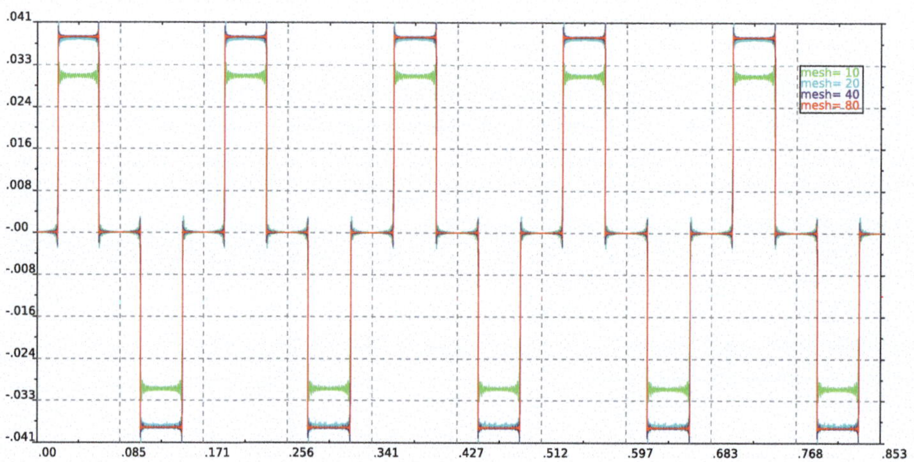

Fig. 5.8 Axial displacement at the center of the rod as computed for succesively refined meshes using the script of code of Listing 5.10

5.9   Elastic Wave Propagation

A basic reference on this topic is the book by Meirovitch [6] (Sects. 8.12 and 8.13), while the previously mentioned book by Graff [3] also contains relevant material. In what follows, we focus on axial vibrations that were previously discussed in Sect. 5.1 while we numerically solve the classical problem presented in Sect. 2.3.4 of [3]. Results are shown in Fig. 5.8 for the axial motion of a rod, following the source code of Listing 5.10.

```
1  import courses.structuraldynamics.prismaticRod
2  hsec=0.5; bsec=0.3
3  rod=new prismaticRod(len:5.0, id:1, Elast:10300000.0,
       Area:bsec*hsec, mass:750.0*hsec*bsec, bcl:1, bcr
       :0)
4  theUniverse.putContraption(rod)
5  c=sqrt(rod.Elast*rod.Area/rod.mass)
6
7  mesh=40
8  P0=1000.0d;
9  Tot = 20.0*rod.len/c
10 Nt=mesh*200
11 rod.dt=Tot/Nt as double
12 rod.Nmodes=4*mesh
13 rod.Nx=200*mesh
14 println rod.dt
15
16 fs=new mathman.functions.Dirac(rod.len) //
17 sigma=0.0001
18 ft=new mathman.functions.GaussianFunction(P0/(sqrt
       (2.0*pi)*sigma),0.0,sigma)
19 d0= {0.0d} as DoubleFunction
20 v0={0.0d} as DoubleFunction
21 if(true){
22   MDMplot= new PlotFrame()
23   timehistory=rod.response(fs, ft, d0, v0, Tot, rod.
       len/2.0)
24   pf=new plotfunction(rod.dt,timehistory)
25   pf.setName("mesh= "+mesh)
26   MDMplot.addFunction(pf)
27
28   MDMplot.makeLegend()
29   MDMplot.setAutoColor(true)
30   MDMplot.show()
31 }else{
32   snapshotPlot=new PlotFrame()
33   snapshotPlot.setTextfontsize(24)
34   (0..2.5).by(0.05).each{
35     time=it*1.0*rod.len/c
36     defshhot=rod.snapshot(fs, ft, d0, v0, time)
37     pf=new plotfunction(rod.len/rod.Nx,defshhot)
38     pf.setName("time= "+time+"s")
39     snapshotPlot.addFunction(pf)
40   }
41   snapshotPlot.show()
42 }
```

Listing 5.10 Elastic wave motion in a bar

5.10   Impact of a Mass on a Beam

We consider a mass colliding on a beam and assume the excitation force imparted on the beam is described by a force density term in the form $q_y(x,t) = f(t)g(x,x_0)$. This force density term is related to the impact force $F(t)$ as follows:

$$f(t) = \frac{F(t)}{\int\limits_{x_0-\delta x}^{x_0+\delta x} g(x,x_0)dx} = \frac{F(t)}{\int g(x,x_0)dx}.$$

In the following, we are interested in determining the contact force $F(t)$ and the deflection of the beam $w(t)$, at the instant the beam is stricken by a mallet of mass m which has a spherical head and with velocity v_o. Assuming the validity of the Hertz contact law in mechanics, then for the relative approach at the contact point can write,

$$\delta(t) = \kappa \, F^{2/3}(t) \tag{5.38}$$

In the above, $F(t)$ is the contact force between the two bodies and k is the Hertz constant [7] which depends on the elastic modulii and the contact geometries of the two bodies in question, as shown below. The relative displacement formulation is defined as the difference between the displacement w_s of the striking sphere due to force $F(t)$ and the displacement w_o of the beam at the contact location, assuming that the sphere has a mass of M_s:

$$\delta(t) = w_s - w_o \tag{5.39}$$

The displacement w_s is given as

$$w_s = v_o t - \frac{1}{M_s} \int\limits_0^t F(\tau)(t-\tau)d\tau \tag{5.40}$$

and according to Eq. (5.39), the relative displacement becomes

$$\delta(t) = v_o t - \frac{1}{M_s} \int\limits_0^t F(\tau)(t-\tau)d\tau - w_o(t) \tag{5.41}$$

The beam's response to the impulsive force $F(t)$ can to be evaluated as an expansion of normal modes, which depend on the beam boundary conditions and on the impact force location. It therefore takes the form [8],

$$w_o(t) = \sum\limits_n^{\infty} \phi_n(x_o)\frac{\int \phi_n(x)g(x,x_0)dx}{\int g(x,x_0)dx} \int\limits_0^t F(\tau)\sin[\omega_m(t-\tau)]d\tau \tag{5.42}$$

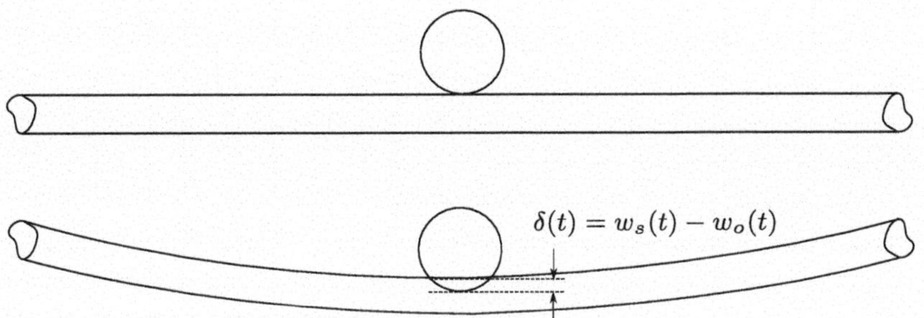

Fig. 5.9 Beam under impact by a mass at the instant of first contact (upper image) and immediately afterwards (lower image)

Combining the above equations leads to the following integral equation, which is classified as a nonlinear, second kind Volterra integral equation (Fig. 5.9).

$$\kappa \, F^{2/3}(t) = v_o \, t - \frac{1}{M_s} \int_0^t F(\tau)(t - \tau)d\tau$$

$$- \sum_n^\infty \phi_n(x_o) \frac{\int \phi_n(x)g(x, x_0)dx}{\int g(x, x_0)dx} \int_0^t F(\tau)\sin[\omega_m(t - \tau)]d\tau \tag{5.43}$$

Following [8], Eq. (5.43) is re-written as

$$\kappa \, F^{2/3}(t) = v_o \, t - \frac{1}{M_s}P(t) - \sum_n^\infty \phi_n(x_o)\frac{\int \phi_n(x)g(x, x_0)dx}{\int g(x, x_0)dx}Q(t) \tag{5.44}$$

where the two force terms are

$$P(t) = \int_0^t F(\tau)(t - \tau)d\tau \tag{5.45}$$

and

$$Q(t) = \int_0^t F(\tau)\sin[\omega_n(t - \tau)]d\tau \tag{5.46}$$

Expressions (5.45) and (5.46) are quite general and depend only on the eigenvalues of the problem at hand. Assuming that the contact force is piecewise linear across a 'small' time step, they take the following form following discretization of the time axis into a total of N steps:

$$P(N\Delta t) = (\Delta t)^2 \left(\frac{F_N}{6} + \sum_{j=1}^{N-1} F_j(N-j) \right), \tag{5.47}$$

$$Q_n(N\Delta t) = \frac{F_N}{\omega_n} \left(1 - \frac{\sin(\omega_n \Delta t)}{\omega_n \Delta t} \right)$$

$$+ \frac{4}{\omega_n^2 \Delta t} \sin^2 \left(\frac{\omega_n \Delta t}{2} \right) \sum_{j=1}^{N-1} F_j \sin[\omega_n \Delta t(N-j)]. \tag{5.48}$$

By combining Eqs. (5.44), (5.47) and 5.48, and furthermore grouping terms that depend only on F_j and terms that depend on the last value F_N of F_j, Eq. (5.44) is condensed as

$$\kappa\, F_N^{2/3} = A_N - B_N F_N \tag{5.49}$$

where

$$A_N = v_o N \Delta t - \frac{(\Delta t)^2}{M_s} \sum_{j=1}^{N-1} F_j(N-j)$$

$$- \frac{4}{\omega_n^2 \Delta t} \sum_n^\infty \frac{a_m g_m(x_o)}{\omega_n^2} \sin^2 \left(\frac{\omega_n \Delta t}{2} \right) \sum_{j=1}^{N-1} F_j \sin[\omega_n \Delta t(N-j)] \tag{5.50}$$

$$B_N = \frac{(\Delta t)^2}{6\,M_s} + \sum_n^\infty \frac{\phi_n(x_o)\dfrac{\int \phi_n(x)g(x,x_0)dx}{\int g(x,x_0)dx}}{\omega_n} \left(1 - \frac{sin(\omega_n \Delta t)}{\omega_n \Delta t} \right) \tag{5.51}$$

Terms B_N remains constant at every time step and F_N can be calculated by transforming Eqs. (5.49) into (5.52), setting $y^3 = F_N$ and finally computing terms A_N that involve all time steps up to $N-1$.

$$y^3 + \frac{\kappa}{B_N} y^2 - \frac{A_N}{B_N} = 0 \tag{5.52}$$

It is mentioned that the java implementation of a cubic equation solution can be found in [9].

5.10.1 Example of a Mass Impacting a Beam

We consider an example of a mass impacting a beam and we calculate, using the developed modules, the contact force between the mass and the beam as well as their response in time. The example is similar to one presented in the literature [10]. The version presented here has been studied by one of the authors (CGP) during a stay at the Hellenic Medditeranena University of Greece, Department of Music Technology and Acoustics, published in conference proceedings [11]. A modification of the published algorithm [8] has been implented here for the case of a continuous beam. Assuming a mass having velocity v0 impacting a beam, with arbitrary boundary conditions, on some specific location impactLoc. For the selection of values for parameters as shown in the code of Listing 5.11 we may observe the result of multiple impacts of the mass on the beam before final seperation. This behaviour can be obsered from the resulted contact force time history shown in the plot of Fig. 5.10 where we see three distinct regimes of contact between the impacting mass and the vibrating beam. The same can be observed in the results of the motion of the impacting mass (red curve) and the vibrating beam (blue curve) shown in the plot of Fig. 5.11 where we can see the three time regimes of contact between the two bodies.

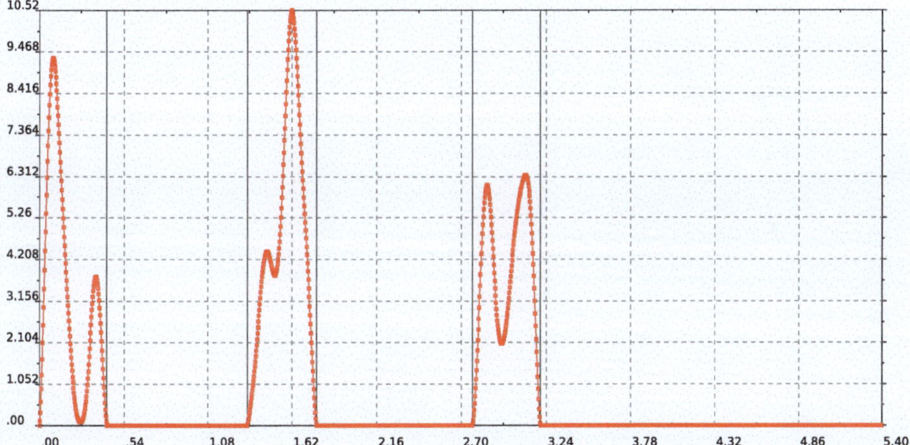

Fig. 5.10 Contact force for a cimple supported beam. It is assumed that a mass identical to the beam's total mass has been impacted the beam in the middle

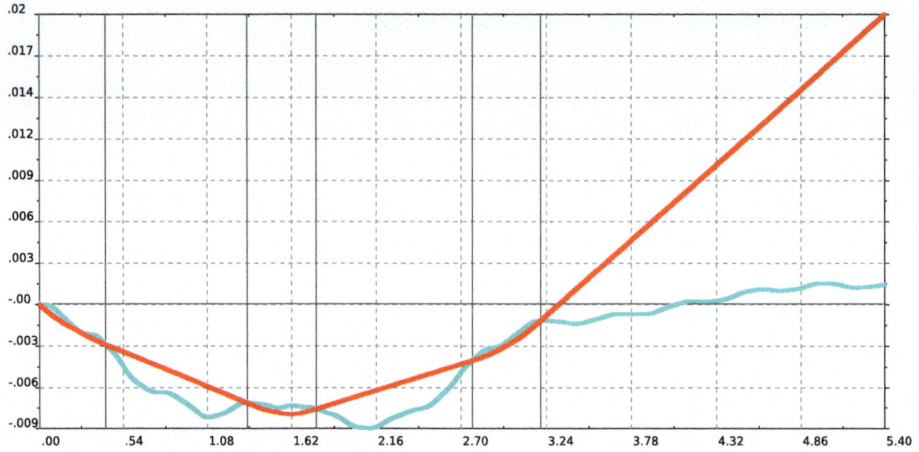

Fig. 5.11 Horizontal motion of the impacting mass (red) and beam's central point (blue)

```
import courses.structuraldynamics.prismaticBeam
theUniverse.cls()

hsec=0.01d; bsec=0.01d
beam=new prismaticBeam(len:0.307d, id:1, Elast:215.82e9,
    Im:bsec*hsec**3/12.0, mass:7960.0*hsec*bsec, bcl:2,
    bcr:2)
theUniverse.putContraption(beam)

mesh=5; beam.Nmodes=1*mesh
beam.Nx=100*mesh; beam.dt=3.0e-6
dt=beam.dt; loadvlc=1.0
numSteps=1800; tot=dt*numSteps
beamMass=beam.mass*beam.len
impactLoc=1.0*beam.len/2.0


//spatialy smoothening function fs
fs={x->
   sz=0.01d//beam.len/beam.Nx
   val=0.0d
   if((x>=impactLoc-sz) && (x<=impactLoc+sz))val=-1.0/(2.0*
     sz)
   return val
} as DoubleFunction

getGenMass={return beam.modalMass(it)}
getGenStif={return beam.modalStiffness(it)}
getGenLoad={return beam.modalForce(it,fs)}

// Ball material data
```

```
29  mass=0.267
30  kh=1.266E-7// Hertz constant
31  v0=0.01d// Impact Velocity
32  dt2=beam.dt*beam.dt
33  F=[0.0]; tcs=[0.0]; contact=true
34  (1..numSteps).each{
35    step=it
36    qi=0.0; pc=0.0; pti=0.0; qc=0.0
37    (1..beam.Nmodes).each{
38      Mn=getGenMass(it); Kn=getGenStif(it)
39      fn=getGenLoad(it); omega_n=sqrt(Kn/Mn)
40      phi_c=beam.eigenfunction(it, impactLoc)
41      qi+=phi_c*(1-sin(omega_n*dt)/(omega_n*dt))*fn/(omega_n
      *omega_n*Mn)
42      cc=(4.0/dt)*phi_c*fn*sin(omega_n*dt/2.0)*sin(omega_n*
      dt/2.0)/(omega_n*omega_n*omega_n*Mn)
43      fcos=0.0; fsin=0.0
44      (1..<step).each{
45        fcos+=F[it]*cos(omega_n*dt*it)
46        fsin+=F[it]*sin(omega_n*dt*it)
47      }
48      qc-=cc*(fcos*sin(omega_n*step*dt)-fsin*cos(omega_n*
      step*dt))
49    }
50
51    (1..<step).each{
52      pc+=F[it]*(step-it)*dt2/mass
53    }
54    pti=(dt2/6.0)/mass
55    Bi=pti-qi;Ai=v0*step*dt-pc-qc
56    cubic = new Cubic();
57    a3=1.0;a2=kh/Bi;a1=0.0;a0=-Ai/Bi
58    cubic.solve(a3, a2, a1, a0);
59    sol=cubic.x1; fc=sol*sol*sol;
60    if(contact){if(fc<=0.0){tcs.add(step*dt*1000.0)}}
61    if(!contact){if(fc>0.0){tcs.add(step*dt*1000.0)}}
62    if(fc>0.0){contact=true}else{fc=0.0; contact=false};F.
      add(fc)
63  }
64
65  thePlot.clear()
66  plot(dt*1000.0, F as double[])
67  thePlot.setTextfontsize(24)
68  thePlot.setMarker(true); thePlot.setAutoColor(true)
69  tcs.each{thePlot.vline(it)};thePlot.show()
70
71  ft={i=(floor(it/dt) as int); return F[i] as double} as
      DoubleFunction
72  if(true){
73    results=beam.compute(fs, ft, {0.0d} as DoubleFunction,
      {0.0d} as DoubleFunction, tot)
74    ws=[0.0];
```

```
75    (1..numSteps).each{
76      N=it
77      val=v0*it*dt-dt*dt*F[N]/(6.0*mass)
78      if(N>2)(1..(N-1)).each{val-=F[it]*(N-it)*dt*dt/(mass)}
79      ws.add(-val*1000.0);
80    }
81    w0=[]
82    (0..<numSteps).each{
83      dx=beam.len/(beam.Nx)
84      int loc=floor(impactLoc/dx)
85      w0.add( beam.u[loc][it]*1000.0 )
86    }
87    dispPlot=new PlotFrame()
88    dispPlot.addFunction(new plotfunction(dt*1000.0,ws as
         double[]))
89    dispPlot.addFunction(new plotfunction(dt*1000.0,w0 as
         double[]))
90    dispPlot.setAutoColor(true)
91    dispPlot.hline(0.0); dispPlot.setMarker(true)
92    tcs.each{dispPlot.vline(it)}
93    dispPlot.setTextfontsize(24);dispPlot.show()
94
95    theGP.stop(); scale=1000.0
96    theGP.plotDeform(scale, 0,numSteps,1)
97  }
```

Listing 5.11 Beam impacted by mass

References

1. Yang B (2005) Stress, strain, and structural dynamics: an interactive handbook of formulas, solutions, and MATLAB toolboxes. Academic
2. Klein Ikkink HA (2018) Groovy goodness: add map constructor with annotation. Cambridge University Press
3. Graff KF (1975) Wave motion in elastic solids. Oxford University Press
4. Clough RW, Penzien J (1975) Dynamics of structures. McGraw-Hill
5. Biggs JM (1964) Introduction to structural dynamics. McGraw-Hill
6. Meirovitch L (2010) Fundamentals of vibrations. Waveland Press
7. Landau L, Lifchitz E (1967) Théorie de l'élasticité. Mir
8. Evans GR, Jones BC, McMillan AJ, Darby MI (1991) A new numerical method for the calculation of impact forces. J Phys D Appl Phys 24(6):854
9. Partheepan S, Sivalingam D (2020) Solution to cubic equation using java programming. Euro J Math Comput Sci 7(1)
10. Stronge WJ (2000) Impact mechanics. Cambridge University Press
11. Kouzoupis S, Panagiotopoulos C, Kontos-Pantazis A (2024) Beam auralization in the time domain using the finite element method. In of Acoustics [74], pp 325–332

Numerical Solutions 6

6.1   Introduction

In the case of complex problems in engineering, analytical solutions are difficult and often impossible to obtain. Use must then be made of numerical methods for the determination of approximate solutions, which in turn are be used in engineering practice, i.e., in structural design. It must be kept in mind, however, that even the rigorous mathematical idealization of an engineering system is itself an approximation of reality.

A numerical method is either of the differential or of the integral type, depending on whether the analysis does or does not a priori integrate the governing differential equation of the problem at hand. The most widely known and used method at present for structural mechanics problems in general and for structural dynamics in particular is the Finite Element Method (FEM), with the Finite Difference Method (FDM) being popular in fluid mechanics and the Boundary Element Method (BEM) used primarily for infinite or semi-infinite media. From a chronological viewpoint, the FDM precedes the other two as it was developed for air flow problems in the 1920s. Specifically, the FDM replaces the governing differential equation(s) by a system of algebraic equations valid at a set of nodes (or nodal points) within the domain of interest, through an approximation of the differential equation derivatives by finite differences. On the other hand, the FEM replaces the domain of interest by a set of sub-domains (known as elements) and then approximates the behavior of the field variable(s) within each element so a to satisfy the governing differential equation(s) locally. This is done either by using equilibrium statements, variational principles or error minimization techniques. The elements are then interconnected at the nodes and comprise the entire domain.

The BEM is in essence the numerical treatment of the integral equation representations of the governing differential equations. There is an intermediate step, however, whereby the integral equation is reformulated so that actual boundary-value problems (BVP) can

© The Author(s), under exclusive license to Springer Nature Switzerland AG 2026   153
G. Manolis and C. Panagiotopoulos, *Vibrations of Structural Systems*, Synthesis Lectures
on Mechanical Engineering, https://doi.org/10.1007/978-3-032-12279-7_6

be solved, a process that results in what is known as a Boundary Integral Equation (BIE) formulation. Then, the BEM is but one of a number of techniques used to numerically evaluate the BIE formulation. For cases involving linear elastic homogeneous media, the BEM requires a surface-only discretization of the problem at hand, in contrast to both FEM and FDM that require full domain discretization. As previously mentioned, the FDM predates the other two, with FEM and BEM developments dating from the 1950 and 1960s, respectively.

Of course, hybrid numerical schemes are possible through combinations of the three basic classes of methods previously mentioned, not to mention variations such as the finite strip method for FEM and meshless methods for the BEM. The question of which numerical approach is best for a given problem is, in many ways, an open question to be answered by the practicing engineer and/or the researcher based on professional experience.

In this introductory-type chapter on numerical methods, we will first present Rayeleigh' method followed by material on Finite Elements as a generalization of this method. Emphasis is given to rod elements with some reference to beam elements. Some additional material will be given on presenting the basic concepts underlying the BEM and show how it can be applied in elastodynamics. Of course, since the turn of the new century, the BEM has been used across the entire spectrum of engineering science (e.g., fluid mechanics, wave propagation, electromagnetic theory, acoustics, fracture mechanics, corrosion, etc.) as a specialized method, as can be ascertained by perusing the literature on the BEM that have appeared in recent years. It should be noted that some early books on structural dynamics, e.g., Hurty and Rubinstein [4], presented integral equation methods (the basis for the BEM) as a means to solve transient problems for both SDOF and MDOF systems.

6.2 Rayleigh–Ritz Method

The Rayleigh–Ritz method downsizes an originally infinite DOF system representation of a continuous medium to one with a finite number of DOF. At first, we consider axial vibrations and use the general virtual work statement for the derivation of the equations of motion. Consider a rod of length L, elasticity modulus E, mass per unit length $m(x)$ and variable cross-section area $A(x)$ that is subjected to a distributed axial load $q_x = q_0(x)f(t)$ along its length. In the Rayleigh–Ritz method, the axial displacement $u(x, t)$ is built as a linear combination of trial functions $\psi_i(x)$ with the generalized coordinates $y_i(t)$ acting as time-dependent weights:

$$u(x, t) = \sum_{i=1}^{N} \psi_i(x) y_i(t) \tag{6.1}$$

The number of DOF N retained in this representation is crucial for a good quality representation of the exact solution, to which $u(x, t)$ converges. The generalized coordinates $y_i(t)$

are the unknown functions to be determined, while the trial functions are chosen randomly as long as they satisfy the following criteria:

i The kinematic boundary conditions are satisfied;
ii Form a set of linearly independent functions;
iii Are integrable functions.

If the above conditions are satisfied, the trial functions are labeled as admissible. In reference to the third condition, it is sufficient to consider a continuously differentiable function $\psi_i(x)$, which is of lesser degree as compared to functions appearing in the integrals. According to the principle of virtual work, if an elastic body is subjected to a virtual displacement/strain field that does not violate the boundary conditions, then the work produced as it acts on the existing force/stress fields is equal to zero. Note that the principle of virtual work requires a reformulation if the boundary conditions are not homogeneous. In the case of the rod, the acting forces are the external distributed forces $q_x(x, t)$, the axial forces $N(x, t) = EA(x)u'(x, t)$ and the inertia forces $f_i(x, t) = m(x)\ddot{u}(x, t)$. Assuming $\delta u(x, t)$ is a virtual displacement imposed on the rod, then the work produced by all these forces is zero, i.e., $\sum W = 0$. Thus we have

$$\int_0^L m(x)\ddot{u}(x, t)\delta u \, dx + \int_0^L EA(x)u'(x, t)\delta u' dx - \int_0^L q_x(x, t)\delta u \, dx = 0 \tag{6.2}$$

Since the virtual displacement is arbitrary, we can select the following representation:

$$\delta u = \sum_{i=1}^N \psi_i(x)\delta y_i(t) \tag{6.3}$$

with the first spatial derivative yielding

$$\delta u' = \sum_{i=1}^N \psi_i'(x)\delta y_i(t) \tag{6.4}$$

Substituting Eqs. (6.3) and (6.4) in the statement of virtual work, Eq. (6.2), yields

$$\int_0^L m(x)\ddot{u}(x, t)\sum_{i=1}^N \psi_i(x)\delta y_i(t)dx + \int_0^L EA(x)u'(x, t)\sum_{i=1}^N \psi_i'(x)\delta y_i(t)dx$$

$$-\int_0^L q_x(x, t)\sum_{i=1}^N \psi_i(x)\delta y_i(t)dx = 0 \tag{6.5}$$

Furthermore, by substituting the representation given in Eq. (6.1) results in

$$\int_0^L m(x) \sum_{i=1}^N \psi_j(x) \ddot{y}_j(t) \sum_{i=1}^N \psi_i(x) \delta y_i(t) dx$$

$$+ \int_0^L E A(x) \sum_{i=1}^N \psi_j'(x) y_j(t) \sum_{i=1}^N \psi_i'(x) \delta y_i(t) dx - \int_0^L q_x(x,t) \sum_{i=1}^N \psi_i(x) \delta y_i(t) dx = 0$$

$$(6.6)$$

The virtual displacement term is a common factor, so a rearrangement gives

$$\sum_{i=1}^N \delta y_i \left[\sum_{i=1}^N \ddot{y}_i \int_0^L m \psi_i \psi_j dx + \sum_{i=1}^N y_i \int_0^L E A \psi_i' \psi_j' dx - \int_0^L q_x \psi_i dx \right] = 0 \qquad (6.7)$$

Since the above relation holds for any arbitrary selection $\delta y_i(t)$, the sum of terms in the brackets must be zero:

$$\sum_{i=1}^N \ddot{y}_i \int_0^L m \psi_i \psi_j dx + \sum_{i=1}^N y_i \int_0^L E A \psi_i' \psi_j' dx - \int_0^L q_x \psi_i dx = 0 \qquad (6.8)$$

Recasting Eq. (6.8) in matrix form gives

$$m_{ij} \ddot{y}_j(t) + k_{ij} y_j(t) = p_i(t) \qquad (6.9)$$

where $m_{ij} = \int_0^L m \psi_i \psi_j dx$ are the coefficients of a mass matrix, $k_{ij} = \int_0^L E A \psi_i' \psi_j' dx$ are the coefficients of a stiffness matrix and finally $p_i(t) = \int_0^L q_x \psi_i dx$ are the coefficients of a load vector.

It is obvious that this procedure can be used in conjunction with other types of structural elements, e.g., for beams, plates, etc. For instance, in the case of flexural vibrations, and by ignoring the work produced by shear forces across the beam's cross-section, the stiffness elements of the matrix equation corresponding to Eq. (6.9) are $k_{ij} = \int_0^L E I(x) \psi_i'' \psi_j'' dx$, where $I(x)$ is the moment of inertia. Furthermore, the presence of springs, dashpots and concentrated masses attached at discrete points along a structural element can also be taken into account. The spatial distribution of these discrete elements can be represented by use of the Dirac–δ function as follows:

$$\int_0^L f(x) \delta(x - x_0) dx = \begin{cases} f(x_0), & \text{if } x \in [0, L] \\ 0, & \text{otherwise} \end{cases} \qquad (6.10)$$

As an example, consider the bending of a beam with a translational spring k_1 at location x_1, plus a rotational spring k_2 at location x_2. Furthermore, attached at location x_3 is a lumped mass m_1 with a mass moment of inertia J_{m1}. The work produced by these spring and inertia elements correspond to W_{S1}, W_{S2} and W_{I1}, W_{I2}, respectively, and are given below as:

$$W_{S1} = \int_0^L k_1 \delta(x - x_1) \psi_i(x) \psi_j(x) dx = k_1 \psi_i(x_1) \psi_j(x_1)$$

$$W_{S2} = \int_0^L k_2 \delta(x - x_2) \psi_i'(x) \psi_j'(x) dx = k_2 \psi_i'(x_2) \psi_j'(x_2)$$

$$W_{I1} = \int_0^L m_1 \delta(x - x_3) \psi_i(x) \psi_j(x) dx = m_1 \psi_i(x_3) \psi_j(x_3)$$

$$W_{I2} = \int_0^L J_{m1} \delta(x - x_3) \psi_i'(x) \psi_j'(x) dx = J_{m1} \psi_i'(x_3) \psi_j'(x_3).$$

6.2.1   Computation of Eigenvalues and Eigenvectors

In order to compute an approximation to the eigenvalue problem of a continuous dynamic system with the Rayleigh–Ritz method, the following algebraic matrix equation

$$(k_{ij} - \lambda_i m_{ij}) z_i = 0 \tag{6.11}$$

must be solved to recover the eigenvalues λ_i and their corresponding eigenvectors z_i. Note that the eigenvectors are essentially the discrete form of the eigenfunctions, which occurs when a continuous system is approximated by a discrete parameter one. Following solution, the approximate eigenvalues (natural frequencies or eigenfrequencies) and their corresponding eigenfunctions (modal shapes) are given as

$$\tilde{\omega}_i = \sqrt{\lambda_i} \tag{6.12}$$

and

$$\tilde{\phi}_i(x) = \sum_{j=1}^{N} z_{ij} \psi_j(x) \tag{6.13}$$

where z_{ij} is the jth component of the ith eigenvector z_i of the aglebraic Eq. (6.11).

The selection of trial functions for implementation within the Rayleigh–Ritz method is crucial for an accurate evaluation of the dynamic system's eigenproperties. These functions must satisfy the kinematic boundary conditions of the problem, but if in addition they satisfy force boundary conditions, the convergence of the method will be faster. One simple way for this selection is to define a polynomial $P(x)$ whose degree matches the number of terms used in the polynomial expansion. An alternative way is to select the displacement function which satisfies the static equivalent of the dynamic problem, i.e., the one defined by the absence of inertia forces. This has the advantage that it guarantees satisfaction of force boundary conditions. Of course, the trial functions must be linearly independent and this is guaranteed by the selection functions such as trigonometric and Bessel among others.

6.2.2   Rayleigh–Ritz Method Using Prismatic Beam Eigenfunctions

The second derivative $\phi_n''(x)$ of eigenfunction $\phi_n(x)$ with respect to x for the pinned–pinned combination of boundary conditions for a beam element is

$$\phi_n''(x) = -\lambda_n^2 \sin(\lambda_n x)$$

while for cases where the left side boundary is clamped,

$$\phi_n''(x) = \cosh(\lambda_n x) - \cos(\lambda_n x) - k_n(\sinh(\lambda_n x) - \sin(\lambda_n x))$$

For the case where the right side boundary is clamped one should replace x with $L-x$ to use the correspond expression for $\phi_n(x)$ eigenfunction.

It is noted that the implementation of class beamRRM given in following section integrates the expressions $EI(X)\psi_n''(x)\psi_m''(x)$ and $\varrho A(X)\psi_n(x)\psi_m(x)$ in order to consider beams of variable cross-sections.

6.3   Class beamRRM

Description: The   beamRRM   appearing   in   the   software   package courses.structuraldynamics is an implementation of the Rayleigh–Ritz Method using as test functions the eigenfunctions of the prismatic cross–section case for any boundary conditions combination. This way we will we are able to estimate the eigensystem solution and even solve the transient problem for the non–prismatic case with increased accuracy.

> *Attributes*
> Class beamRRM attributes include the length of the beam (double len), the elasticity
> modulus (double Elast) as well the mass density (double dens) together with the
> definition of left (Integer bcl) and right (Integer bcr) boundary conditions. The later
> two attributes signal boundary condition types, namely the value of 0 implying a free
> end, the value of 1 implying a fixed end and finally the value of 2 for a pinned end.
> Furthermore, Nx stands for the spatial discretization over the length of the beam in order
> for integrals to be computed using the trapezoidal rule.
>
> *Instance*
> No specific constructors have been considered, so a Groovy's map constructor with
> annotation might be used [5].
>
> *Methods*
>
> - void calcEigen (DoubleFunction Im, DoubleFunction Area, int n)
> - void calcEigen (DoubleFunction Im, DoubleFunction Area, DoubleFunction Spring,
>   int n)
> - double eigenfrequency (int n)
> - DoubleFunction eigenfunction (int n)
> - double eigenfunction (int n, double x)
> - double norm (int n)
>
> Note that methods similar to those found for the case of prismaticBeam such as the
> response, dresponse, ddresponse, snapshot and compute, can be easily
> set, however have not been implemented in the current version.

6.3.1   Winkler–Type Elastic Foundation Supported Beam

We use the beamRRM in order to approximate the eigensystem of a beam on elastic foundation. Analytial solution for the eigenfrequencies considering a Winkler-type elastic foundation with a spring distribution, for a prismatic beam may be found in the literature [6].

As can be seen frome the source code of Listing 6.1, after definition of the beamRRM,
using the Groovy's map constructor with annotation, then uppon eigensystem's calculation
we should specify distribution's over the length of the beam for the moment of inertia and
mass density, controling this way the beam's elastic and inertial properties, as well as the
elastic constant of springs for a Winkler–type foundation. The script can easily used also
for the case of prismatic beam, where we consider a constant distribution for moment of
inertia and mass density, as well for absence of elastic foundation by defining a zero valued

distribution of the ealstic constant or by simply omitting the value of it (see first version of `calcEigen` in Sect. 6.3).

```
import courses.structuraldynamics.beamRRM

YModul=20.0e9;Imp=0.00423d
Ar=0.69d;p=27395.51
L=4.0;k=150000000.0d
println "elastic foundation spring's constant=$k"
eig=6

npPlot = new PlotFrame()
npPlot.setAutoColor(true)
npPlot.makeLegend()
npPlot.setTextfontsize(24)
npbeam=new beamRRM(len:L, Elast:YModul, dens:p, bcl:2, bcr
    :2)
Im={Imp}; Area={Ar}; Spring={k}
npbeam.Nx=100
npbeam.calcEigen(Im as DF, Area as DF, Spring as DF, eig)
(1..eig).each{
  println "Rayleigh-Ritz omega_"+it+":\t"+npbeam.
    eigenfrequency(it)
  pf = new plotfunction(linspace(0.0, npbeam.len,1001),
    npbeam.eigenfunction(it))
  pf.setName(" _"+it+"="+String.format("%.3f", npbeam.
    eigenfrequency(it))+"rad/s")
  npPlot.addFunction(pf)
}
npPlot.title="Rayleigh-Ritz method"
npPlot.show()
println "========================="
(1..eig).each{println "analytical omega_"+it+":\t"+sqrt((it
    *it*it*it*pi*pi*pi*pi*YModul*Imp/(L*L*L*L)+k)/(p*Ar))}
```

Listing 6.1 Simple example using the Rayleigh–Ritz method beam implementation object beamRRM

6.4 Finite Element Method

This section is an introductory presentation of the Finite Element Method, including as example an implementation of the Rayleigh–Ritz method previously discussed. It is possible to say that the FEM can be seen as a variant of the Rayleigh–Ritz method, the difference being that in the former method the base functions are locally defined, while in the latter method they are globally defined. Such a situation arises when the trial functions for the simple case of the one–dimensional rod are described in the literature as hat functions [7].

We start by dividing the axis of the rod into N_e non-overlapping sub-intervals, each of length L_i, see Fig. 6.1. We therefore refer to element i of length L_i having the appropriate number of nodes at specific locations, see Fig. 6.2. We will retain the convention under which

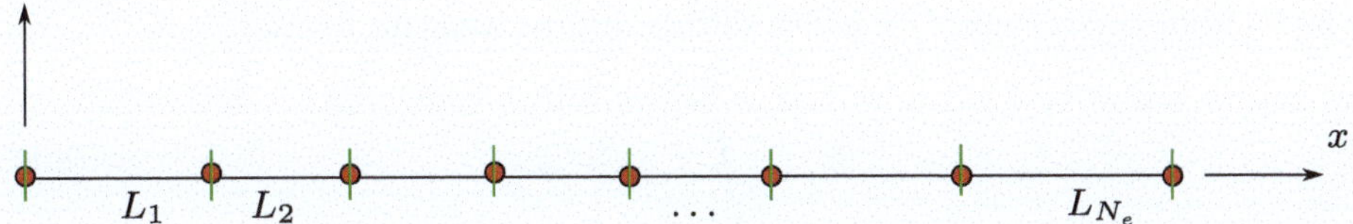

Fig. 6.1 Finite element mesh for discretization in space

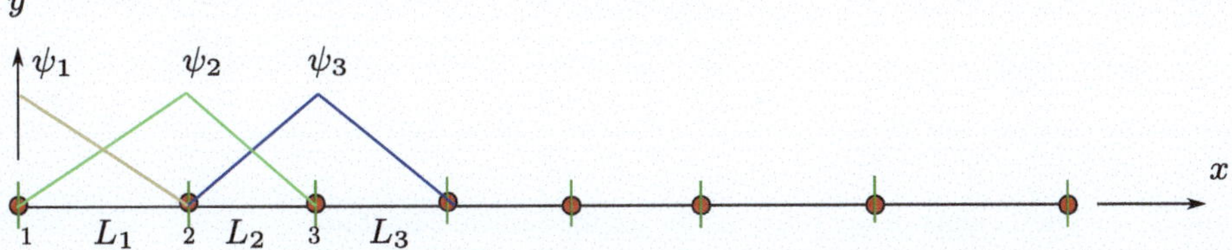

Fig. 6.2 Hat functions related to the first three nodes of the rod starting from the left

element i starts from node i and ending at node $i + 1$. The x coordinate of node i is written as x_i, while the interpolation functions applicable to element i are $\psi_i(x)$ and $\psi_{i+1}(x)$. This is illustrated in Fig. 6.2 which shows how the hat functions relate to the first three rod nodes from the left. These basis functions can be explicitly written as follows:

$$
\psi_i(x) = \begin{cases}
0, & x < x_{i-1}, \\
(x - x_{i-1})/(x_i - x_{i-1}), & x_{i-1} \le x < x_i, \\
1 - (x - x_i)/(x_{i+1} - x_i), & x_i \le x < x_{i+1}, \\
0, & x \ge x_{i+1}.
\end{cases}
\tag{6.14}
$$

In the case of the FEM, the number of basis functions employed are not in general equal to the number of nodes, but for the case of 1D rod presented here this holds true. In more general cases, the number of basis functions used is equal to the number of degrees of freedom of the problem. In reference to Eq. (6.1), the displacement field can be simply written as follows:

$$
u(x, t) = \sum_{i=1}^{N} \psi_i(x) u_i(t)
\tag{6.15}
$$

In the above, $u_i(t)$ is the axial displacement registered at the location of node x_i and at time t. From here on we follow a similar approach to the one introduced in Sect. 6.2. The coefficients of the stiffness matrix of Eq. (6.9) are defined as

$$
k_{ij} = \int_0^L EA\psi_i'\psi_j' dx = \sum_{e=1}^{N_e} \int_{x_e}^{x_{e+1}} EA\psi_i'\psi_j' dx
\tag{6.16}
$$

with k_{ij} being coefficients of the global stiffness matrix, while the terms

$$k_{ij}^e = \int_{x_e}^{x_{e+1}} E A \psi_i' \psi_j' dx \tag{6.17}$$

refer to the stiffness matrices for the individual eth element. In a similar way, the mass matrix for the eth element is defined as

$$m_{ij}^e = \int_{x_e}^{x_{e+1}} \rho A \psi_i \psi_j dx \tag{6.18}$$

Finally, the coefficients of the force vector for the same element are

$$p_i^e = \int_{x_e}^{x_{e+1}} q_x \psi_i dx \tag{6.19}$$

with $q_x(x)$ the distribution of the external load along along the x-axis of the rod. By considering the case of constant material properties E^e, ρ^e and cross section area A^e along the length of element e, and then following integration of the stiffness terms listed above, Eq. (6.17) results in

$$K^e = \frac{E^e A^e}{L^e} \begin{bmatrix} 1 & -1 \\ -1 & 1 \end{bmatrix}. \tag{6.20}$$

In a similar fashion, integration of the mass terms yields the mass matrix of element e, as

$$M^e = \frac{\rho^e A^e L^e}{6} \begin{bmatrix} 2 & 1 \\ 1 & 2 \end{bmatrix}. \tag{6.21}$$

There is an advantage that accrues when implementing explicit time integration schemes if the mass matrices of the rod elements are diagonal. To this end, there exist certain techniques of mass matrix diagonalization [8]. For the case examined here, it suffices to consider the assumption of masses lumped at the location of the nodes [9]. This results in the following form for the mass matrix of the eth element,

$$M^e = \frac{\rho^e A^e L^e}{2} \begin{bmatrix} 1 & 0 \\ 0 & 1 \end{bmatrix}. \tag{6.22}$$

6.5   Class beamFEM

Description: The beamFEM appearing in the software package courses.structuraldynamics is an implementation of the Finite Element Method for thcase of the Eulre–Bernouli beam theory.

Attributes

Class beamFEM attributes include the length of the beam (double totlen), the elasticity modulus (double Elast_ref), the moment of inertia (double Im_ref) together with their distributions per length given by the ImDist and ArDist as DoubleFunction variables. Furtermore, the mass per unit length (double mass_ref) together with the definition of left (Integer bcl) and right (Integer bcr) boundary conditions. The later two attributes signal boundary condition types, namely the value of 0 implying a free end, the value of 1 implying a fixed end and finally the value of 2 for a pinned end. Finally, the restrainedLocations a Double ArrayList serves to define locations on the length of the beam where the transverse displacement is restrained. Furthermore, Nx stands for the spatial discretization over the length of the beam while the boolean variable consistentMass for a full, see Eq.(6.21), rather than a diagonal, see Eq. (6.22), mass matrix.

Instance

- beamFEM (double len, double Elast, double Im, double mass)

Methods

- void calcEigen()
- void calcEigen(DoubleFunction Im, DoubleFunction Area)
- double eigenfrequency(int n)
- DoubleFunction eigenfunction(int n)
- double[] eigenVector(int n)
- double[] eigenDeform(int n)

Note that methods for the the transient response of the beam can be easily set, however have not been implemented in the current version.

6.5.1 FEM Introductory Example

A FEM introductory example is given here. A simple supported prismatic beam also restrained, as regards the transverse deformation, at specfic locations over each length is considered. These location are defined by setting the `restrainedLocations` ArrayList which each component l_i refers to some location onto the length L of the beam $x_i = l_i * L$ where transverse displacement set to zero.

```
import courses.structuraldynamics.beamFEM

YModul=20.0e9;Imp=0.00423d
Ar=0.69d;p=27395.51
L=4.0

femBEAM=new beamFEM(L, YModul, Imp, p*Ar)
femBEAM.bcl=2; femBEAM.bcr=2;
femBEAM.consistentMass=true
femBEAM.Nx=100
femBEAM.restrainedLocations=[0.25d,0.5d,0.75d]

femBEAM.calcEigen()

plot()
thePlot.hline(0.0)
thePlot.setAutoColor(true)
femBEAM.restrainedLocations.each{thePlot.vline(it*L,Color.
    blue)}
(1..3).each{
  println "FEM omega_"+it+":\t"+femBEAM.eigenfrequency(it)
  plot(L/femBEAM.Nx,femBEAM.eigenDeform(it))
}
thePlot.getPlotFunctions().each{it.linewidth=3}
```

Listing 6.2 Eigensystem of a simply supported beam being restrained to several points over each length as computed using the finite element method beam implementation object `beamFEM`

The first three eigenvectors computed using the FEM are depicted in Fig. 6.3 and correspond to eigenfrequencies of 660.269, 770.279 and 1031.466 rad/s, respectivey.

6.6 Boundary Element Method

The basic theory behind integral equations was developed by Fredholm [10], followed by their use in potential theory by Kellogg [11] and in elastostatics by Muskhelishvili [12]. A comprehensive treatise on the theory of singular integral equations is that of Mikhlin [13], while the earliest numerical method for the solution of engineering problems based on integral equations is attributed to Trefftz [14]. It was only after the introduction of the electronic computer as a computational tool in the mid-1950s that a new impetus was given to

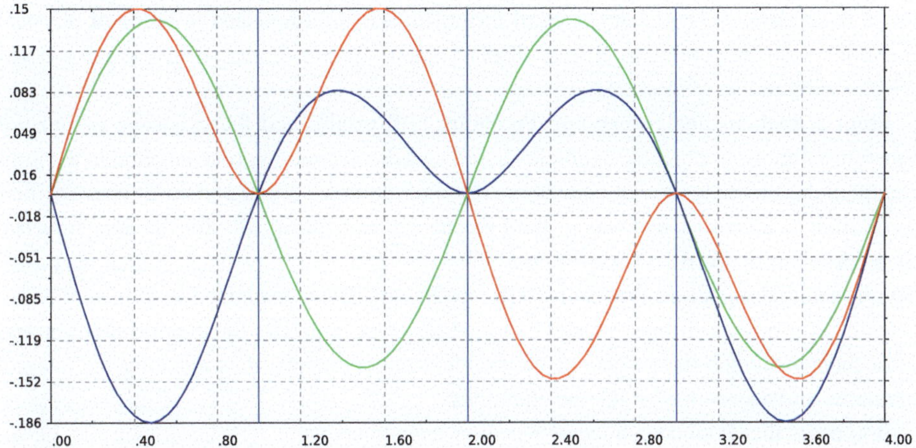

Fig. 6.3 The three first eigenfunctions for a simply supported prismatic beam with restrained points

the subject of computational methods based on integral equation formulations. Specifically, we have the emergence of indirect boundary integral equation methods in potential theory [15], in acoustics [16] and in elastostatics [17]. At about the same time, boundary integral methods of the direct type, which are better suited for the solution of BVP, were introduced in elastostatics by Rizzo [18] and in elastodynamics by Cruse and Rizzo [19]. Also, the relation between integral methods and the general weighed residual statement was identified by Zienkiewicz et al. [20].

Numerous articles have since appeared on the use of numerical methods based on boundary integral equation formulations, which result in either the indirect or the direct BEM, in all fields of engineering science such as potential theory, potential fluid flow, acoustics, torsion, electric and magnetic field theories, elastostatics, elastodynamics, structural mechanics, heat conduction, viscoelasticity, thermoelasticity, fracture mechanics, plasticity, water waves, viscous flow, ground water flow, corrosion, contact problems, coupled-field problems, etc., see for instance the book by [3].

In the BEM, the problem is formulated in terms of integral equations relating boundary values of the dependent variables and solution is then found numerically. If values in the interior domain are needed, they can be calculated as a second-tier approach from the boundary data. Thus, the dimension of the problem at hand is reduced by one, while the resulting system of algebraic equations is invariably smaller in size compared to similar systems obtained from use of the FEM or the FDM. Only in certain cases (e.g., nonlinearities, inhomogeneities, certain types of body forces) does the BEM require discretization of part or all of the interior domain. As mentioned, there are two basic categories of BEM, namely indirect and direct. In the former case, which is also known as the source method, the boundary integral formulation is accomplished in terms of fictitious source densities from which the physical quantities (i.e., the dependent variables) of the problem can be recovered.

In the latter case, the boundary integral formulation is accomplished in terms of physical quantities such as potentials, fluxes, displacements, tractions, etc., which are computed directly without use of intermediate steps.

From a mathematical viewpoint, Fredholm [10] established the existence of integral equation solutions to potential problems based on a discretization procedure and in the limit identified the Fredholm integral equations (FIE) of the first, second and third type. The first application of a direct BEM was in fluid mechanics and specifically for the axisymmetric air-jet problem by Trefftz [14], who used the method of successive approximations to satisfy the integral equation. Prager in 1928 (see Prager, 1961) examined doubly symmetric potential flow past an elliptic cylinder using a direct-type boundary integral formulation, and subsequently divided the surface of the problem into elements, thus reducing the integral equations into a system of algebraic equations.

It appears that boundary integral equation techniques were well known by the 1950s, but not popular due to the high computational effort involved. Among the first to use the electronic computer to solve axisymmetric potential fluid flow problems was Hess [15]. Martensen (1959) (see Martensen 1968) extended Prager's method to 2D fluid flow problems, while Hess and Smith (1964) solved 3D potential fluid flow problems using an indirect BEM called the 'panel' method because the surface of the body in question was discretized into flat quadrilateral elements. At about the same time, Jawson (1963) and Symm (1963) presented a numerical technique for solving Fredholm integral equations of the first kind resulting from use of the indirect BEM in 2D potential problems. They also presented a direct BEM based on Green's third identity, which was capable of relating physical quantities of the general mixed BVP in integral form.

Since the mid-1960s, a number of both direct and indirect BEM formulations for elastostatic problems appeared. The most notable of these is the direct formulation of Rizzo [18] relating boundary displacements and tractions, which allowed for the solution of general, mixed-boundary value problems. This formulation was extended to elastodynamics a year later by Cruse and Rizzo [19], which proved seminal in constructing BEM schemes for solving a variety of problems in structural dynamics. Next, the 1970s witnessed an extension of the BEM to many engineering fields, followed by the introduction of higher–order elements for discretization purposes, which brought about improved accuracy and efficiency. This was followed by the coupling of the BEM with other methods such as the FEM in an effort to exploit the best features of different categories of numerical techniques. Nowadays, the BEM has become an established computational tool whose efficiency for certain specialized categories of problems is well established and complements the most popular numerical method to date, namely the FEM. The main pros and cons of BEM usage are summarized below:

Advantages:

1. It is a numerical technique founded on a firm mathematical basis and is therefore applicable to a wide range of problems in engineering science

2. It usually requires surface discretization only, which reduces the dimensionality of the problem at hand by one.

3. The method is capable of yielding a high degree of accuracy, since the underlying integral equation is an exact statement of the problem at hand. Approximations are introduced at the second stage, i.e., when the surface of the problem is discretized.

4. The fundamental singular solutions (or Green's functions) used in the BEM satisfy radiation-type boundary conditions. As a result, infinite or semi-infinite domains can be treated without the artificial boundaries introduced in the FEM and the FDM.

5. Values of the unknowns at points inside the domain can be selectively found from the boundary solution.

6. The BEM can be used for the construction of influence matrices for large homogeneous regions or regions containing semi-infinite boundaries (super elements), which in turn can be incorporated within FEM analyses.

Disadvantages:

1. The BEM generates non-symmetric, fully populated influence matrices. This is in contrast to FEM, which generates symmetric and sparse (diagonally banded) system matrices. The order of the BEM influence matrices, however much smaller than that of those generated by the FEM.

2. For certain classes of problems, it is either very difficult or impossible to obtain the appropriate fundamental solutions. It is also possible that the fundamental solution is an approximate one, or is known in discrete form only, thus giving rise to rather inefficient computational schemes.

3. The BEM is inefficient as compared to the FEM for problems where one spatial direction is small compared to the remaining ones, and for problems with rapidly varying material properties.

4. The presence of known distributed body forces in linear problems or pseudo-incremental body forces in nonlinear problems requires volume integrals, which in turn necessitate internal domain discretization.

6.6.1   Implementation of BEM

The BEM formulation implemented in here was previously presented in [21] and follows the conventional treatment outlined in [22]. While it is possible to work with two and three dimensional BEM formulations (see also [23]), for structural dynamics purposes we choose to develop a one-dimensional Euler–Bernoulli beam formulation following the steps outlined in [24]. A parallel BEM formulation for Timoshenko beams is also used, see [25], but not implemented here. Starting with the free vibration case where the transverse displacement is $w(x, t) = W(x) \sin\omega t$, substitution in Eq. (5.3) yields the reduced form in the freqeuncy domain as

$$\frac{\partial^4 W}{\partial x^4} - \frac{\varrho A \omega^2}{EI} W = 0, \tag{6.23}$$

with ω the circular frequency. Multiplying both sides by the displacement fundamental solution $G(x, \xi, \omega)$ (yet to be determined) and integrating by parts four times we arrive at

$$W(\xi) = EI \left[-Y'''G + Y''G' - Y'G'' + YG''' \right]_0^l \tag{6.24}$$

where l the length of the beam and primes indicate differentiation with respect to x. Moreover, Eq. (6.24) can be written in the form,

$$W(\xi) = [sG - mF + \theta E - WD]_0^l \tag{6.25}$$

where

$$\theta = \theta(x) = \frac{dW}{dx}, \qquad\qquad F = F(x, \xi) = \frac{dG}{dx}$$
$$m = m(x) = -EI\frac{d^2W}{dx^2}, \qquad E = E(x, \xi) = -EI\frac{d^2G}{dx^2}$$
$$s = s(x) = -EI\frac{d^3W}{dx^3}, \qquad D = D(x, \xi) = -EI\frac{d^3G}{dx^3}$$

In the above, dependence on ω has been dropped, but tacitly implied. Furthermore, the two kinematic field variables are the amplitude of the transverse displacement $W(x)$ and the corresponding slope $\theta(x)$, while the two stress field variables are the bending moment $m(x)$ and the shear force $s(x)$. Given that the Green's function is a displacement influence coefficient between two field points x, ξ along the beam's length, the quantities that can be defined in parallel to the physical variables besides G are F, E, and D. Differentiating Eq. (6.25) with respect to ξ and considering that $dW(\xi)/d\xi = \theta(\xi)$ we get,

$$\theta(\xi) = \left[sG' - mF' + \theta E' - WD' \right]_0^l \tag{6.26}$$

where

$$D' = D'(x, \xi) = \frac{dD}{d\xi}, \quad E' = E'(x, \xi) = \frac{dE}{d\xi}, \quad F' = F'(x, \xi) = \frac{dF}{d\xi}.$$

Equations (6.25) and (6.26) provide the transverse displacement and slope of the beam at any interior point ξ due to boundary values of shear s, moment m, slope θ and deflection W. Explicit expressions for the fundamental solution G and its various derivatives are given in the Appendix of Ref. [24], where the typographical error in Eq. (A.3) suggests that the cosine function has to be replaced by a sine function. Here we reproduce the expression for the fundamental solution as follows:

$$G(x, \xi) = \frac{1}{4\lambda^3 EI} \left[\sec(\lambda l) \sin(\lambda(l - |x - \xi|)) - \mathrm{sech}(\lambda l) \sinh(\lambda(l - |x - \xi|))\right],$$

$$(6.27)$$

In the above, sinh and cosh the hyperbolic sine and cosine functions, respectively, and coefficient $\lambda^4 = \varrho A \omega^2 / EI$. Next, by moving the field point ξ to the boundaries, Eqs. (6.25) and (6.26) now form an 8×8 $H(\omega)$ homogeneous matrix equation which in compact form reads as follows:

$$H(\omega)\, x = 0 \qquad\qquad (6.28)$$

The system matrix $H(\omega)$ incorporates the boundary equations (6.25) and (6.26), as well as the appropriate equations for a given set of boundary conditions. Vector x comprises eight entries, namely the displacements, slopes, moments and shears at the boundaries (i.e., the endpoints) of the beam. Each separate domain i has its own coefficient matrix $H^i(\omega)$ and degrees of freedom at the boundaries. Following imposition of the interface conditions, i.e., kinematic compatibility and equilibrium, see also [26], successive subdomains of the total coefficient matrix and of the respective vector of unknowns will have the same form as that given in Eq. (6.28). Nontrivial solutions of equation (6.28) for x are then recovered by setting the system matrix determinant equal to zero, i.e.,

$$det[H(\omega)] = 0. \qquad\qquad (6.29)$$

The beam's natural frequencies are provided as the real roots ω of the frequency equation (6.29), while solution of equation (6.28) in conjunction with Eqs. (6.25) and (6.26) provide the modal shapes of the beam (see also [24]). However, numerical problems might arise since the dense, non-symmetric linear systems obtained from the discretization process of the BEM are often ill-conditioned, especially when large-scale problems are considered [27]. In order to overcome this problem we use as mode indication index the lower singular value $\sigma_1(\omega)$ of the matrix H. Zero or numerically near zero values of $\sigma_1(\omega)$ indicates that ω is a natural frequency of the beam. Using the singular value matrix decomposition (SVD) technique, that is $H = USV^H$, we can also identify the respective eigenmodes as the right-hand singular vectors corresponding to the lowest singular value (the last column of V) for an ω value that gives

$$\sigma_1(\omega) = 0. \tag{6.30}$$

Performing a frequency sweep analysis in order to find the roots of equation (6.29) or Eq. (6.30) is a computationally demanding task. To that purpose, we set up the following procedure, in order to have a first estimation of a potential eigenfrequency. We commence by searching a particular frequency range of interest $\hat{\omega} \in [\hat{\omega}_{min}, \hat{\omega}_{max}]$ using a rather coarse frequency grid defined by step $\Delta\omega$, as shown in Fig. 6.4, plus the usual automatic peak-picking method to identify an estimate $\tilde{\omega}_i$ of the ith natural frequency ω_i. After obtaining a first estimate of an eigenfrequencys ω_i, we then use the standard golden–section search technique in order to zero into a more accurate value. Also, the search for each eigenfrequency ω_i is done within its corresponding range $[\tilde{\omega}_i - \Delta\omega, \tilde{\omega}_i + \Delta\omega]$, see also Fig. 6.4. It should be noted that the above method for the case of a beam with piecewise constant cross-section provides the exact values of all (theoretically infinite in number) natural frequencies and corresponding eigenmodes.

An interesting framework for application of the present BEM to shape optimization of beams can be found in the literature [21].

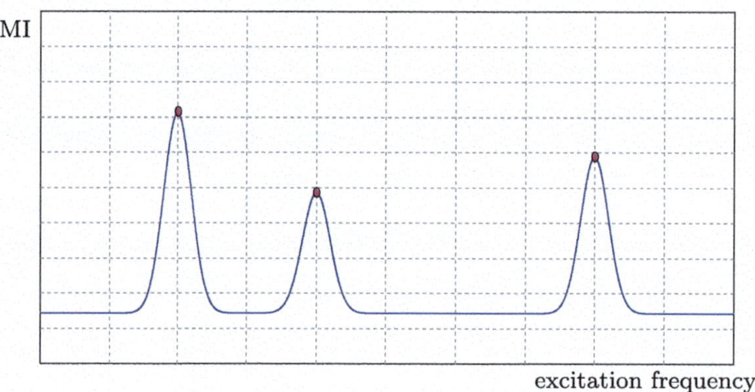

Fig. 6.4 An initially coarse frequency grid for an eigenfrequencies search in conjunction with a mode indicator (MI) selected as the inverse of the argument of the determinant $1/det[H(\omega)]$ or $1/\sigma_1(\omega)$

6.7   Class beamBEM

Description: The beamBEM appearing in the software package courses.structuraldynamics is an implementation of the Boundary Element Method for the case of the Eulre–Bernouli beam theory under harmonic excitation as described in [24].

Attributes
Class beamBEM attributes include the length of the beam (double totlen), the elasticity modulus (double Elast_ref), the moment of inertia (double Im_ref) together with their distributions per length given by the ImDist and ArDist as DoubleFunction variables. Furtermore, the mass per unit length (double mass_ref) together with the definition of left (Integer bcl1 and bcl1) and right (Integer bcr1 and bcr2) boundary conditions. The later four attributes signal boundary condition types, namely the value of 0 implying imposition of displacement, that of 1 imposition of rotation, next a 3 signifies force while a value of 4 stands for a moment. Furthermore, Nx defines the division of the beam length into subregions.

Instance
- beamBEM(double len, double Elast, double Im, double mass)

Methods
- void solve(double omega, double val_l1, double val_l2, double val_r1, double val_r2)
- double deflection(double xloc)
- double slope(double xloc)
- double moment(double xloc)
- double shear(double xloc)

6.7.1 Introductory BEM Example

As an introductory example we consider a prismatic beam with prescribed boundary conditions imposed on specfic frequency. Homogeneous boundary conditions are considered for the displacments of bot edges while non-zero values values are considered for the slopes at both boundary points. Output of script given in source code of Listing 6.3 is given in the plots of Fig. 6.5.

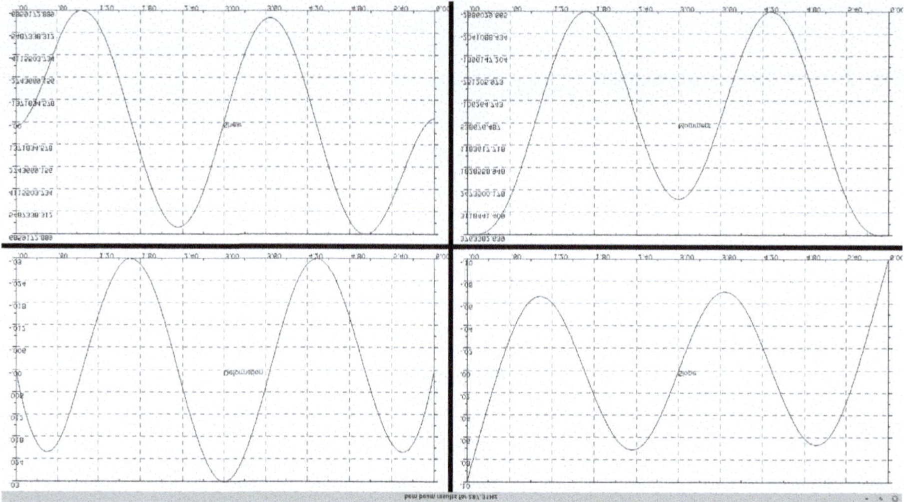

Fig.6.5 (top-left) Displacement, (top-right) slope, (bottom-left) shear force and (bottom-right) bending moment (in c-w sense)

```
import courses.structuraldynamics.beamBEM
elem=new beamBEM(6.0, 50.0e9, 0.25**4/12.0,
    2500.0*0.25*0.25)
elem.Nx=1
elem.bcl1=0; elem.bcl2=1
elem.bcr1=0; elem.bcr2=1
omega=451.30137848901313*4
elem.solve(omega, 0.0, 0.1, 0.0, -0.1)

Npoints=200
defPlot=new PlotFrame(); defPlot.lock();
defPlot.addFunction(new plotfunction(elem.totlen/(Npoints
    -1),elem.deflection(Npoints)))
defPlot.text("CENTER","Deformation")
slpPlot=new PlotFrame(); slpPlot.lock();
slpPlot.addFunction(new plotfunction(elem.totlen/(Npoints
    -1),elem.slope(Npoints)))
slpPlot.text("CENTER","Slope")
shrPlot=new PlotFrame(); shrPlot.lock();
shrPlot.addFunction(new plotfunction(elem.totlen/(Npoints
    -1),elem.shear(Npoints)))
shrPlot.text("CENTER","Shear")
momPlot=new PlotFrame(); momPlot.lock();
momPlot.addFunction(new plotfunction(elem.totlen/(Npoints
    -1),elem.moment(Npoints)))
momPlot.text("CENTER","moment")

fm = new javax.swing.JFrame();
```

```
24  fm.setExtendedState(javax.swing.JFrame.MAXIMIZED_BOTH);
25  fm.setLayout(new java.awt.GridLayout(2,2,10,10))
26  fm.getContentPane().setBackground(Color.BLACK);
27  fm.getContentPane().add(defPlot.getPlotPanel())
28  fm.getContentPane().add(slpPlot.getPlotPanel())
29  fm.getContentPane().add(shrPlot.getPlotPanel())
30  fm.getContentPane().add(momPlot.getPlotPanel())
31  fm.setTitle("bem beam results for "+String.format("%.2f",
        omega/(2*pi))+"Hz")
32  fm.setSize(new java.awt.Dimension(500, 300));
33  fm.setVisible(true);
```

Listing 6.3 Response of a fixed-fixed beam under harmonic conditions

6.8   Beam with Variable Cross Section

Here we investigate the problem of dynamical features of continuous beam of non–constant cross–section. We present results of a tapered beam and compare with analytical solution for the eigenfrequencies [28] (Fig. 6.6).

For the case we present here the cross–section area and $A(x)$ and the second mement of inertia $I(X)$ vary as follows:,

$$A(x) = A_g \left(1 + c\frac{x}{L}\right)^n ; \quad I(x) = I_g \left(1 + c\frac{x}{L}\right)^{n+2} \tag{6.31}$$

The characterisation of a tapered beam in Eq. (6.31) clearly indicates that for positive values of c the beam tapers upward from the thin end (g) at the left and to the thick end (h) at the right so that the area and second moment of area at the right-hand end are $A_h = A_g (1 + c)^n$ and $I(x) = I_g (1 + c)^{n+2}$, respectively. If the beam tapers downward with the thick end (h) at the left and thin end (g) on the right, the alternative form of Eq. (6.31) is given by

$$A(x) = A_g \left(1 - \tilde{c}\frac{x}{L}\right)^n ; \quad I(x) = I_g \left(1 - \tilde{c}\frac{x}{L}\right)^{n+2} \tag{6.32}$$

where $\tilde{c}$ is a new taper ratio of the downwards tapering beam which has a positive value. Clearly, the area and second moment of area at the right-hand end of the tapered beam corresponding to Eq. (6.32) are respectively $A_h = A_g (1 - \tilde{c})^n$ and $I(x) = I_g (1 - \tilde{c})^{n+2}$. It can be shown with the help of Eqs. (6.31) and (6.32) that the taper ratios c and $\tilde{c}$ for the tapering upward and tapering downward beams can be related as follows.

$$\tilde{c} = \frac{c}{1+c}; \quad c = \frac{\tilde{c}}{(1-\tilde{c})}. \tag{6.33}$$

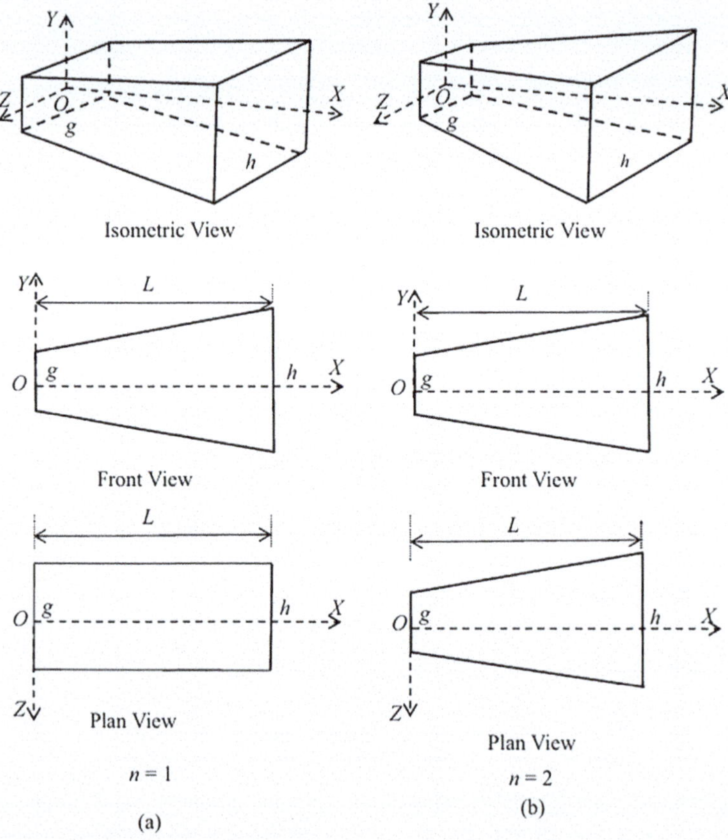

Fig. 6.6 A tapered beam of solid rectangular cross–section with **a** a constant width and a linearly varying depth for which the variations of the cross-sectional area and the second moment of area along the length are respectively linear and cubic ($n = 1$) and **b** a linearly varying width and depth for which the variations of the cross–sectional area and the second moment of area along the length are respectively second and fourth order ($n = 2$). Reprinted from [28]

The scripts for the Rayleigh-Ritz solution of tapered beams are shown in Listing 6.4.

```
import courses.structuraldynamics.beamRRM

YModul=20.0e9;Imp=0.00423d
Ar=0.69d;p=27395.51
L=4.0;k=0.0d
matcoef=sqrt(p*Ar*L*L*L*L/(YModul*Imp))
eig=20

RRMeigPlot = new PlotFrame()
RRMeigPlot.setAutoColor(true)
RRMeigPlot.makeLegend()
```

```
12  RRMeigPlot.setTextfontsize(24)
13  npbeam=new beamRRM(len:L, Elast:YModul, dens:p, bcl:2, bcr
        :2)
14  gc=0.2; n=1
15  Im={Imp*(1.0-gc*it/L)**(n+2)}; Area={Ar*(1.0-gc*it/L)**n};
        Spring={k}
16  npbeam.calcEigen(Im as DoubleFunction, Area as
        DoubleFunction, Spring as DoubleFunction, eig)
17  (1..5).each{
18    println "tapered lambda_"+it+": "+npbeam.eigenfrequency(
        it)*matcoef
19    pf = new plotfunction(linspace(0.0, npbeam.len,1001),
        npbeam.eigenfunction(it))
20    pf.setName("omega_"+it+"="+String.format("%.3f", npbeam.
        eigenfrequency(it))+"rad/s")
21    RRMeigPlot.addFunction(pf)
22  }
23  RRMeigPlot.show()
```

Listing 6.4 Rayleigh-Ritz method for tapered beam

For the results we present we have used up to fifteen test functions ψ_i in order to have a good approximation (accuracy to the third decimal place) for the first five eigenvalues and eigenvectors. One can easily see that the more functions used the better approximation for larger value of eigenvalues. That is shown in results of Table 6.2 that produced using code of Listing 6.4. We give the eigenfrequencies in a non-dimensionalized form

$$\lambda_i = \omega_i \sqrt{\frac{\rho A_h L^4}{E I_h}}.$$

as presented in [28] to compare also with exact values given there and reproduced here in Table 6.1.

For the case of FEM a similar script developed and is given in Listing 6.5. Results for the non-dimensional eigenfrequencies are presented in Table 6.3 for number of elements from 10 up to 160. It is obvious that convergence is more difficult to be established in that case since even with the finer mesh (that of 160 equal size elements) we did not make to have accuracy to the decimal places. However it is evident the results are slowly converge to the correct solution.

Table 6.1 Exact values for non-dimensional natural frequencies for the case $\tilde{c} = 0.2$ and $n = 1$ as given in the literature [28]

λ_1	λ_2	λ_3	λ_4	λ_5
8.846	35.445	79.730	141.721	221.420

Table 6.2 Non-dimensional natural frequencies and their convergence for tappered beam using the Rayleigh-Ritz

N	λ_1	λ_2	λ_3	λ_4	λ_5
5	8.846	35.446	79.739	141.755	223.701
10	8.846	35.445	79.730	141.722	221.421
15	8.846	35.445	79.730	141.721	221.420

Table 6.3 Non-dimensional natural frequencies and their convergence for tappered beam using the Finite Element Method

N	λ_1	λ_2	λ_3	λ_4	λ_5
10	8.647	34.653	77.980	138.765	217.300
20	8.747	35.049	78.840	140.148	218.996
40	8.797	35.247	79.285	140.929	220.185
80	8.821	35.346	79.508	141.325	220.801
160	8.834	35.396	79.619	141.523	221.111

```
import  courses.structuraldynamics.beamFEM

YModul=20.0e9;Imp=0.00423d
Ar=0.69d;p=27395.51
L=4.0
matcoef=sqrt(p*Ar*L*L*L/(YModul*Imp))

femBEAM=new beamFEM(L, YModul, Imp, p*Ar)
femBEAM.bcl=2; femBEAM.bcr=2;
femBEAM.consistentMass=true
femBEAM.Nx=10
gc=0.2
Im={(1.0-gc*it/L)*(1.0-gc*it/L)*(1.0-gc*it/L)}; Area={(1.0-
    gc*it/L)};
femBEAM.calcEigen(Im as DF, Area as DF)



plot()
thePlot.hline(0.0)
thePlot.setAutoColor(true)
(1..5).each{
   println femBEAM.eigenfrequency(it)*matcoef
   plot(L/femBEAM.Nx, femBEAM.eigenDeform(it))
}
```

Listing 6.5 FEM approach for solving the tapered beam eigenproblem

Table 6.4 Non-dimensional natural frequencies and their convergence for tappered beam using the Boundary Element Method. Here N refers to the number of subdomains considered in order to succeed an approximation of the geometry beams' section

N	λ_1	λ_2	λ_3	λ_4	λ_5
5	8.841	35.423	79.676	141.552	221.940
10	8.845	35.441	79.718	141.702	221.381
20	8.847	35.448	79.729	141.708	221.409

Moving to the case BEM it shown to be evern more dificult to get good results for the eigenrequencies. The code for the script developed for that purpose is giben in the Listing 6.6, while results are reported in Table 6.4. Similar, a script for a solution using a BEM approximation appears below. It is shown that while BEM can not be considered the more suitable method for solving eigenvalue problems it makes to gain good accuracy with increased number of subdomains.

```
import courses.structuraldynamics.beamBEM
YModul=20.0e9;Imp=0.00423d
Ar=0.69d;p=27395.51
L=4.0;
matcoef=sqrt(p*Ar*L*L*L*L/(YModul*Imp))
elem=new beamBEM(L, YModul, Imp, Ar*p)
elem.Nx=40
elem.bcl1=0; elem.bcl2=3
elem.bcr1=0; elem.bcr2=3
gc=0.2
elem.ImDist={(1.0-gc*it/elem.totlen)*(1.0-gc*it/elem.totlen
    )*(1.0-gc*it/elem.totlen)} as DoubleFunction
elem.ArDist={(1.0-gc*it/elem.totlen)}  as DoubleFunction
elem.refine=true
a=elem.eigen(10.0, 1200.0, 100)
elem.tolfinder=1.0e-12
elem.tolrefine=1.0e-12;
elem.maxIter=30
(0..<a.size()).each{println( (it+1)+": "+a[it])}
svPlot=new PlotFrame()
svPlot.addFunction(new plotfunction(elem.oms as double[],
    elem.svs as double[]))
a.each{svPlot.vline(it,Color.blue)}
svPlot.setAutoColor(true)
svPlot.setMarker(true)
svPlot.show()

np=200;
BEMeigPlot = new PlotFrame()
BEMeigPlot.setAutoColor(true)
BEMeigPlot.makeLegend()
BEMeigPlot.setTextfontsize(24)
(0..<a.size()).each{
```

```
32   pf = new plotfunction(linspace(0.0, elem.totlen,np),elem.
       eigenvec(a[it],np))
33   pf.setName(" _"+(it+1)+"="+String.format("%.3f", a[it])+"
       rad/s")
34   println a[it]*matcoef
35   BEMeigPlot.addFunction(pf)
36 }
37 BEMeigPlot.title="bem eigenvectors"
38 BEMeigPlot.show()
```

Listing 6.6 BEM approach for solving the tapered beam eigenproblem

References

1. Zienkiewicz OC, Taylor RL (2005) The finite element method for solid and structural mechanics. Elsevier
2. Bathe K-J, Wilson E (1976) Numerical methods in finite element analysis
3. Brebbia CA, Telles JCF, Wrobel LC (eds) (1984) Boundary element techniques: theory and applications in engineering. Springer
4. Rubinstein MF, Hurty WC (1964) Dynamic analysis of a rocket model by the component modes method. Technical report, SAE Technical Paper
5. Klein Ikkink HA (2018) Groovy goodness: add map constructor with annotation. Cambridge University Press
6. Radeş M (1970) Steady-state response of a finite beam on a Pasternak-type foundation. Int J Solids Struct 6(6):739–756
7. Langtangen HP (2013) Introduction to finite element methods
8. Paraskevopoulos EA, Talaslidis DG (2004) A rational approach to mass matrix diagonalization in two-dimensional elastodynamics. Int J Numer Meth Eng 61(15):2639–2659
9. Katsikadelis JT (2020) Dynamic analysis of structures. Academic
10. Fredholm I (1903) Sur une classe d'équations fonctionnelles. Acta Mathematica Sweden 27:365–390
11. Kellogg OD (1953) Foundations of potential theory, vol 31. Courier Corporation
12. Muskhelishvili N (1953) Some basic problems of the mathematical theory of elasticity, vol 15. Noordhoff Groningen
13. Mikhlin SG (2014) Multidimensional singular integrals and integral equations. Elsevier
14. Trefftz E (1926) Ein gegenstuck zum ritzschen verfahren. In: Proceedings of the 2nd international congress for applied mechanics
15. Hess JL (1962) Calculation of potential flow about bodies of revolution having axes perpendicular to the free-stream direction. J Aerosp Sci 29(6):726–742
16. Friedman MB, Shaw R (1962) Diffraction of pulses by cylindrical obstacles of arbitrary cross section. J Appl Mech 29(6):40–46
17. Massonet CE (1966) Stress analysis, chapter numerical use of integral procedures. Wiley, Chichester
18. Rizzo FJ (1967) An integral equation approach to boundary value problems of classical elastostatics. Q Appl Math 25(1):83–95
19. Cruse TA, Rizzo FJ (1968) A direct formulation and numerical solution of the general transient elastodynamic problem. i. J Math Anal Appl 22(1):244–259

20. Zienkiewicz OC, Kelly DW, Bettess P (1977) The coupling of the finite element method and boundary solution procedures. Int J Numer Meth Eng 11(2):355–375
21. Panagiotopoulos CG, Kouzoupis S (2024) Tuning of idiophones using shape optimization on finite and boundary element 1d models. Appl Acoust 217:109850
22. Manolis GD, Beskos DE (1988) Boundary element methods in elastodynamics. Taylor & Francis
23. Dominguez J (1993) Boundary elements in dynamics. WIT Press
24. Providakis CP, Beskos DE (1986) Dynamic analysis of beams by the boundary element method. Comput Struct 22(6):957–964
25. Antes H, Schanz M, Alvermann S (2004) Dynamic analyses of plane frames by integral equations for bars and Timoshenko beams. J Sound Vib 276(3–5):807–836
26. de Mesquita NE, Barretto SFA, Pavanello R (2000) Dynamic behavior of frame structures by boundary integral procedures. Eng Anal Boundary Elem 24(5):399–406
27. Xiao H, Chen Z (2007) Numerical experiments of preconditioned Krylov subspace methods solving the dense non-symmetric systems arising from BEM. Eng Anal Boundary Elem 31(12):1013–1023
28. Banerjee JR, Ananthapuvirajah A (2019) Free flexural vibration of tapered beams. Comput Struct 224:106106

7.1 Random Processes

If the outcome of a conceptual experiment assigns a real value to variable x, then x is known as a random variable. Furthermore, if x assumes only a finite number of values, it is called a discrete random variable. Finally, if x assumes a continuous range of values, it is called a continuous random variable. Probabilities associated with a random variable are conveniently described by a distribution function such that the probability of x assuming a value less than a fixed number X is $P(x \leq X)$. Note that $P(x < -\infty) = 0$ and $P(x < +\infty) = 1$, where zero denotes impossibility and unity denotes certainty. The probability for x to lie within interval $(a, b]$ is

$$P(x \leq b) - P(x \leq a) = P(a \leq x \leq b) \geq 0. \tag{7.1}$$

From the above equation it is seen that if $b \geq a$, $P(x \leq b) > P(x \leq a)$ and hence the distribution function is a monotonically, non-decreasing function of X. Figure 7.1 shows the distribution function for both discrete and continuous random variables. By differentiating the distribution function $P(x \leq X)$ in the regions where the derivative exists, we obtain the probability density function (PDF) as

$$p(x) = \lim_{\Delta X \to 0} \frac{P(X - \Delta X < x \leq X)}{\Delta X} \tag{7.2}$$

In the case of a discrete random variable, the PDF can be represented by a series of impulses (or Dirac delta functions) at the location of each jump, see Fig. 7.2a. Each impulse represents an area equal to the magnitude of the corresponding jump in P, so that the probability that $X - \Delta X < x \leq X$ is approximated as $p(X)\Delta X$. Figure 7.2b shows the PDF as a smooth function corresponding to a continuous random variable.

© The Author(s), under exclusive license to Springer Nature Switzerland AG 2026
G. Manolis and C. Panagiotopoulos, *Vibrations of Structural Systems*, Synthesis Lectures
on Mechanical Engineering, https://doi.org/10.1007/978-3-032-12279-7_7

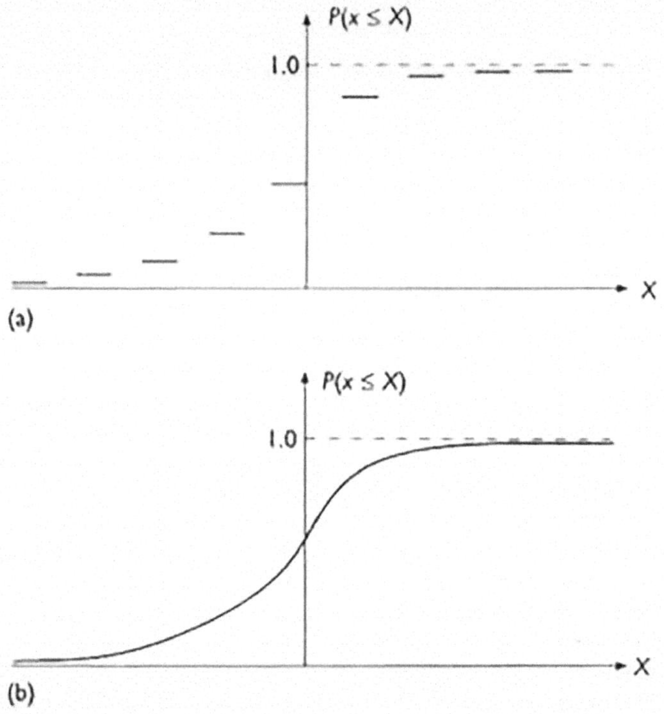

Fig. 7.1 Distribution function for **a** discrete and **b** continuous random variables

7.2　Random Functions

Consider a random process that generates an infinite ensemble of sample functions $x(t)$ so that a series of probabilities can be defined. As shown in Fig. 7.3, at any time t, a first-order distribution function and a first-order PDF may be defined across the ensemble as a limiting process in the form

$$p(X, t)dX = P(X - dX < x(t) < X) \tag{7.3}$$

Similarly, a joint PDF can be defined as

$$p(X_1, t_1, \ldots, X_n, t_n)dX_1 \ldots dX_n =$$
$$P(X_1 - dX_1 < x(t_1) \le X_1, \ldots, X_n - dX_n < x(t_n) \le X_n) \tag{7.4}$$

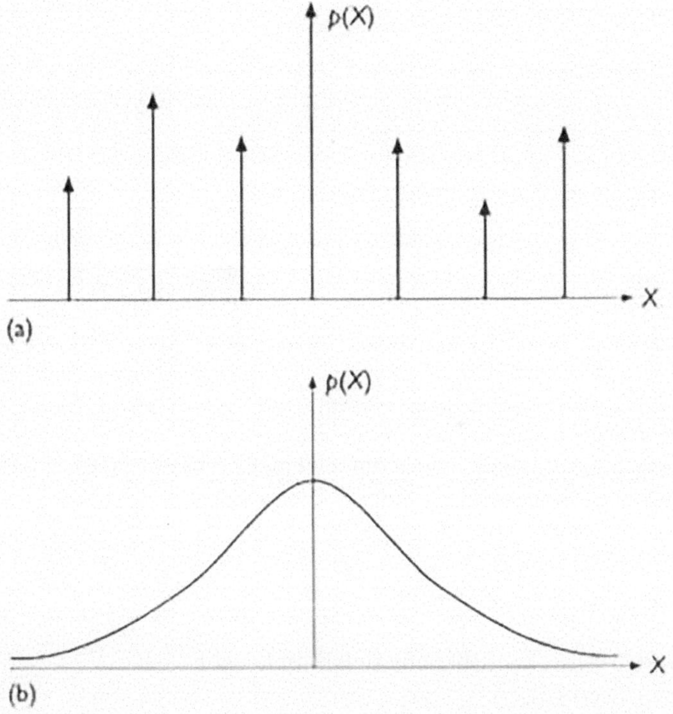

Fig. 7.2 Probability density function (PDF) for **a** discrete and **b** continuous random variables

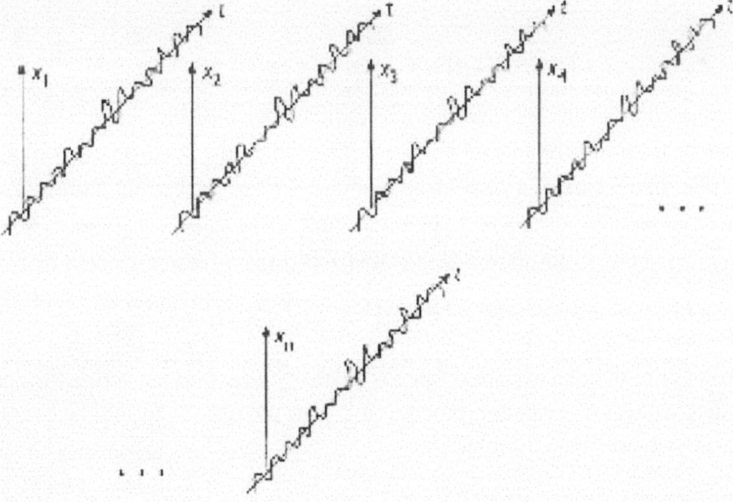

Fig. 7.3 Ensemble of records $x(t)$

with the following properties:

$$p(x_1, t_1, \ldots, x_n, t_n) \geq 0,$$

$$\int_{-\infty}^{\infty} \cdots \int_{-\infty}^{\infty} p(x_1, t_1 + T, \ldots, x_n, t_n + T)dx_1 \ldots dx_n = 1 \tag{7.5}$$

In terms of notation, X_1 has been replaced by x_1 and so on. A lower order joint PDF (index m) can be found from a higher order one (index n), where $m < n$, by integrating across $x_{m+1}, \ldots, x_n$ as in Eq. (7.5). A joint PDF which is invariant to shifts in the time axis is said to be stationary:

$$p(x_1, t_1, \ldots, x_n, t_n) = p(x_1, t_1 + T, \ldots, x_n, t_n + T) \tag{7.6}$$

Since it is not possible in practice to determine all joint probabilities required for completely defining a random process, one has to settle for a few easily obtainable averages that partially specify a random process. First, the mean (or statistical average or ensemble average) of $x(t)$ is

$$E[x(t)] = \int_{\infty}^{\infty} X \, p(X, t)dX \tag{7.7}$$

where E is the expected value (or expectation) of $x(t)$. Of particular interest is a set of averages called central moments, which do not vary with time for stationary cases:

$$E[x_t - E(x_t)]^n = \int_{\infty}^{\infty} (x_t - E(x_t))^n p(x_t)dx_t \tag{7.8}$$

In the above, n is an integer and subscript t indicates that the averages refer to a particular instant of time. In the case of a zero-mean random process, the central moments are simply referred to as moments. The second central moment is of importance and is known as the variance σ_x^2 (or square of the standard deviation):

$$\sigma_x^2 = E[x_t - E(x_t)]^2 = E[x_t^2] - [E(x_t)]^2 \tag{7.9}$$

The operation of finding an expected value is basically an averaging across the ensemble of sample functions $x(t)$ of a random process. It is also possible to form time averages along a particular member of the ensemble as

$$\bar{x}(t) = \lim_{T \to \infty} \frac{1}{2T} \int_{-T}^{T} x(t)dt \tag{7.10}$$

where the overbar indicates a time-averaged value. If the time averages and the ensemble averages are identical, then the random process is ergodic. Obviously, this property holds for stationary processes only, since for a nonstationary process the ensemble average varies with time. Ergodicity is a desirable property and a stationary process is assumed to be ergodic unless there are reasons to the contrary.

The correlation coefficient xy between two random variables x and y with joint PDF $p(x, y)$ is defined as

$$\rho_{xy} = \frac{E[(x - E[x])(y - E[y])]}{\rho_x \rho_y} \tag{7.11}$$

It is common practice to normalize both x and y such that their means are zero and their variances are equal to unity. In such cases, $\rho_{xy} = \rho_{yx} = \rho = E[xy]$, where ρ is the slope of a straight line that best fits, by mean-square error minimization, the data of a normalized (x, y) scatter plot, as shown in Fig. 7.4. Also, $\rho \leq 1$ and intermediate values measure the degree of statistical dependence between x and y. For a random process $x(t)$, we express the correlation between $x(t_1)$ and $x(t_2)$ by the autocorrelation function

$$R_x(t_1, t_2) = E[x(t_1)x(t_2)] \tag{7.12}$$

In the stationary case, it is only the time lag τ between t_1 and t_2 is important, i.e.,

$$R_x(\tau) = E[x(t)x(t + \tau)] = R_x(-\tau) \tag{7.13}$$

Also note that $R_x(0) = E[x^2] = \sigma_x^2 + (E[x])^2$. For ergodic cases, R_x can be found by averaging any sample function of the ensemble across time as

$$R_x(\tau) = \lim_{T\infty} \frac{1}{2T} \int_{-T}^{T} x(t)x(t + \tau)dt \tag{7.14}$$

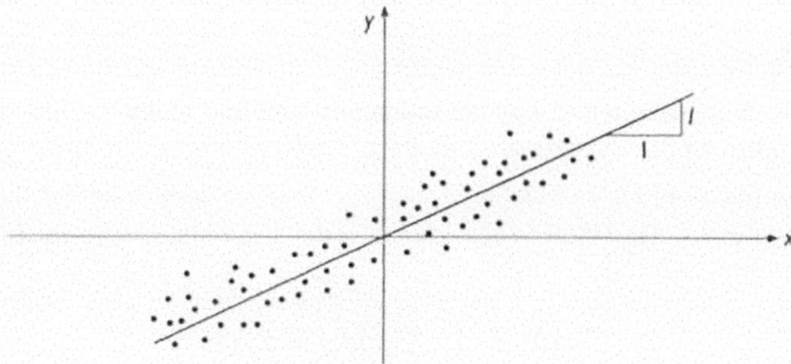

Fig. 7.4 Correlation coefficient for a scatter of sample values

Next, the correlation between two samples from random processes $x(t)$ and $y(t)$ is described by the cross-correlation functions

$$R_{xy}(t_1, t_2) = E[x(t_1)y(t_2)], \ R_{yx}(t_1, t_2) = E[y(t_1)x(t_2)] \tag{7.15}$$

For stationary cases, $R_{xy}(t_1, t_2) = R_{xy}(\tau)$, $R_{yx}(t_1, t_2) = R_{yx}(\tau)$, $R_{xy}(\tau) = R_{yx}(-\tau)$. It is observed that although the autocorrelation is an even function of τ, the cross-correlation functions are not. Finally, by using the fact that differentiation and expectation are linear operators and as such commute, the time derivatives of a stationary autocorrelation function are

$$\frac{d R_x(\tau)}{d\tau} = E[x(t)\dot{x}(t + \tau)] = E[\dot{x}(t - \tau)x(t)]$$

$$\frac{d^2 R_x(\tau)}{d\tau^2} = -E[\dot{x}(t - \tau)\dot{x}(t)] = -E[\dot{x}(t)\dot{x}(t + \tau)] \tag{7.16}$$

where dots indicate derivatives with respect to t. The above two equations indicate that the second derivative of the autocorrelation function of x is the negative of the autocorrelation function of $\dot{x}$. Finally, both $R_x(\tau)$ and $R_{xy}(\tau)$ tend to zero as τ, provided x is non-periodic.

7.3 Spectral Analysis

Spectral analysis is not only a standard method in random vibrations, but is also essential in signal processing. We define the autocorrelation function $R_x(\tau)$ for a stationary process and the power spectral density function (PSDF) $S_x(\omega)$ as being a Fourier transform (FT) pair:

$$S_x(\omega) = \frac{1}{2\pi} \int_{-\infty}^{\infty} R_x(\tau) \exp(-i\omega\tau)d\tau,$$

$$R_x(\tau) = \int_{-\infty}^{\infty} S_x(\omega) \exp(i\omega\tau)d\omega \tag{7.17}$$

In the above, ω is the frequency, i is the imaginary unit and factor $\frac{1}{2\pi}$ may be associated with either member of the above pair or may be evenly split between them. Since $R_x(\tau)$ is a real and even function, Fourier cosine transforms may be used in lieu of the exponential transform shown above. The PSDF is also known as the mean square spectral density because

$$R_x(0) = \int_{-\infty}^{\infty} S_x(\omega)d\omega = E[x^2] \tag{7.18}$$

This implies that $S_x(\omega)d\omega$ can be interpreted as the power or mean square density contained in an infinitesimal band of complex exponentials (sines and cosines) into which the random function is resolved. The PSDF is a positive, real-valued function and even in ω. Since physical meaning can only be assigned to positive frequencies, an experimentally obtained spectrum is plotted by halving the measured $S_x(\omega)$ at each frequency and plotting the result for positive ω. A spectrum $S_{xy}(\omega)$ for the cross-correlation function $R_{xy}(\tau)$ can also be defined for the stationary case as in Eq. (7.17).

As expected, the PSDF of an ergodic process and the FT of a sample function $x(t)$ of the random process are related. If $x(t)$ is a non-periodic function, its Fourier transform $X_T(\omega)$ is given as

$$X_T(\omega) = \int_0^T x(t) \exp(-i\omega t)dt \tag{7.19}$$

where $x(t)$ is assumed zero outside the integration interval. An energy density spectrum for $x(t)$ and its power density spectrum are respectively given as

$$\Phi_x(\omega, T) = |X_T(\omega)|^2$$

$$S_x(\omega, T) = \frac{|X_T(\omega)|^2}{2\pi T} \tag{7.20}$$

The power density spectrum is now a random variable dependent on both $x(t)$ and T. Although it can be shown that the $\lim_{T \to \infty} E[S_x(\omega, T)] = S_x(\omega)$, the manner in which the power density spectrum approaches the PSDF needs to be investigated in each case. For a normal (Gaussian) process, it is known that the variance of $S_x(\omega, T)$ does not approach zero as $T \to \infty$, and hence measurements of $S_x(\omega, T)$ provide questionable estimates for the PSDF.

7.4   SDOF System Response to Random Input

Consider the equation of motion of a single degree-of-freedom (SDOF) linear system

$$(\ddot{x})(t) + 2\omega_0\zeta\dot{x}(t) + \omega_0^2 x(t) = \frac{\tilde{f}(t)}{m} = f(t), \qquad t > t_0 \tag{7.21}$$

shown in Fig. 7.5, where the natural frequency is $\omega_0^2 = k/m$, the damping ratio is $\zeta = c/2m\omega_0$, and f(t) is the scaled forcing function. Also, m, c, k is the mass, damping and stiffness, while $x(t)$ is the displacement response of the system to a Gaussian stochastic input $f(t)$, which is a member function of a stochastic process $f(t)$. Equation (7.21) is accompanied by initial conditions $x(t_0) = x_0, \dot{x}(t_0) = \dot{x}_0$. As previously mentioned, the probability law for a random process cannot be determined only from the mean and covariance of that

Fig. 7.5 SDOF representation with $h(t)$ the unit impulse response and $H(\omega)$ the complex frequency response

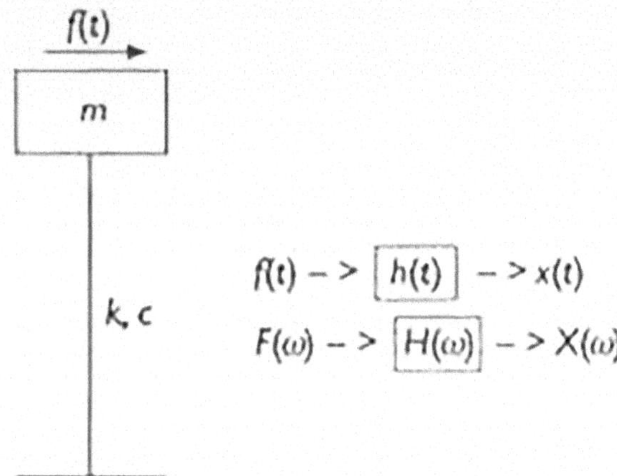

process. The exception is when the functional form of the probability law is known and utilizes parameters which are simply related to the mean and covariance, as in the case of a normal (Gaussian) distribution. In what follows, it is assumed that the input process $f(t)$ is Gaussian, and so is the output. First, the mean-square (deterministic) response of the SDOF system is given by Duhamel's integral as

$$x(t) = x_0 g(t - t_0) + \dot{x}_0 h(t - t_0) + \int_0^{t-t_0} h(\xi) f(f - \xi) d\xi$$

$$g(t) = \exp\left(-\zeta \omega_0 t\right)\left(\cos \bar{\omega} t + \zeta \omega_0 \frac{\sin \bar{\omega} t}{\bar{\omega}}\right)$$

$$h(t) = \exp\left(-\zeta \omega_0 t\right)\frac{\sin \bar{\omega} t}{\bar{\omega}} \tag{7.22}$$

with $\bar{\omega} = \omega\sqrt{1 - \zeta^2}$ the damped natural frequency and $0 < \zeta < 1$ the damping ratio. As before, function $h(t)$ is the unit impulse response of a linear SDOF system. The mean $m_x(t)$ of the output process is obtained by averaging across the ensemble as

$$m_x(t) = E[x(t)] = x_0 g(t - t_0) + \dot{x}_0 h(t - t_0) + \int_0^{t-t_0} h(\xi) m_f(t - \xi) d\xi \tag{7.23}$$

For a system with infinite operating time ($t_0 = \infty$), the mean simplifies to

$$m_x(t) = \int_0^\infty h(\xi) m_f(t - \xi) d\xi \tag{7.24}$$

The covariance $K_{xx}(t_1, t_2)$ of the output process, which is the autocorrelation function taken about the mean, is given as

$$K_{xx}(t_1, t_2) = E[(x(t_1) - m_x(t_1))(x(t_2) - m_x(t_2))] =$$

$$= \int_{t_0}^{t_1} \int_{t_0}^{t_2} h(t_1 - \tau_1)h(t_2 - \tau_2)E[(f(\tau_1) - m_f(\tau_1))(f(\tau_2) - m_f(\tau_2))]d\tau_1 d\tau_2 =$$

$$= \int_{0}^{t_1-t_0} \int_{0}^{t_2-t_0} h(\xi_1)h(\xi_2)K_{ff}(t_1 - \xi_1, t_2 - \xi_2)d\xi_1 d\xi_2 \tag{7.25}$$

with a change of variables as $\xi_1 = t_1 \tau_1$, $\xi_2 = t_2 \tau_2$. As before, the upper limits are replaced by for a system with infinite operating time. Furthermore, the variance $\sigma_x^2(t)$ of the output process is obtained by setting $t_1 = t_2 = t$ in the expression for the covariance, i.e., $\sigma_x^2(t) = K_{xx}(t, t)$. Finally, given a normal (Gaussian) input, the PDF of $x(t)$ is given as

$$p(x) = \frac{\exp\left(-\frac{(x-m_x(t))^2}{2\sigma_x^2(t)}\right)}{\sqrt{2\pi}\sigma_x(t)} \tag{7.26}$$

so that the probability of x lying in the interval $(x, x + dx)$ at time t is given by $p(x)dx$. Note that the above development is for nonstationary processes, but for stationary processes, the above averages no longer vary with time.

7.5   MDOF System Response to Random Input

Consider the response of a multiple degree-of-freedom (MDOF) system to nonstationary random input. The development follows along the lines of the SDOF system, expect for the introduction of matrix notation. At first, the governing equation of motion of an N-DOF system and the initial conditions are as follows:

$$M\ddot{x}(t) + C\dot{x}(t) + Kx(t) = f(t), \quad t > t_0$$

$$x(t_0) = x_0, \ \dot{x}(t_0) = \dot{x}_0 \tag{7.27}$$

In the above, M, C and K are symmetric $N \times N$ matrices and x and f are $N \times 1$ vectors denoting the input and output processes, respectively. In particular, the mass matrix M is positive definite, while the damping C and K stiffness matrices are non-negative definite. Also, $f(t)$ is a vector of Gaussian random variables whose mean and covariance matrix are respectively given by

$$m_f(t) = E[f(t)]$$

$$K_{ff}(t_1, t_2) = E[(f(t_1) - m_f(t_1))(f(t_2) - m_f(t_2))^T] \tag{7.28}$$

with superscript T denoting transposition.

We examine the case where the system of Eq. (7.27) possesses real eigenvectors, else known as classical normal modes. In that case, the matrix of normalized eigenvectors A defines a set of modal coordinates

$$y = A^{-1}x = A^T Mx \tag{7.29}$$

These are used for uncoupling the system of governing equations of motion as follows:

$$I\ddot{y} + \overline{C}\dot{y} + \overline{K}y = A^T f(t) = q(t) \tag{7.30}$$

In the above, I is the identity matrix $\overline{C}$ and $\overline{K}$, are diagonal matrices. The ith row of the above equation gives an SDOF equation for the ith mode, whose solution for a Gaussian input results in Gaussian output:

$$\ddot{y}_i + 2\omega_i \zeta_i \dot{y}_i + \omega_i^2 y_i = q_i(t) = \sum_{j=1}^{N} A_{ij} f_j \tag{7.31}$$

Thus, the mean-square (deterministic) response of the MDOF system in modal coordinates is

$$y = U(t - t_0)y(t_0) + H(t - t_0)\dot{y}(t_0) + \int_{t_0}^{t} H(t - \tau)q(\tau)d\tau, \quad \text{where}$$

$$U_{ii}(t) = \exp\left(-\zeta_i \omega_i t\right)\left(\cos\overline{\omega}_i t + \zeta_i \omega_i \frac{\sin\overline{\omega}_i t}{\overline{\omega}_i}\right)$$

$$H_{ii}(t) = \exp\left(-\zeta_i \omega_i t\right)\frac{\sin\overline{\omega}_i t}{\overline{\omega}_i}$$

$$\overline{\omega}_i = \omega_i \sqrt{1 - \zeta_i^2}, \quad 0 < \zeta_i < 1 \tag{7.32}$$

Reverting back to the physical coordinates via the transformation of Eq. (7.29) gives

$$x = AU(t - t_0)A^T Mx_0 + AH(t - t_0)A^T M\dot{x}_0 + \int_{t_0}^{t} AH(t - \tau)A^T f(\tau)d\tau \tag{7.33}$$

Given the above solution for the mean-square (deterministic) response, the stochastic means are

$$m_x(t) = AU(t - t_0)A^T M x_0 + AH(t - t_0)A^T M \dot{x}_0 + \int_{t_0}^{t} AH(t - \tau)A^T m_f(\tau)d\tau$$

$$(7.34)$$

Next, the covariance matrix is given by the stochastic average of the outer product of the zero-mean response vector evaluated at two different times, i.e.,

$$K_{xx}(t_1, t_2) = E[(x(t_1) - m_x(t_1))(x(t_2) - m_x(t_2))^T] =$$

$$\int_0^{t_1-t_0} \int_0^{t_2-t_0} AH(\xi_1)A^T K_{ff}(t_1 - \xi_1, t_2 - \xi_2)AH(\xi_2)A^T d\xi_1 d\xi_2 \qquad (7.35)$$

Finally, the PDF for the ith component of the response $x(t)$ is given as

$$p(x_i) = \frac{\exp \frac{-(x_i(t)-m_{xi}(t))^2}{2\sigma_{xi}^2(t)}}{\sqrt{2\pi}\sigma_{xi}(t)} \qquad (7.36)$$

where variance σ_{xi}^2 is the ith diagonal component of the covariance matrix evaluated at $t_1 = t_2 = t$, i.e., $\sigma_{xi}^2 = K_{ii}(t, t)$. If the components of the input f are jointly normally distributed, so are the components of the output x with a joint PDF given by

$$p(x_1, x_2, \ldots, x_N) = \exp \frac{(-0.5(x(t) - m_x(t))^T K_{xx}(t)^{-1}(x(t) - m_x(t)))}{(2\pi)^{N/2}(det(K_{xx})(t))^{1/2}} \qquad (7.37)$$

As before, for stationary processes all statistical averages are time-independent.

If the damping and stiffness matrices are non-symmetric, then the classical normal mode approach fails and a more general approach must be sought. The key idea is to convert the second order matrix differential equation (7.27) into a first order matrix differential equation by defining

$$z(t)^T = \begin{bmatrix} \dot{x}(t)^T & x(t)^T \end{bmatrix} \qquad (7.38)$$

By combining the above equation with the matrix equation of motion that has been pre-multiplied by M^{-1}, the following $2N \times 2N$ matrix differential equation system is obtained

$$\dot{z} = Bz + b(t),$$

$$B = \begin{bmatrix} -M^{-1}C & -M^{-1}K \\ I & 0 \end{bmatrix},$$

$$b(t)^T = \begin{bmatrix} -M^{-1}f(t)^T & 0^T \end{bmatrix} \qquad (7.39)$$

with $z(t - t_0) = z_0$ as initial condition. It is assumed that the input vector $b(t)$ is Gaussian with mean $m_b(t)$ and covariance $K_{bb}(t_1, t_2)$. It is beyond the scope of this chapter to give further details on the derivation of statistical moments for this case.

7.6　Nonstationary Processes

As we saw, Fourier transforms play a central role in the analysis of stationary random processes by relating the autocorrelation (or autocovariance) to the PSDF and vice-versa. These relations can be extended to nonstationary processes starting with function $f(t)$, a member of a real-valued, nonstationary process, defined as follows:

$$f_T = \begin{cases} f(t) & \text{for } |t| < T \\ 0 & \text{for } |t| \geq T \end{cases} \tag{7.40}$$

Assume for simplicity that $f(t)$ is a zero-mean process and subsequently define the Fourier transform pair as

$$F_T(\omega) = \frac{1}{2\pi} \int_{-T}^{T} f(t) \exp(-i\omega t)dt$$

$$f_T(t) = \int_{-\infty}^{\infty} F_T(\omega) \exp(i\omega t)d\omega \tag{7.41}$$

Since f_T is a real function, it is equal to its complex conjugate f_T^* so that

$$f_T(t) = \int_{-\infty}^{\infty} F_T^*(\omega) \exp(-i\omega t)d\omega$$

The above equations can now be used in conjunction with definition of the covariance, i.e.,

$$K_{ff}(t_1, t_2, T) = E[f_T(t_1)f_T(t_2)] =$$

$$= \int_{-\infty}^{\infty} \int_{-\infty}^{\infty} E[F_T^*(\omega_1)F_T(\omega_2)] \exp(-i(\omega_1 t_1 - \omega_2 t_2))d\omega_1 d\omega_2 \tag{7.42}$$

The covariance of $f(t)$ is therefore given by

$$K_{ff}(t_1, t_2) = \lim_{T \to \infty} K_{ff}(t_1, t_2, T) =$$

$$\int_{-\infty}^{\infty} \int_{-\infty}^{\infty} S_{ff}(\omega_1, \omega_2) \exp(-i(\omega_1 t_1 - \omega_2 t_2))d\omega_1 d\omega_2 \tag{7.43}$$

where $S_{ff}(\omega_1, \omega_2) = \lim\limits_{T\to\infty} E[F_T^*(\omega_1)F_T(\omega_2)]$ is the generalized PSDF for random process $f(t)$. Applying the inverse FT to Eq. (7.42) yields

$$S_{ff}(\omega_1, \omega_2) = \frac{1}{4\pi^2} \int\limits_{-\infty}^{\infty} \int\limits_{-\infty}^{\infty} K_{ff}(t_1, t_2) \exp\left(i(\omega_1 t_1 - \omega_2 t_2)\right) dt_1 dt_2 \qquad (7.44)$$

Next, for a linear system with infinite operating time, the response $x_T(t)$ can be determined as

$$x_T(t) = \int\limits_0^{\infty} h(\xi) f_T(t - \xi) d\xi \qquad (7.45)$$

where $h(\xi)$ is the unit impulse response, see Eq. (7.22). The FT of the above equation yields

$$X_T(\omega) = \frac{1}{2\pi} \int\limits_{-\infty}^{\infty} x_T(\omega) \exp\left(-i\omega t\right) dt = H(\omega) F_T(\omega)$$

$$H(\omega) = \int\limits_{-\infty}^{\infty} h(t) \exp\left(-i\omega t\right) dt = \frac{1}{(\omega_0^2 - \omega^2) + i(2\omega\omega_0\zeta)} \qquad (7.46)$$

Function $H(\omega)$ is known as the complex frequency response of a SDOF system. If the above equation is multiplied by its complex conjugate to obtain

$$X_T^*(\omega_1) X_T(\omega_2) = H^*(\omega_1) H(\omega_2) F_T^*(\omega_1) F_T(\omega_2),$$

then the generalized power spectrum of $x(t)$ is

$$S_{xx}(\omega_1, \omega_2) = \lim\limits_{T\to\infty} E[X_T^* X_T] = H^* H \lim\limits_{T\to\infty} E[F_T^* F_T] =$$

$$H^*(\omega_1) H(\omega_2) S_{ff}(\omega_1, \omega_2) \qquad (7.47)$$

The above equation may be regarded as the generalization of the equation given below, namely

$$S_{xx}(\omega) = |H(\omega)|^2 S_{ff}(\omega) \qquad (7.48)$$

that holds true for stationary processes. As before, the inverse FT gives the covariance of the output process as

$$K_{xx}(t_1, t_2) = \int\limits_{-\infty}^{\infty} \int\limits_{-\infty}^{\infty} H^*(\omega_1) H(\omega_2) S_{ff}(\omega_1, \omega_2) \exp\left(-i(\omega_1 t_1 - \omega_2 t_2)\right) d\omega_1 d\omega_2 \qquad (7.49)$$

Finally, the variance of the response is

$$\sigma_x^2(t) = E[x^2(t)] = K_{xx}(t,t) = \int_{-\infty}^{\infty} \int_{-\infty}^{\infty} S_{ff}(\omega_1, \omega_2) \exp\left(-i(\omega_1 - \omega_2)t\right) d\omega_1 d\omega_2$$

(7.50)

7.7 Stochastic Design

7.7.1 Stationary Case Example

Consider an SDOF system defined in terms of its complex frequency response function $H(\omega) = 1/[(\omega_0^2 - \omega^2) + i(2\omega_0^2\omega^2\zeta)]$, where $\omega_0 = \sqrt{k/m}$ is the natural frequency, ζ is the damping coefficient and ω is the running frequency, see Fig. 7.6. The description of the random loading $f(t)$ is through its input PSDF function $S_f(\omega)$, which is labelled as white noise S_0 if it is independent of the frequency. Based on the previous development, the variance of the response $x(t)$ of the SDOF system for a white noise input is given as

$$\sigma_x^2 = \frac{S_0}{2\pi} \int_0^{\infty} |H(\omega)|^2 \, d\omega = \frac{S_0}{2\pi} \int_0^{\infty} \frac{1}{\left(1 - (\frac{\omega}{\omega_0})^2\right)^2 + \left(4\zeta^2(\frac{\omega}{\omega})^2\right)} \, d\omega$$

(7.51)

The above formula is valid even if $S_f(\omega)$ is a slowly varying function of frequency, see Fig. 7.6, since the form of the complex frequency response function is a highly pronounced

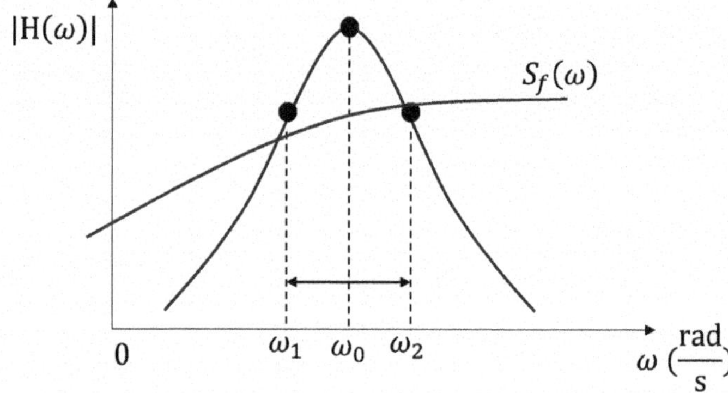

Fig. 7.6 The complex frequency response function H of an SDOF system showing the half power points ω_1, ω_2 and the PSDF of the load S_f

spike centered at the natural frequency ω_0. For a damping ratio $\zeta \ll 1.0$, the above integral can be evaluated analytically and is equal to

$$\sigma_x^2 = \frac{S_0}{2\pi} \frac{\pi \omega_0}{4\zeta} = \frac{S_0 \omega_0}{8\zeta} \tag{7.52}$$

An engineering approach to the evaluation of the variance σ_x^2 is the half-power method based on the identification of the half-power points ω_1, ω_2 centered about the natural frequency and tracing a bandwidth of $\omega_2 - \omega_1 = 2\zeta\omega$, see Fig. 7.6. The half-power points are the locations where the square of the complex frequency response function (an indirect energy measure) has one-half the value, as compared with the value at the natural frequency, i.e.,

$$|H(\omega_1)|^2 = |H(\omega_2)|^2 = \frac{1}{2}|H(\omega_0)|^2.$$

Thus, the value of the integral of $|H|^2$ is computed as $(\pi/2)\max H(\omega)^2$ times the bandwidth, yielding the same answer as before, i.e.,

$$\int_0^\infty |H(\omega)|^2 \, d\omega = \frac{\pi}{2} \left(\frac{1}{2\zeta}\right)^2 2\zeta\omega = \frac{\pi\omega}{4\zeta}$$

7.7.2   An SDOF Design Example

In this example, we will investigate a re-design case for a structure represented by a SDOF system originally designed with a natural frequency $\omega_0 = 8$ rad/s and a damping ratio $\zeta = 0.01$. The design specification is that the lateral displacement of the SDOF system should not exceed 3 cm. The PSDF of the forcing function $f(t)$ is slowly decreasing with frequency and has a value of $S_f(\omega = 8) = 0.01$ cm^{2}s/rad, dropping to a value of $S_f(\omega = 16) = 0.0025$ cm^{2}s/rad, see Fig. 7.7. Starting with the original SDOF design, we have that the variance of its response is approximately

$$\sigma_x^2 \approx \frac{S_{ff}(\omega)}{2\pi} \int_0^\infty |H(\omega)|^2 \, d\omega = \frac{S_{ff}(\omega_0)\omega_0}{8\zeta} = 1 \text{ cm}^2$$

From statistical tables [5] and noting that we have a zero-mean response, the probability of exceedance of the design value is

$$Pr(|x(t)| > 3) = 0.0027 = 0.27\%$$

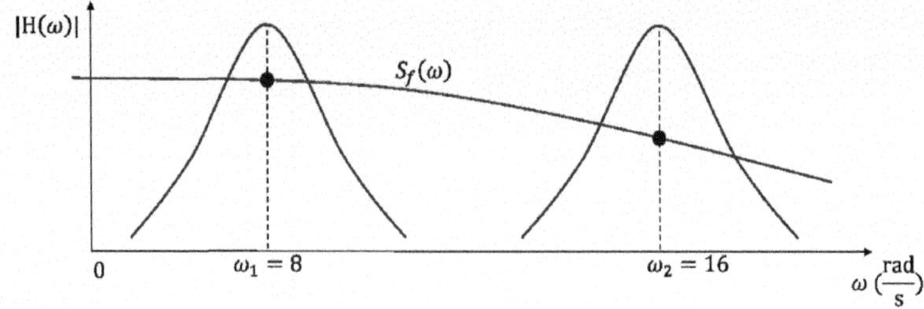

Fig. 7.7 Redesign of a SDOF system by increasing the natural frequency: Both the complex frequency response function H and the slowly varying PSDF of the load S_f are depicted

The next step is to redesig the structure by making it stiffer so that the new natural frequency is $\omega_0 = 16$ rad/s. Now we have a variance $\sigma_x^2 \approx 0.5$ cm^2. This gives a $Pr(|x(t)|>3) = 0.02\%$, which is now a negligible value.

We now briefly discuss the normal distribution for a random variable X, with values realized in the domain $-\infty < x < \infty$ and defined by a probability density function $p(x) = (1/\sqrt{2\pi}\sigma)\exp\left(-(x-\mu)^2/2\sigma^2\right)$. A random variables can be normalized as follows: $y = (x-\mu)/\sigma$, yielding $m_y = 0$, $\sigma_y = 1$, thus making standard statistical tables [5] accessible for computing probabilities. The normal distribution is defined as one with mean $m_x = 0$ and standard deviation $\sigma_x = 1$. Thus, for the first case examined above, the probability:

$$Pr(|x(t)|>3) = \int_{-\infty}^{3} p(x)\,dx + \int_{3}^{\infty} p(x)\,dx = 0.00027.$$

References

1. Crandall SH, Mark WD (2014) Random vibration in mechanical systems. Academic
2. Nigam NC, Saunders H (1986) Introduction to random vibration
3. Hurty WC, Rubinstein MF (1964) Dynamics of structures
4. Vanmarcke EMSGI, Shinozuka M, Nakagiri S, Schueller GI, Grigoriu M (1986) Random fields and stochastic finite elements. Struct Saf 3(3–4):143–166
5. Zwillinger D, Kokoska S (1999) CRC standard probability and statistics tables and formulae. CRC Press

Structural Control

8

8.1 Passive Control

As previously mentioned, passive structural implies minimization of vibrations in a primary dynamic system system by the addition of a secondary system, as would be the case of a tuned mass damper.

8.1.1 Vibration Absorbing System Without Damping

Figure 8.1 shows the simple combination of a primary SDOF system with mechanical properties comprising a mass and a stiffness (m_p, k_p), to which a secondary SDOF system is attached, with mass and stiffness (m_s, k_s). The combined dynamic system is a 2-DOF oscillator whose equations of motion in matrix are simple to derive, see below:

$$\underbrace{\begin{bmatrix} m_p & 0 \\ 0 & m_s \end{bmatrix}}_{M} \underbrace{\begin{Bmatrix} \ddot{u}_p \\ \ddot{u}_s \end{Bmatrix}}_{\ddot{u}(t)} + \underbrace{\begin{bmatrix} k_p + k_s & -k_s \\ -k_s & k_s \end{bmatrix}}_{K} \underbrace{\begin{Bmatrix} u_p \\ u_s \end{Bmatrix}}_{u(t)} = \underbrace{\begin{Bmatrix} 1 \\ 0 \end{Bmatrix}}_{B_f} f_0 \sin \omega t \tag{8.1}$$

In the above, index p is used for the primary system and index s for the secondary system. The displacement response of the combined 2-DOF system to a harmonic input at the level of the mass of the primary system is

$$\begin{Bmatrix} u_p \\ u_s \end{Bmatrix} = \begin{Bmatrix} U_p \\ U_s \end{Bmatrix} \sin \omega t$$

© The Author(s), under exclusive license to Springer Nature Switzerland AG 2026
G. Manolis and C. Panagiotopoulos, *Vibrations of Structural Systems*, Synthesis Lectures on Mechanical Engineering, https://doi.org/10.1007/978-3-032-12279-7_8

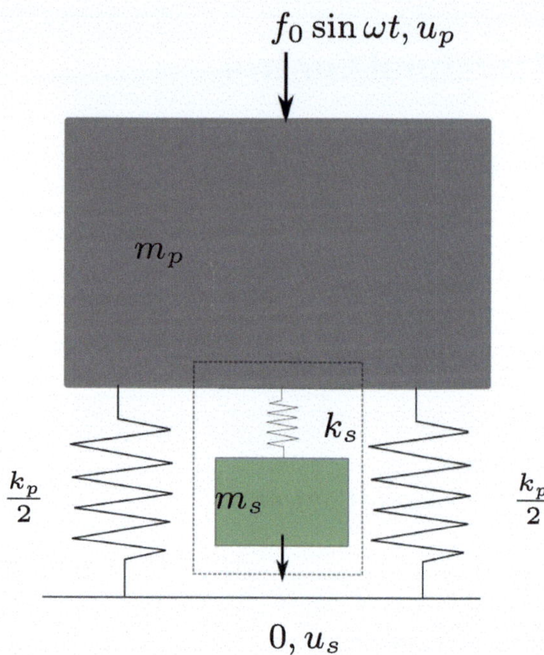

Fig. 8.1 An SDOF passive vibration minimization system without damping

The amplitude of the response can be obtained from Eq. (3.42) as follows:

$$\begin{Bmatrix} U_p \\ U_s \end{Bmatrix} = \frac{f_0}{(k_p + k_s - m_p\omega^2)(k_s - m_s\omega^2) - k_s^2} \begin{Bmatrix} k_s - m_s\omega^2 \\ k_s \end{Bmatrix} \qquad (8.2)$$

From the above equation, we observe that in the case where $k_s = m_s\omega^2$, the residual response of the system is ($u_p = 0$), while the vibrations of the control system are $u_s = -f_0 \sin \omega t / k_s$. This leads to the conclusion that we can adjust the properties of the secondary system to completely suppress the vibrations of the primary system, albeit for a harmonic input. For the purpose of a concise notation, we define the following parameters:

$$r = \frac{\omega}{\omega_p} \qquad \text{Input to primary system frequency ratio}$$

$$\mu = \frac{m_s}{m_p} \qquad \text{mass ratio}$$

$$\omega_p = \sqrt{\frac{k_p}{m_p}} \qquad \text{Primary system natural frequency}$$

$$\omega_s = \sqrt{\frac{k_s}{m_s}} \qquad \text{Secondary system natural frequency}$$

$$\alpha = \frac{\omega_s}{\omega_p} \qquad \text{Primary to secondary system natural frequency ratio}$$

Next, Eq. (8.2) can be written in dimensionless form, similar to that used for the dynamic load factor in Sect. 2.3,

$$\begin{Bmatrix} \dfrac{U_p}{f_0/k} \\[2mm] \dfrac{U_s}{f_0/k} \end{Bmatrix} = \dfrac{1}{(1 + \mu\alpha^2 - r^2)(\alpha^2 - r^2) - \mu\alpha^4} \begin{Bmatrix} \alpha^2 - r^2 \\[2mm] \alpha^2 \end{Bmatrix} \tag{8.3}$$

From Eq. (8.3) we can extract the natural frequencies of the coupled, 2-DOF system, and by setting the denominator of the resulting expression equal to zero, we have

$$r_1, r_2 = \frac{1}{\sqrt{2}}\sqrt{1 + (1+\mu)\alpha^2 \mp \sqrt{(1 + (1+\mu)\alpha^2)^2 - 4\alpha^2}}$$

Therefore, the natural frequencies of the combined system are

$$\omega_i = r_i \omega_p, \quad i = 1, 2$$

and furthermore we have

$$\omega_1 < \omega_p < \omega_2.$$

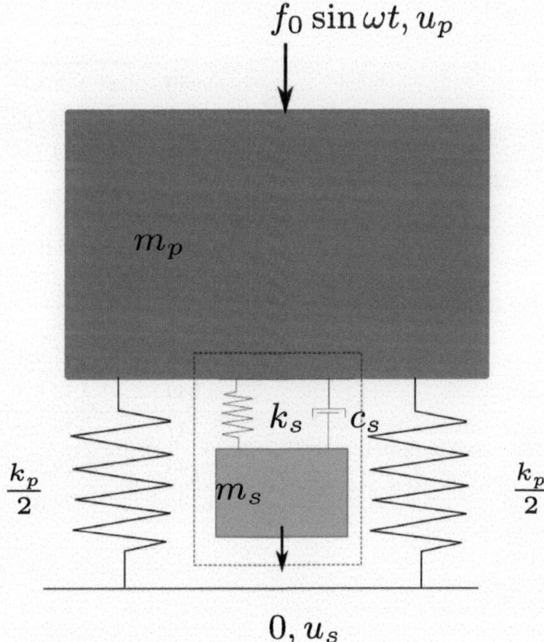

Fig. 8.2 An SDOF passive vibration minimization system with a damping system

8.1.2 Energy Dissipation with Damping

he previously presented passive control system can be extended as a dissipation mechanism due to the inclusion of a damper c_s in the secondary system attachment. This way, the frequency spectrum over which this secondary system is effective can be increased, see Sect. 8.3.2. This coupled, 2-DOF system is illustrated in Fig. 8.1, and the equations of motion are given below as follows (Fig. 8.2).

$$\underbrace{\begin{bmatrix} m_p & 0 \\ 0 & m_s \end{bmatrix}}_{M} \underbrace{\begin{Bmatrix} \ddot{u}_p \\ \ddot{u}_s \end{Bmatrix}}_{\ddot{u}(t)} + \underbrace{\begin{bmatrix} c_s & -c_s \\ -c_s & c_s \end{bmatrix}}_{C} \underbrace{\begin{Bmatrix} \dot{u}_p \\ \dot{u}_s \end{Bmatrix}}_{\dot{u}(t)} + \underbrace{\begin{bmatrix} k_p + k_s & -k_s \\ -k_s & k_s \end{bmatrix}}_{K} \underbrace{\begin{Bmatrix} u_p \\ u_s \end{Bmatrix}}_{u(t)} = \underbrace{\begin{Bmatrix} 1 \\ 0 \end{Bmatrix}}_{B_f} f_0 \sin \omega t \qquad (8.4)$$

For this case, we can now derive the solution for the displacement response of the forced vibration of the system as

$$u_{p,ss}(t) = U_p(\omega) \sin(\omega t + \phi_p)$$
$$u_{s,ss}(t) = U_s(\omega) \sin(\omega t + \phi_s)$$

with the amplitudes of vibration of the two DOF as

$$U_p(\omega) = \frac{f_0}{k_p} \sqrt{\frac{(r^2 - \alpha^2)^2 + 4\xi^2 \alpha^2 r^2}{\Delta(r)}}$$

$$U_s(\omega) = \frac{f_0}{k_p} \sqrt{\frac{\alpha^4 + 4\xi^2 \alpha^2 r^2}{\Delta(r)}}$$

The denominator in the above expressions is

$$\Delta(r) = \left[(r^2 - 1)(r^2 - \alpha^2) - \mu \alpha^2 r^2 \right]^2 + 4\xi^2 \alpha^2 r^2 \left[1 - (1 + \mu)r^2 \right]^2$$

and the phase angles are

$$\phi_p = \tan^{-1}\left(\frac{2\xi \alpha r}{\alpha^2 - r^2} \right), \quad \phi_s = \tan^{-1}\left(\frac{2\xi r}{\alpha} \right)$$

Note that parameters α, μ, r were all defined in the previous sub-section, with the damping ratio in the SDOF control system equal to

$$\xi = \frac{c_s}{2 m_s \omega_s} = \frac{c_s}{2 \sqrt{m_s k_s}}$$

8.2   Active Control

Consider a structural or a mechanical system described by N DOF, which is augmented with a network of N_r sensors (or receivers) and with N_a actuators (or sources). The function of this system is to activate the actuators so as to control the response of the dynamic system with feedback (closed loop concept) or without feedback (open loop concept) , as shown in the Fig. 8.3.

The equation of motion of this MDOF system is well known, i.e.,

$$M\ddot{u}(t) + C\dot{u}(t) + Ku(t) = f(t) + B_a q(t) \tag{8.5}$$

where the extra term of the RHS of the equation is the active control system, Specifically, matrix B_a of size $N \times N_a$ represents the distribution of the actuators, whose action is controlled by the $N_a \times 1$ vector $q(t)$. Furthermore, the sensors act as receivers, collect data and form response vector $y(t)$, which might contain kinematic variables such as displacements, velocities and accelerations.

$$y(t) = R_u u(t) + R_v \dot{u}(t) + R_a \ddot{u}(t) \tag{8.6}$$

where R_u, R_v and R_a are matrices of size $N_r \times N$. A retroactive controller is one which adjusts the action of the actuators according to the response recorded in the sensors. This action can be stated in matrix form as follows:

$$q(t) = -K_a y(t) = -K_a \left(R_u u(t) + R_v \dot{u}(t) + R_a \ddot{u}(t) \right)$$

In the above, K_a is the gain matrix that must be adjusted according the expected result. Thus, a closed loop control system can be written as given below, once the control system parameters have been incorporated in the equation of motion of the dynamic system:

Fig. 8.3 Dynamic system enriched with sensors and actuators

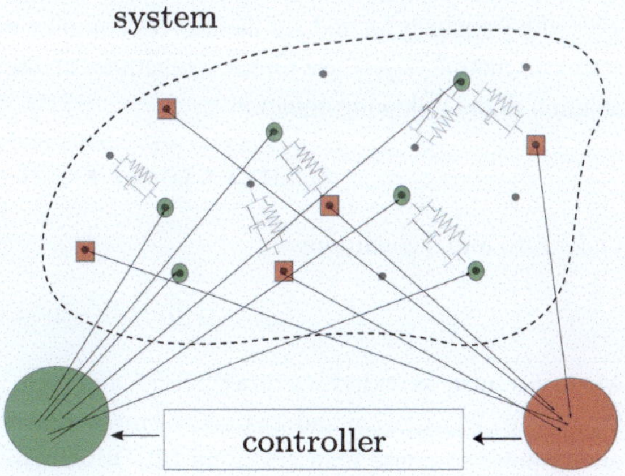

$$(M + B_a K_a R_a)\ddot{u}(t) + (C + B_a K_a R_v)\dot{u}(t) + (K + B_a K_a R_u)u(t) = f(t)$$

At this point, the equation of motion of the controlled system can be written in the state space (a system of first-order differential equations) as

$$\dot{x}(t) = Ax(t) + Bq(t) + F(t) \tag{8.7}$$

where the state vector is equal to

$$x(t) = \begin{Bmatrix} u(t) \\ \dot{u}(t) \end{Bmatrix}$$

Also, matrices A, B and vector F are given as

$$= \begin{bmatrix} 0_{N \times N} & I_{N \times N} \\ -M^{-1}K & -M^{-1}C \end{bmatrix}, \; B = \begin{bmatrix} 0_{N \times N} \\ M^{-1}B_a \end{bmatrix}, \; F(t) = \begin{Bmatrix} 0_{N \times 1} \\ M^{-1}f(t) \end{Bmatrix}$$

$$\dot{x}(t) = (A - BK_a R)x(t) + F(t).$$

8.3 Application Examples

8.3.1 Frequency and Impulse Response of MDOF Systems

An important component in structural control applications are the matrices which contain the frequency response and the impulse response of MDOF systems. These matrices form a Fourier transform pair. More specifically, for a dynamic system with-DOF, the elements of the frequency response matrix $H_{ij}(\omega)$ are the steady-state to a complex harmonic excitation $e^{i\omega t}$ with frequency ω, and are themselves complex numbers. The elements of the impulse response matrix $G_{ij}(t)$ are Green's functions of the MDOF system and derive from the solution of the following equation

$$M\ddot{G}(t) + C\dot{G}(t) + KG(t) = I_{N \times N}\delta(t)$$

under zero initial conditions

$$G(0) = 0\dot{G}(0) = 0.$$

Modules such as `mdof`, `shearframe,`and `sdofSeries` all include the methods `TransferFunction` and `ImpulseFunction` for computing the frequency response and impulse response matrices of an MDOF system in the frequency and time domains, respectively. We begin by computing the frequency and impulse response matrices for the dynamic system presented in Sect. 3.9.1, where the mass and stiffness matrices were computed in Sect. 4.6.3 and are given below as

$$K = \begin{bmatrix} k_1 + k_0 & -k_0 \\ -k_0 & k_2 + k_0 \end{bmatrix} = \begin{bmatrix} 26517.925 & -13140.145 \\ -13140.145 & 26042.655 \end{bmatrix}$$

and

$$M = \begin{bmatrix} m_1 & 0 \\ 0 & m_2 \end{bmatrix} = \begin{bmatrix} 7.95 & 0 \\ 0 & 9.25 \end{bmatrix}$$

In reference to damping, there are a number of assumptions for modal damping, the most common assigning the same percentage of critical damping ξ to all eigenmodes In our case, these damping values are $\xi \in [0, 2, 4, 8, 16]\%$, so as to suppress the high frequency vibrations associated with higher-order modes. The elements of the frequency and impulse response matrices were computed by using module `mdof` and are plotted in Fig. 8.4.

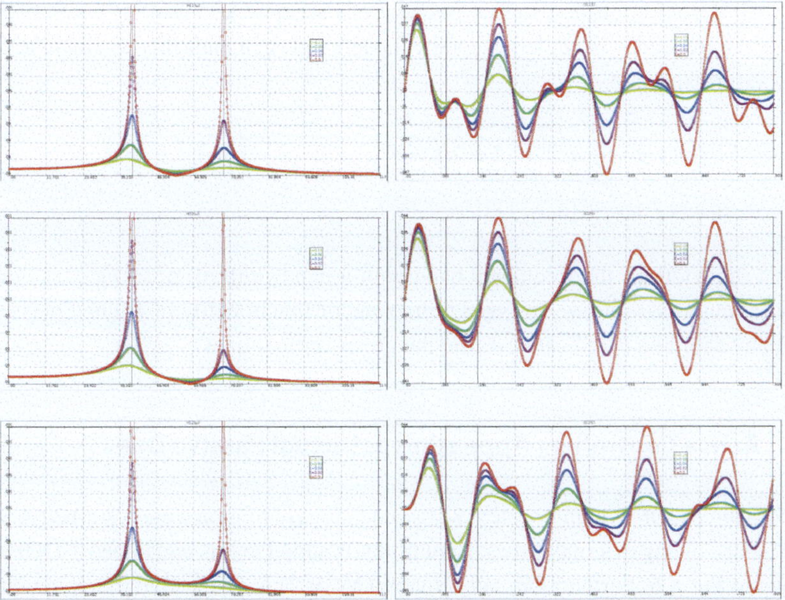

Fig. 8.4 Frequency response matrix $|H_{ij}(\omega)|$ (left column) and impulse response matrix $G_{ij}(t)$ (right column). From top to bottom are matrix elements (i, j) corresponding to $(1, 1)$, $(2, 2)$ $(1, 2)$

8.3.2 Passive Mass Damper

When designing systems such as the passive mass damper for absorbing the energy released due to vibrations, see also Sect. 8.1.1, the natural frequency of the secondary system mass must be selected so as to minimize the force vibration of the primary system due to harmonic input of frequency ω. To achieve this, it must be that $\omega_s = \omega$ or alternatively $\alpha = r$. In practice, however, it is not possible to assume that the forcing frequency ω will be close to the secondary system frequency ω_s, which negates the role of energy absorption and may lead to large motions or even to resonance. If the forcing frequency does fall within the range of the natural frequency ω_s, such that the relation given in Eq. (8.3) holds true,

$$\left| \frac{U_p}{f_0/k} \right| = \left| \frac{\alpha^2 - r^2}{(1 - r^2)(\alpha^2 - r^2) - \mu\alpha^2 r^2} \right| \leq 1 \tag{8.8}$$

then the passive mass damper will offer a measure of protection to the primary system.

$$r_L \leq r \leq r_R$$

and a small deviation of the external frequency is acceptable. From the equality given in Eq. (8.8), the frequency band delineated by the left and right ends that results in acceptable system performance are

$$r_L = \frac{1}{\sqrt{2}}\sqrt{2 + (1 + \mu)\alpha^2 - \sqrt{(2 + (1 - \mu)\alpha^2)^2 - 8\alpha^2}}$$

$$r_R = \alpha\sqrt{1 + \mu}$$

Note that the width of the frequency band with a bandwidth of $\Delta r = r_R - r_L$) increases with increasing mass ratio μ. In Fig. 8.5 the zone of functionality of a tuned mass damper without any damping is indicated within the green vertical lines.

8.3.3 Active Controllers Appearing as Virtual Passive Ones

Consider a three DOF dynamic system whose equations of motion were described in Sect. 8.2 in the absence of any external forces, a controller is added to this undamped 3-DOF system, as shown in Fig. 8.6. Furthermore, an actuator ($N_a = 2$) is attached to the system that acts at the two end masses, namely the DOF u_1 and u_3. The system's governing equation of motion is

$$M\ddot{u}(t) + C\dot{u}(t) + Ku(t) = B_a q(t) \tag{8.9}$$

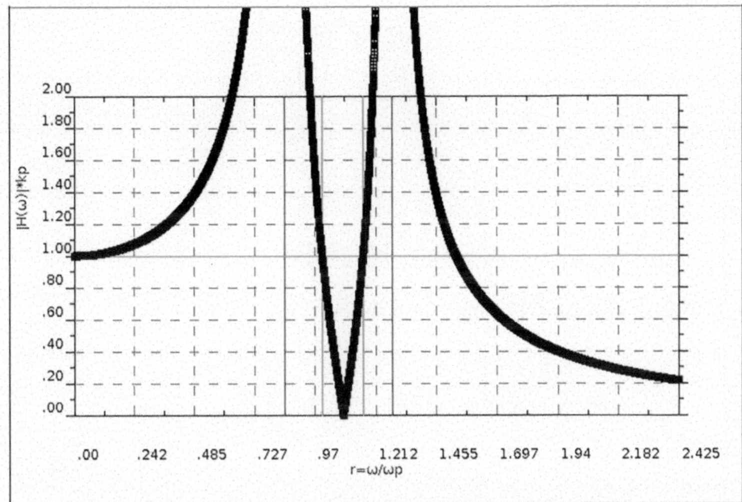

Fig. 8.5 Functionality zone for a tuned mass damper: The vertical green lines demarcation is at points r_L and r_R, while for the red lines these points are r_1 and r_2

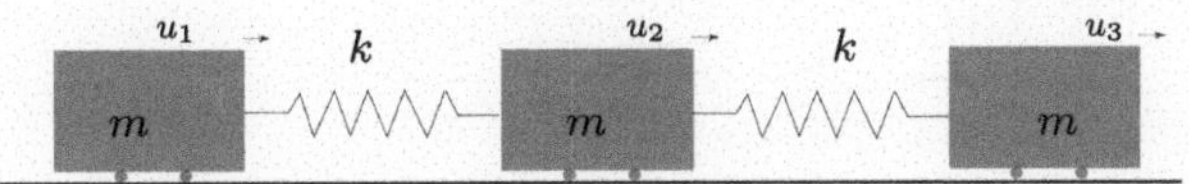

Fig. 8.6 A 3-DOF system coupled through springs without any dampers

The mass matrix of this 3-DOF system is

$$M = \begin{bmatrix} m & 0 & 0 \\ 0 & m & 0 \\ 0 & 0 & m \end{bmatrix}$$

and the stiffness matrix is

$$K = \begin{bmatrix} k & -k & 0 \\ -k & 2k & -k \\ 0 & -k & k \end{bmatrix}$$

while vector B_a defining the placement of the actuators is

$$B_a^T = \begin{bmatrix} 1 & 0 & -1 \end{bmatrix}.$$

Furthermore, the control system is equipped with two sensors ($N_r = 2$) which monitor the difference in velocities $\dot{u}_1$ $\dot{u}_3$ between the two end masses, relative to the velocity $\dot{u}_2$ of the centrally-placed mass. This information can be written in matrix form as follows:

$$y(t) = \underbrace{\begin{bmatrix} -1 & 1 & 0 \\ 0 & -1 & 1 \end{bmatrix}}_{R_v} \underbrace{\begin{bmatrix} \dot{u}_1 \\ \dot{u}_2 \\ \dot{u}_3 \end{bmatrix}}_{\dot{u}(t)}$$

Note that we have set $R_u = R_a = 0$. We now construct the gain matrix as follows:

$$K_a = \begin{bmatrix} -\zeta & -\zeta \end{bmatrix}.$$

where ζ is a positive constant. For checking the response, we have that

$$q(t) = -K_a y(t) = -\begin{bmatrix} -\zeta & -\zeta \end{bmatrix} \begin{bmatrix} -1 & 1 & 0 \\ 0 & -1 & 1 \end{bmatrix} \begin{bmatrix} \dot{u}_1 \\ \dot{u}_2 \\ \dot{u}_3 \end{bmatrix}$$

The force produced by the actuators in response to the system's motion is

$$B_a q(t) = -\underbrace{\begin{bmatrix} 1 \\ 0 \\ -1 \end{bmatrix}}_{B_a} \underbrace{\begin{bmatrix} -\zeta & -\zeta \end{bmatrix}}_{K_a} \underbrace{\begin{bmatrix} -1 & 1 & 0 \\ 0 & -1 & 1 \end{bmatrix}}_{R_v} \underbrace{\begin{bmatrix} \dot{u}_1 \\ \dot{u}_2 \\ \dot{u}_3 \end{bmatrix}}_{\dot{u}(t)} = -\begin{bmatrix} \zeta & 0 & -\zeta \\ 0 & 0 & 0 \\ -\zeta & 0 & \zeta \end{bmatrix} \begin{bmatrix} \dot{u}_1 \\ \dot{u}_2 \\ \dot{u}_3 \end{bmatrix}$$

This matrix operates as a damping matrix in a closed-loop control system, i.e.,

$$\ddot{u}(t) + B_a K_a R_v \dot{u}(t) + K u(t) = 0$$

$$\ddot{u}(t) + B_a K_a R_v \dot{u}(t) + K u(t) = 0$$

Viewed another way, the force that acts on the dynamic system as provided by the actuators in a closed-loop system, is equivalent to the addition of damping. This alludes to an equivalent passive control system which dampens out the vibrations of the dynamic system under study, see Fig. 8.7. For more information on this control equivalence, see Juang and Phan [2].

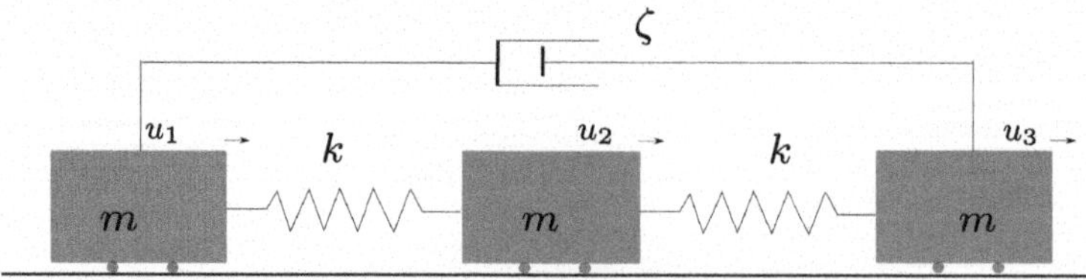

Fig. 8.7 Equivalent passive damping system for control of a 3-DOF dynamic system

References

1. Yang B (2005) Stress, strain, and structural dynamics: an interactive handbook of formulas, solutions, and MATLAB toolboxes. Academic
2. Juang J-N, Phan MQ (2001) Identification and control of mechanical systems. Cambridge University Press

Structural Identification 9

9.1 Introduction

Structural identification alludes to the fact that it should be possible to determine the dynamic characteristics of a structure from either field measurements or from experimentally obtained data. The importance of this activity cannot be overstated, given the categories of external loads structures are subjected to [1] over their lifetime. Furthermore, structural ageing requires comparing currently obtained structural data with existing benchmarks so as to decide if repairs and/or strengthening measures are required.

The dynamic properties of a structure are defined by three sets of parameters, namely the natural frequencies, the damping ratios and the modal shapes. These are independent of the loads acting on the structure and change only if the geometric properties, the material properties and the boundary conditions change. Furthermore, if the frequency content of an external load falls within the regions of the natural frequencies, then resonance develops that causes the structure to vibrate with dangerously high amplitudes. Modal analysis is perhaps the most comprehensive method for analyzing structural behavior, with further applications in numerical modelling validation, structural health monitoring, load estimation, sensitivity analysis and damage detection.

There are currently three main branches in modal analysis techniques, namely experimental modal analysis (EMA), operational modal analysis (OMA) and the newer impact synchronous modal analysis (ISMA). EMA is the oldest modal analysis technique with a history spanning over fifty years. It is performed by artificially exciting a structure, usually by an impact hammer or a shaker, and is often performed under controlled laboratory conditions. By way of contrast, the other two methods, namely OMA and ISMA, are conducted while the structure is in operation. The difference between them is that the former uses as input environmental loads (i.e., ambient vibrations) which are unknown, while the latter uses artificially generated excitations for input, which are known.

9.2   Operational Modal Analysis

Operational Modal Analysis (OMA) is also known as output-only modal analysis, or as ambient modal identification, and is used to identify the modal parameters of a structure based on vibration data collected when the structure is under its operating conditions [2]. OMA is employed in situations where the structure is too large to respond to an artificial excitation or the system cannot be shutdown. Furthermore, either ambient vibrations or the forces resulting from cyclic loading of the structure are used as excitation. As these excitations are unknown, OMA uses only the measured response data and different algorithms are then employed to extract the modal parameters. Since the input excitation is not needed, it is best to analyze the dynamics of structures that are subjected to random excitation generated by their ambient environment, as is often the case with tall buildings and bridges. Processing data in OMA is challenging and may often give erroneous results without the possibility of establishing error bounds. This is especially true if the structure exhibits a nonlinear response. To overcome this problem, different OMA techniques have been developed over the last three decades [3] which are presented next.

9.2.1   OMA in the Frequency Domain

Although modal analysis has been in use for a long time, a major issue in modal testing is the availability of computational power and the implementation of efficiency, as hundreds (or even thousands) of records must be processed simultaneously. OMA has become an active research area since the late 1980s due to the advances in the field of electronic computation and OMA techniques can be divided into frequency and time domain methods. The former techniques are based on the relation between the input and output power spectrum density (PSD) functions, while the latter ones are based on the analysis of response time histories and on producing correlation functions.

The earliest (and simplest) OMA technique is the peak picking (PP) method, identification of the natural frequencies is simply based on identifying peaks in the power spectrum. Obviously, the basic assumption is that the natural modes are well separated, and damping is low. Although the method is simple to use, see Listing 9.1, misleading results can be obtained if a structural system has closely spaced modes. Unfortunately, in real-life structures, closely spaced modes are almost always encountered. To overcome this serious limitation, a new technique called the frequency domain decomposition (FDD) was developed, which remains one of the most popular OMA techniques to date. More specifically, it uses the singular value decomposition (SVD) of the power spectral density (PSD) matrix in order to detect mode multiplicity. Through use of the SVD, the spectral matrix is decomposed into a set of auto-spectral density functions, each corresponding to a single-degree-of-freedom system. Thus, closely spaced (and even repeated modes) can be identified accurately. The major limitation of FDD method is that it cannot estimate damping ratios.

```groovy
// response file can be downloaded from:
// \url{https://symplegma.civil.auth.gr/docs/resp.txt
   }
dt=1.0/256.0

basepath=\.. folder location where response file
   located..\
fileinp=basepath+"resp.txt"

z726=[]; z728=[]; z729=[]; count=0
new File(fileinp).withReader('UTF-8') { reader ->
def line
    while (((line = reader.readLine()) != null) &&
   count<86016) {
       //println line
       lsp=line.split(" "); count++
  z726.add(Double.parseDouble(lsp[0])+1.0)
  z728.add(Double.parseDouble(lsp[1])+1.0)
  z729.add(Double.parseDouble(lsp[2])+1.0)
     }
}

hann={inp->
   out=[]
   if(inp.getClass().isArray()){L = inp.length-1 }else
   {L = inp.size()-1}
   (0..<inp.size()).each{out.add(inp[it]*(0.5*(1.0-cos
   (2*pi*it/L))))}
   return out
}

resp=[z726,z728,z729]

plot()
thePlot.setAutoColor(true)
resp.each{plot(dt,it)}

// make FFT
fftSPlot=new PlotFrame()
fftSPlot.setAutoColor(true)
fftresp=[]
Fs=1.0/dt
resp.each{
   insignal=it
   fft = new FastFourierTransformer(DftNormalization.
   STANDARD);
   Sfwd = fft.transform(pad(hann(insignal)),
   TransformType.FORWARD);
```

```
42    freq=[]
43    Lp=Sfwd.size();
44
45    fftSource=[]
46    (0..(Lp)/2).each{
47      freq.add(it*Fs/Lp)
48      fftSource.add(Sfwd[it])
49    }
50    fftSPlot.addFunction(new plotfunction(freq as
       double[], fftSource.collect{it.abs()} as double
       []))
51    fftresp.add(fftSource)
52 }
53 fftSPlot.show()
54
55 println "Peak picking extreme values"
56 println fftSPlot.getPlotPanel().getMax(8.0,12.0)[0]+"
       Hz"
57 println fftSPlot.getPlotPanel().getMax(34.0,41.0)[0]+
       " Hz"
58 println fftSPlot.getPlotPanel().getMax(80.0,90.0)[0]+
       " Hz"
```

Listing 9.1 Peak picking

9.2.2 OMA in the Time Domain

The time domain decomposition (TDD) method is based on a decomposition of the structural response into a series single degree of freedom (SDOF) system signals in the time domain. Using this method, the undamped mode shapes are extracted using the singular value decomposition of the output correlation matrix relative to the locations of sensors. After extracting mode shapes, natural frequencies and damping ratios are computed from the individual SDOF signals using the PP method.

TDD is a computationally efficient method when a large number of sensors are involved in the structure analysis procedure. The mode separation task is achieved by using digital band-pass filters that reduce the operator interaction during the analysis process, thus making the method suitable for automated online health structural monitoring applications. On the other hand, since this method is based on SDOF approach, it is difficult to extract modal parameters from closely spaced modes.

9.2.3 Modal Damping

When studying structures subjected to dynamic loads, knowledge of the amount of damping available is of paramount importance in interpreting field measurements and in calibrating analytical/numerical models [4]. Broadly speaking, damping is classified as either viscous or hysteretic. The former is a linearized version of energy dissipation, proportional to the velocity and a function of the construction materials used. The latter is a nonlinear phenomenon and often has to do with the integrity of the connections and supports of the structure in question. Viscous damping is commonly computed from the free-vibration regime of a structure using the logarithmic decrement concept. This procedure is unambiguous for a structure modelled as a single degree-of-freedom (SDOF) system, but for multiple degree-of-freedom (MDOF) systems, viscous damping thus computed reflects superposition of all modes of vibration. It is therefore imperative when processing transient signals to transform them to the frequency domain and isolate the peaks that correspond to the eigenfrequencies of the system. If a Fourier transformed (FT) signal is cleared of all eigenfrequencies expect one, then the inverse FT transformation to the time domain contains the contribution of that specific mode only. Therefore, the logarithmic damping concept will yield the percentage damping associated with that particular mode only. In terms of analysis, this necessitates the use of generalized (or modal) coordinates so as to separate the modes of the structure. In reality, structural systems have a continuous mass distribution, but when conducting measurements, data is produced at discrete locations, and hence have to be modelled as discrete MDOF systems.

9.3   OMA Application Example

Consider the structural system presented in Chap. 3, where two coupled shear frames were modelled as a two-DOF system. For a given external force input, the challenge is to determine the dynamic properties of this system from its output. At this point, we need to discuss the difference between the two basic categories of what was previously called "modal testing", namely (a) experimental modal analysis (EMA) that utilizes controlled input forces and (b) the operational modal analysis (OMA) that utilizes operational forces. During the last few decades, both pre-determined and in-operation vibrations have been used, and both are effective in determining the dynamic characteristics of structures. As previously mentioned, the concept behind OMA testing techniques is that the structure to be investigated is subjected to excitations from different sources such as wind, traffic, machinery, and live loads in the building, which often contain white noise characteristics. These kinds of excitations produce energies distributed over a wide range of frequencies that depend on the capacity of the recording instruments to capture. At the very least, the recordings should cover the frequency range of the fundamental mode shapes of the structure that is being tested. Form a mechanical viewpoint, the most crucial issue in OMA is that the required modes shapes are adequately excited during the testing procedure so that the resulting measurements method may record their energy contribution.

9.4   EMA Application Example

As previously discussed, accurate determination of damping from measurements in the laboratory and in the field, but without using expensive equipment or complicated setups, is possible but requires careful interpretation of the recorded data. As a first step, we examine two methods used for eigenvalue-eigenvector extraction, namely the Frequency Domain Decomposition (FDD) and Principal Component Analysis (PCA), which are expanded in a second step to compute modal damping from measurements of the response of a simply supported beam under impact. The case study is shown in Fig. 9.1 and a dynamic model

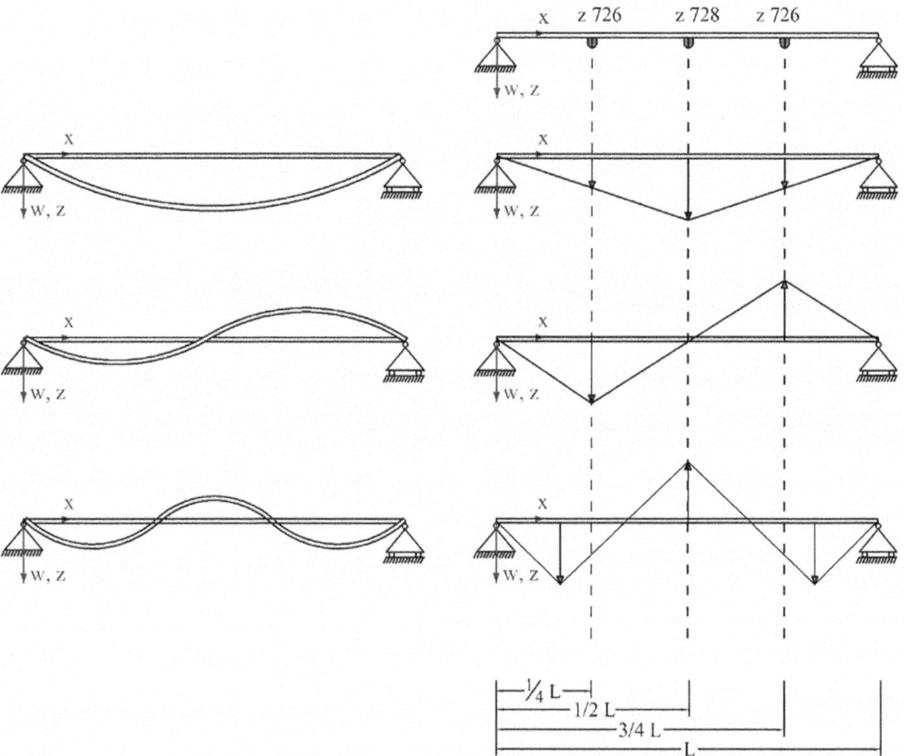

Fig. 9.1 The first three eigenfunctions of the continuous beams (left column) and the location of acceleration sensors labelled as Z726, Z728 and Z729 showing the eigenvectors for an MDOF representation (right column)

was developed for a point load moving across the span of a continuous, simply-supported beam [5]. The acceleration generalized coordinates as the basic kinematic variables (in lieu of the displacement generalized coordinates) to process the transient accelerations recorded by sensors, followed by use of the logarithmic decrement method in the free vibration regime to compute modal damping values.

9.4.1   Mechanical Model for a Continuous Beam

Figure 9.1 shows a supported, Euler-Bernoulli beam the length with a continuous mass distribution. If accelerometers are used for recording motion, then separation of variables for the accelerations gives $\ddot{w}(x, t) = \Phi_n(x)\ddot{q}_n(t)$, where x are the locations of the sensors, $\Phi_n = sin(n\pi x/L)$ are the eigenfunctions for a simply-supported beam and are the generalized coordinates. Furthermore, overdots indicated time derivatives and the summation convention is implied for a repeated index $n = 1, 2, \ldots, \infty$. Assuming that three sensors are placed at nodes L_i, $i = 1, 2, 3$, then the first three terms are given as follows:

$$
\begin{Bmatrix} \ddot{w}(L_1, t) \\ \ddot{w}(L_2, t) \\ \ddot{w}(L_3, t) \end{Bmatrix} = \begin{bmatrix} \Phi_1(L_1) & \Phi_2(L_1) & \Phi_3(L_1) \\ \Phi_1(L_2) & \Phi_2(L_2) & \Phi_3(L_2) \\ \Phi_1(L_3) & \Phi_2(L_3) & \Phi_3(L_3) \end{bmatrix} \begin{Bmatrix} \ddot{q}_1 \\ \ddot{q}_2 \\ \ddot{q}_3 \end{Bmatrix}
\tag{9.1}
$$

The location of the sensors is arbitrary, and in this example are placed at $L_1 = L/4$, $L_2 = L/2$ and $L_3 = 3L/4$. The discrete eigenfunctions values are now eigenvectors given as

$$
\Phi_1 = \begin{Bmatrix} \Phi_1(L_1) \\ \Phi_1(L_2) \\ \Phi_1(L_3) \end{Bmatrix} = \begin{Bmatrix} 1/2 \\ \sqrt{2}/2 \\ 1/2 \end{Bmatrix}, \quad \Phi_2 = \begin{Bmatrix} \Phi_2(L_1) \\ \Phi_2(L_2) \\ \Phi_2(L_3) \end{Bmatrix} = \begin{Bmatrix} \sqrt{2}/2 \\ 0 \\ -\sqrt{2}/2 \end{Bmatrix},
$$

$$
\Phi_3 = \begin{Bmatrix} \Phi_3(L_1) \\ \Phi_3(L_2) \\ \Phi_3(L_3) \end{Bmatrix} = \begin{Bmatrix} 1/2 \\ -\sqrt{2}/2 \\ 1/2 \end{Bmatrix}.
\tag{9.2}
$$

Since the eigenvectors are normalized, the inverse of the eigenvector matrix is $\Phi^{-1} = \Phi^T$, where T is the transpose operator, yielding the three generalized coordinates as $\ddot{q}(t) = \Phi^T \ddot{w}(t)$. Once the generalized coordinates are extracted from the experimental data, then the logarithmic decrement method can be applied to their free vibration part yielding the critical damping ratios ξ_i, $i = 1, 2, 3$, for each mode. The accuracy of this computation is obviously dependent on the quality of the eigenvector data from the experiment.

9.4.2   Extraction of Eigenfrequencies and Eigenvectors

Two methods will be examined for evaluating the eigenproperties of the simply supported beam model of Fig. 9.1, as will be discussed in the next sections. The beam's transient response due to impact was measured at three locations where sensors were placed [5]. More specifically this was a low–cost experiment involving three wireless acceleration sensors and a plastic impact hammer for striking a HEB 100 steel beam of length at select locations, thus imparting a pseudo white noise excitation, with results shown in Fig. 9.2.

A Fourier transforms (FT) of the above records (see next section) will easily show the eigenfrequencies of the simply supported beam by simply using the peak picking (PP) method. These are given below in Table 9.1 and agree to within three significant digits with the analytic results [6].

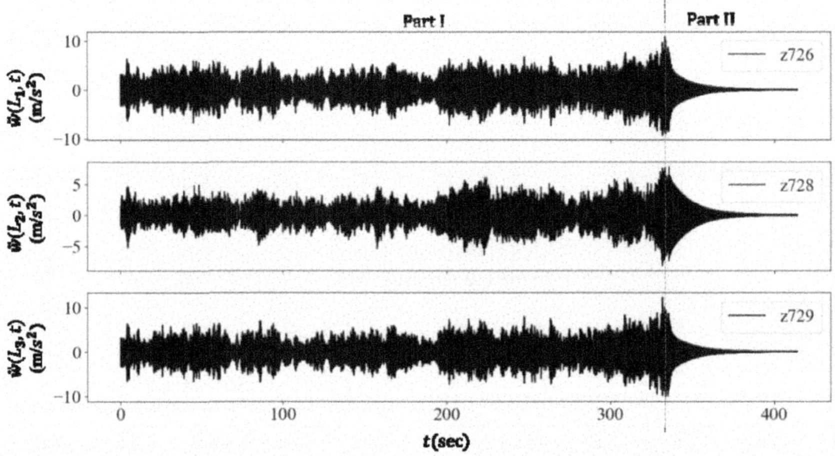

Fig. 9.2 Transient acceleration records recovered at three locations along the beam under impact. Each record comprises Part I, a quasi-white noise vibration and Part II, free vibrations

Table 9.1 Eigenfrequencies of the simply-supported beam

f_1 (Hz)	f_2 (Hz)	f_3 (Hz)
9.745	36.61	83.97

9.5   **Frequency Domain Decomposition (FDD)**

The FDD requires a white-noise input to the structure, but the impact hammer used produces an imperfect white noise input, and this undoubtedly affects the quality of the results. Both parts of the transient accelerations shown in Fig. 9.3, which were obtained for a sampling rate 256 Hz, 128 Hz as the highest frequency value, were used. This frequency range covers the first three eigenfrequencies of the beam used in the experiments. At first, a standard baseline correction is applied by subtracting the gravity effect. Next, Part I of each transient record is split into 21 segments, with each segment length containing 2^{12} (or 86016) data points covering 336 s of response. A Hanning window function [2] is applied to these segments to reduce leakage in the ensuing Fourier transforms (FT). Furthermore, the FT's of all 21 segments are averaged out for better accuracy and the final frequency plots are given in Fig. 9.3. We then compute the Power Spectral Density (PSD) matrix $\hat{G}_{yy}(j\omega_i)$ of the FT plots, where $j = \sqrt{-1}$ and ω_i the ith eigenfrequency. This matrix is formed by multiplying the FT acceleration at ω_i by its complex conjugate, followed by a Singular Value Decomposition (SVD) yielding $\hat{G}_{yy}(j\omega_i) = U_i S_i U_i^H$. According to FDD theory, the first column in matrix U_i is an estimate of the eigenvector of the structure $\hat{\Phi}_i$. The first three eigenvector estimates of the beam are as follows:

$$\Phi_1 = \left\{ \begin{array}{c} 0.49165975 \\ 0.71104878 \\ 0.50267308 \end{array} \right\}, \ \Phi_2 = \left\{ \begin{array}{c} 0.67759235 \\ 0.02828831 \\ -0.73485401 \end{array} \right\}, \ \Phi_3 = \left\{ \begin{array}{c} 0.46162609 \\ -0.69121135 \\ 0.54818159 \end{array} \right\}.$$

A comparison between the experimentally determined eigenvectors with the analytically derived ones from Eq. (9.2) is given in Table 9.2 and is carried out using the Modal Assurance

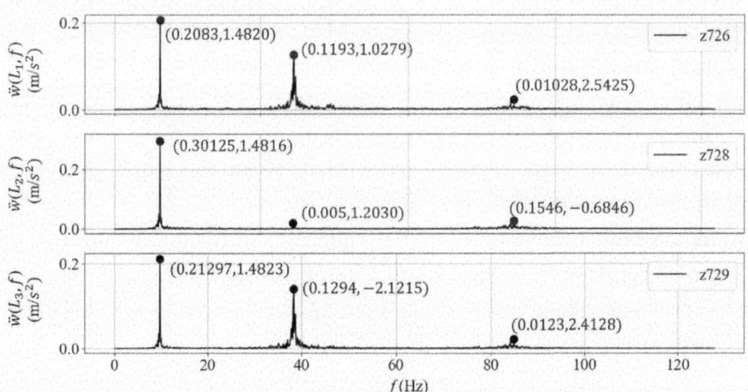

Fig. 9.3 FT of three acceleration records with each peak giving the magnitude and phase angle of the complex-valued eigenfrequencies. Note the phase angles are either 0 or π (rad)

Table 9.2 MAC index for comparing analytical (Φ) and experimentally ($\hat{\Phi}$) derived eigenvectors using the FDD

	Φ_1	Φ_2	Φ_3
Φ_1	0.99990776	0.00007445	0.00026288
Φ_2		0.99756018	0.00377852
Φ_3			0.99595860

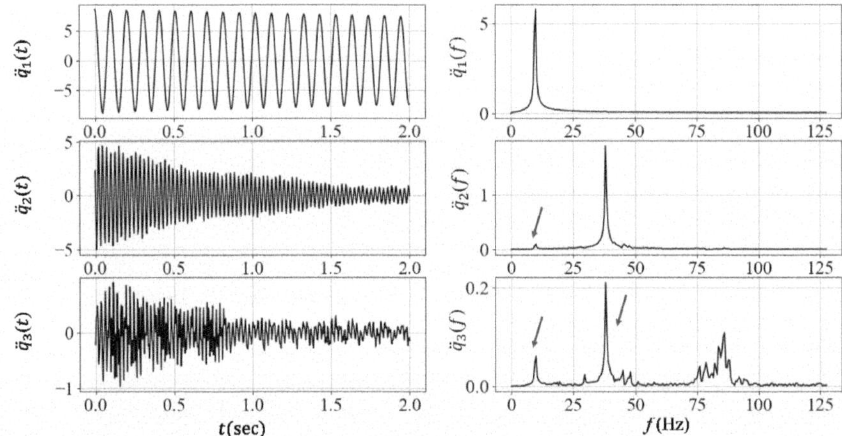

Fig. 9.4 Generalized coordinates $\ddot{q}_i$ by the FDD method: **a** Transient response versus **b** FT of the free vibration part with the arrows pointing the intrusion of other modes

Criterion $\mathrm{MAC}(\Phi_i^T, \hat{\Phi}_i) = |\Phi_i^T \hat{\Phi}_i|^2 / ((\Phi_i^T \Phi_i)(\hat{\Phi}_i^T \hat{\Phi}_i))$, which is equal to $\cos^2(\theta)$, where θ is the angle between two eigenvectors. As the MAC index approaches unity, $\theta \rightarrow 0$, while as it approaches zero, $\theta \rightarrow \pi/2$.

Once the eigenvectors are known, the transformation from the physical coordinates $\ddot{w}(x = L_i, t)$ to the generalized coordinates $\ddot{q}(t)$ is possible. This is shown in Fig. 9.4, where the generalized coordinates associated with the beam's acceleration do not uncouple properly, with the exception of the first one ($\ddot{q}_1$). The arrow in the FT of the generalized coordinates indicate the leakage of the eigenfrequencies from other modes into the current one, which can be construed as a measure of insufficient accuracy.

Listing 9.2 provides the algorithm for computing the eigenvectors of this example.

```
import org.apache.commons.math3.linear.*

SV0=[]
nrec=fftresp.size()
(0..(Lp)/2).each{
  q=it
  da=new double[nrec][nrec]
  (0..<nrec).each{
    i=it
    (0..<nrec).each{
      j=it
      da[i][j]=(fftresp[i][q]*fftresp[j][q].conj()).
  abs();
    }
  }

  A =new Array2DRowRealMatrix(da)
  SVD=new SingularValueDecomposition(A)
  SV0.add(SVD.getSingularValues()[0].abs())
}

SVplots = new PlotFrame()
SVplots.addFunction(new plotfunction(freq as double
    [], SV0 as double[]))
SVplots.setAutoColor(true)
SVplots.setMarker(true)
SVplots.show()

println "SVD extreme values"
println SVplots.getPlotPanel().getMax(8.0,12.0)[0]+"
    Hz"
println SVplots.getPlotPanel().getMax(34.0,41.0)[0]+"
    Hz"
println SVplots.getPlotPanel().getMax(80.0,90.0)[0]+"
    Hz"

getSingularVector={
  q=it
  nrec=fftresp.size()
  da=new double[nrec][nrec]
  (0..<nrec).each{
    i=it
    (0..<nrec).each{
      j=it
      da[i][j]=(fftresp[i][q]*fftresp[j][q].conj()).
  re();
    }
  }
```

```
43    A =new Array2DRowRealMatrix(da)
44    SVD=new SingularValueDecomposition(A)
45    return SVD.getV().getColumn(1)
46 }
47
48 loc=floor(SVplots.getPlotPanel().getMax(8.0,12.0)[0]/
      freq[1]) as int
49 println SV0[loc]
50 println getSingularVector(loc)
51 loc=floor(SVplots.getPlotPanel().getMax(34.0,41.0)
      [0]/freq[1]) as int
52 println SV0[loc]
53 println getSingularVector(loc)
54 loc=floor(SVplots.getPlotPanel().getMax(80.0,90.0)
      [0]/freq[1]) as int
55 println SV0[loc]
56 println getSingularVector(loc)
```

Listing 9.2 Modal analysis using FDD, script given in Listing 9.1 should be loaded at the beginning

9.6   Principal Component Analysis (PCA)

This method is defined in the time domain and does not require any specific type of input. The free vibration regime, Part II, from Fig. 9.1, which contains minimal noise, can be used to determine the eigenvectors of the beam. We start by computing the covariance matrix of the acceleration records as $\text{cov}\left(\ddot{w}(L_i, t), \ddot{w}(L_j, t)\right) = \ddot{w}(L_i, t)\ddot{w}^T(L_j, t)/(N - 1)$ from the three points along the beam's span $i, j = 1, 2, 3$, where N is the number of time samples. The covariance is the dot product between two vectors, which is close to zero if two vectors are perpendicular and close to unity if they are parallel. A diagonalization of the covariance matrix yields the eigenvectors $\hat{\Phi}$, followed by the Modal Assurance Criterion in Table 9.3.

$$\Phi_1 = \begin{Bmatrix} 0.4885036 \\ 0.7112993 \\ 0.5053885 \end{Bmatrix}, \; \Phi_2 = \begin{Bmatrix} 0.68694459 \\ 0.04362915 \\ -0.7015339 \end{Bmatrix}, \; \Phi_3 = \begin{Bmatrix} 0.5053885 \\ -0.72539894 \\ 0.46731021 \end{Bmatrix}.$$

In PCA, the matrix containing the eigenvectors $\Phi^T = \ddot{W}\ddot{W}^T$ is the Principal Component base, and every column of Φ^T corresponds to an eigenvector. This can be proved by observing that Φ^T is Hermitian and thus has real eigenvalues. Furthermore, there exists an orthonormal transformation such that $\Phi^T = K \Lambda K^{-1}$, with Λ a diagonal matrix containing the eigenvalues and matrix K containing the eigenvectors in its columns, where $K^{-1} = K^T$. We can therefore write that $\ddot{q} = \Phi^T \ddot{W} = K^T \ddot{W}$. Next, the covariance of the

Table 9.3 MAC index for comparing analytical (Φ) and experimentally ($\hat{\Phi}$) derived eigenvectors using the PCA

	Φ_1	Φ_2	Φ_3
Φ_1	0.999821238	0.00057444	0.00070677
Φ_2		0.99792146	0.000724978
Φ_3			0.99856825

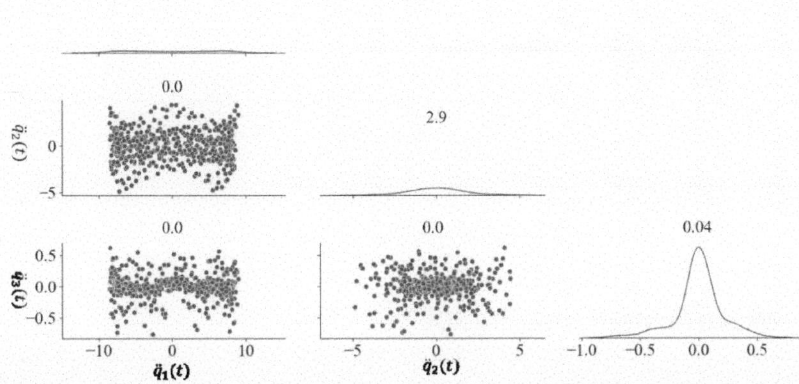

Fig. 9.5 Covariance matrix for the generalized acceleration coordinate time histories recorded at three locations along the beam's span

generalized coordinates $\ddot{q}$ is defined as $(M - 1)C_{\ddot{q}} = \ddot{q}\ddot{q}^T = \Lambda$, where M is now the number of eigenvectors. The covariance matrix of the generalized coordinates $\ddot{q}$ is shown in Fig. 9.5, where we observe a zero cross-correlation, which is a good measure of accuracy. Furthermore, the numbers along the main diagonal reveal the contribution of each mode to the total response, i.e., the participation factors are $\gamma_j = \sum_{i=1}^{j} \lambda_i / \sum_{i=1}^{3} \lambda_i$ and yield a 91.82%, participation for the first mode and a 99.89% participation for the first two modes. Finally, Fig. 9.6 plots the generalized coordinates as function of time and also gives their FT, where in contrast to the FDD, the separation of modes is complete as each coordinate vibrates with its own eigenfrequency.

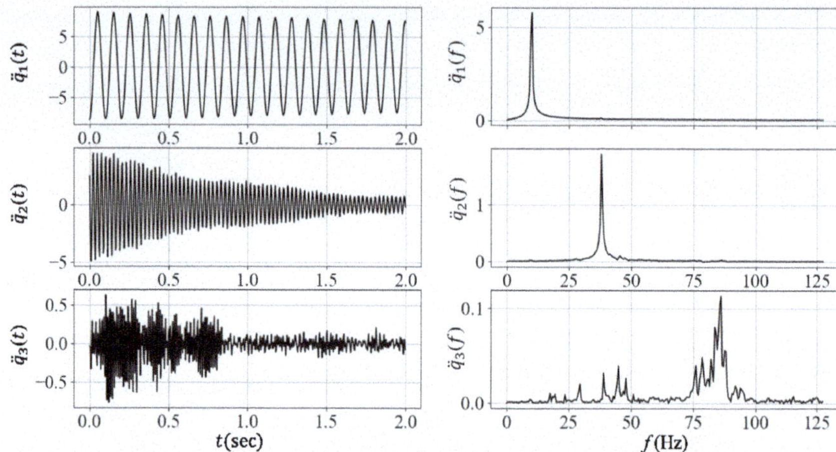

Fig. 9.6 Generalized coordinates $\ddot{q}_i$ by the PCA method: **a** Transient response versus **b** FT response of the free-vibration to check separation of eigenmodes

9.6.1 Modal Analysis Utilizing SDE

The script given in Listing 9.3 performs modal analysis for the continuous beam example using the PCA method

```
import org.apache.commons.math3.stat.correlation.
    Covariance
dt=1.0/256.0

basepath=\.. folder location where response file
    located..\
fileinp=basepath+"resp.txt"

z726=[];z728=[];z729=[];count=0
new File(fileinp).withReader('UTF-8') { reader ->
def line
    while (((line = reader.readLine()) != null) &&
    count<86016) {
        //println line
        lsp=line.split(" "); count++
    z726.add(Double.parseDouble(lsp[0])+1.0)
    z728.add(Double.parseDouble(lsp[1])+1.0)
    z729.add(Double.parseDouble(lsp[2])+1.0)
    }
}

resp=[z726,z728,z729]
```

```
21  cov=new Covariance()
22
23  resp=[z726,z728,z729]
24  N=resp[0].size()
25  COV=new double[resp.size()][resp.size()]
26  (0..<resp.size()).each{
27    i=it
28    (0..<resp.size()).each{
29      j=it
30      COV[i][j]=cov.covariance(resp[i] as double[],
       resp[j] as double[])/(N-1)
31      print COV[i][j]+" "
32    }
33    println()
34  }
35
36  covMat=Matrix(COV)
37  eigen=EigenDecomposition(covMat)
38  println eigen.getRealEigenvalues()
39  eigen.getV().print()
```

Listing 9.3 Modal analysis case using PCA

9.7   Modal Damping Factors

By comparing Figs. 9.4 and 9.6 the PCA method, as developed for the generalized coordinates, is better suited for computing the modal damping factors $\xi_i = 1, 2, 3$. For subcritical damping, the free vibration response at the generalized coordinate level can be written as

$$q_i(t) = \alpha_1 e^{-(\delta_i + j\omega_{d,i}t)} + \alpha_2 e^{-(\delta_i + j\omega_{d,i}t)} \tag{9.3}$$

where α_1, α_2 are constants determined form the initial conditions and $\delta_i = \omega_i \xi_i$. By taking the time second time derivatives of the displacement generalized coordinates we obtain the acceleration generalized coordinates for computing the modal damping ratios as

$$\ddot{q}_i(t) = e^{-\delta_i t}\left(\alpha_1(\delta_i + j\omega_{d,i})^2 e^{-\omega_{d,t}t} + \alpha_2(\delta_i - j\omega_{d,i})^2 e^{-\omega_{d,t}t}\right) \tag{9.4}$$

From Fig. 9.7, the free vibration response is periodic with damped natural period $T_{d,i} = \omega_{d,i}/(2\pi)$, so that the logarithm of the amplitude of two consecutive peaks, or of any points on that graph differing by an integer number N of periods, gives

$$\ln(\ddot{q}(t)/\ddot{q}_i(t + NT_{d,i})) = \delta_i N T_{d,i} = N\Lambda_i. \tag{9.5}$$

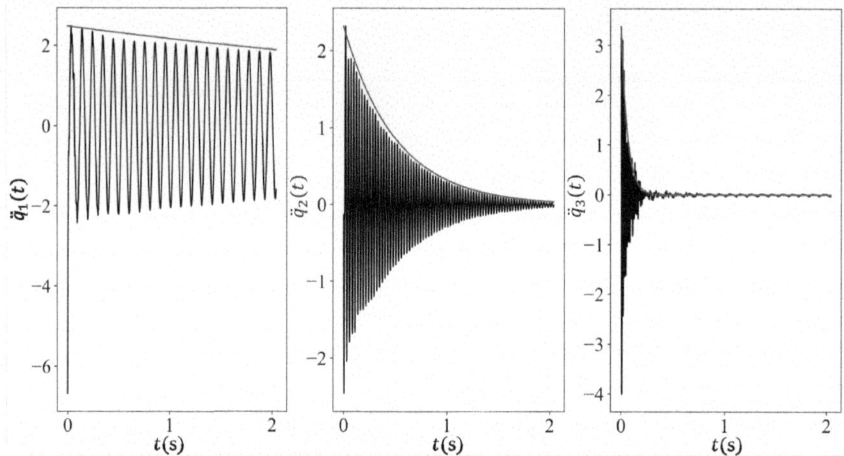

Fig. 9.7 Envelope function (in red) for the generalized coordinate accelerations in the free vibration regime

As shown in Fig. 9.8, the slope of the envelop is a straight line equal to λ_i, so for the three generalized coordinates we have $\Lambda_1 = 0.01353$, $\Lambda_2 = 0.05287$ and $\Lambda_3 = 0.21669$. From Eq. (9.5) we may write $\xi_i = \Lambda_i / (4\pi^2 + \Lambda_i^2)^{-1/2}$. The values recovered for the three modal damping factors are given in Table 9.4.

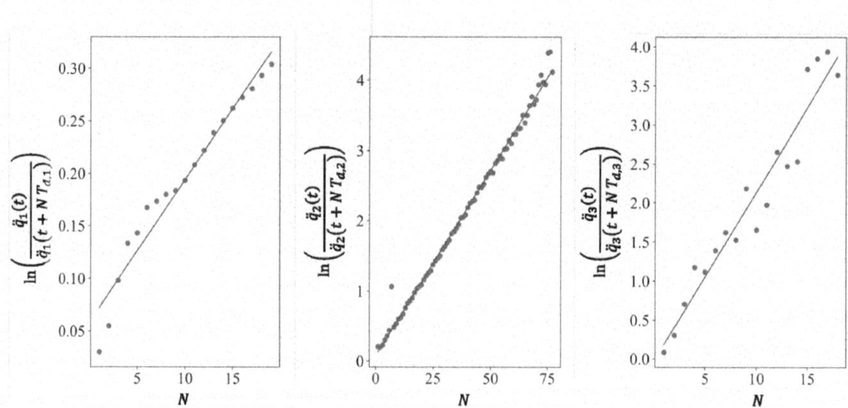

Fig. 9.8 Linear interpolation of the logarithm of two consecutive peaks in the generalized coordinate acceleration plots $\ddot{q}_i(t)$

Table 9.4 Damping factors for the HEB 100 steel simply–supported beam

ξ_1	ξ_2	ξ_3
0.002	0.008	0.034

9.8   Structural Health Monitoring

9.8.1   Overview of the Problem

Vibration tests on civil engineering structures have been carried on for many years now. It has become quite clear over the years that environmental parameters affect the dynamic behavior of a structure. For instance, the elasticity modulus of reinforced concrete decreases with increasing temperature and with high humidity. Since damage detection is one of the main aims of vibration monitoring, the basic premise that holds is that loss of stiffness observed in time shifts the frequencies to the low end of the spectrum. However, changes due to damage can be completely masked by changes due to predictable environmental conditions as for instance the changes of the seasons. The main problem that remains when analyzing vibration measurements as a tool for structural health monitoring (SHM) of infrastructure components is thus separating abnormal changes from normal changes in the dynamic response. Normal changes are caused by conditions such as temperature, humidity, wind and ground motions, or other reasons, while abnormal changes are caused by a loss of stiffness in the bridge due to accumulated damage. It is clear that the normal changes should not trigger a false alarm in the monitoring system, whereas the abnormal changes may be critical for structural safety purposes.

As an example of what SHM entails, the Z24 bridge in Switzerland which was monitored for one year before it was artificially damaged [7]. Black-box models were then built from the undamaged bridge data by applying system identification techniques. These models described the variations of eigenfrequencies as a function of temperature and were used for gauging new incoming data. If a natural frequency exceeds certain confidence intervals established by the model, then it is probable that another cause, perhaps damage, drives the natural frequency variations. More specifically, an automatic modal analysis (AMA) procedure, based on stochastic subspace identification, was proposed to extract the modal parameters from stabilization diagrams without any user interaction. By carefully inspecting the available data, the physical phenomenon behind the typical bilinear relation between frequency and temperature was recovered. Due to the relatively large amount of recorded data, a more detailed data analysis was possible as compared to the classical statistical regression analysis. A unique data set could be used to validate the proposed method, in the sense that measurements were available year-round, and the bridge could be artificially damaged at the end of the monitoring period. It was shown that it is indeed possible to filter out the environmental variations and thus detect true damage.

9.8.2   Damage Detection in Structures

Over the last few decades, many damage detection techniques have been proposed to exploit changes in the modal parameters (primarily natural frequencies and modal shapes) of a structure over time and thus to help identify the extent and location of damage. These techniques typically determine the baseline parameters through acquisition of test data based on forced or ambient vibrations. These are often verified by numerical models based on actual blueprints of the structure in question. Detection of structural damage is based on the assumption that it will be possible to see changes in the measured vibration data with time. Most of these detection techniques, however, overlook the influence of some categories of environmentally induced loads on the modal parameters. More specifically, temperature effects (often accompanied by humidity) induce changes in the overall stiffness of the structure but may also affect the boundary conditions and the way design loads are distributed across the structure. If the moving loads traversing a bridge are substantial, then the eigen-properties will change and become time dependent for as long as the loads remain on the bridge, only to return to their original values once the load have passed through. In the case of excessive humidity, concrete bridges absorb moisture, which changes their mass and as a result alters their natural frequencies. In fact, the changes due to environmental effects can often mask subtle structural changes caused by actual damage, such as the appearance of cracks, deterioration of the integrity of the connections, support settlement, etc.

To give an example, a linear adaptive filter was introduced that discriminates between changes of modal parameters due to temperature from those caused by structural damage by using data from the Alamosa Canyon Bridge in the state of New Mexico [8]. This filter solves the eigenvalue problem using conventional methods but is able to modify its prediction for the values of the natural frequencies of the structure using a pre-determined time-temperature profile. Thus, it becomes possible to discriminate between changes of modal parameters due to temperature from those caused by other environmental factors or by structural damage. Specifically, when the measured frequencies move outside the predicted confidence intervals, the system can provide a reliable indication that structural changes are likely caused by factors other than thermal. Changes in the natural frequencies are found to correlated linearly with temperature readings from different locations on the bridge. The filter uses spatial and temporal temperature distributions to determine changes in the first and second mode frequencies. Furthermore, it is possible to account for non-stationarity in the natural frequencies caused by environmental factors. More specifically, a linear filter with two spatially separated and two temporally separated temperature measurements reproduces the variation of the frequencies from a first data set. Based on the trained filter system, a prediction interval of the frequency for a new temperature profile is computed and the prediction performance is tested using a second data set. Finally, the system defines a confidence interval for future values of modal parameters in order to discriminate between variations caused by temperature changes and those from other sources.

It should be kept in mind that filter systems are developed for specific structures and under particular environmental conditions. Furthermore, filters have been superseded by machine learning techniques such as neural networks and it is possible to construct a dedicated, continuous data collection system that allow the filter coefficients to be more reliably updated, and thus decrease the size of the confidence intervals. Furthermore, the whole process can be placed within an artificial intelligence (AI) environment for the benefit of the practicing engineer, see Fig. 9.9.

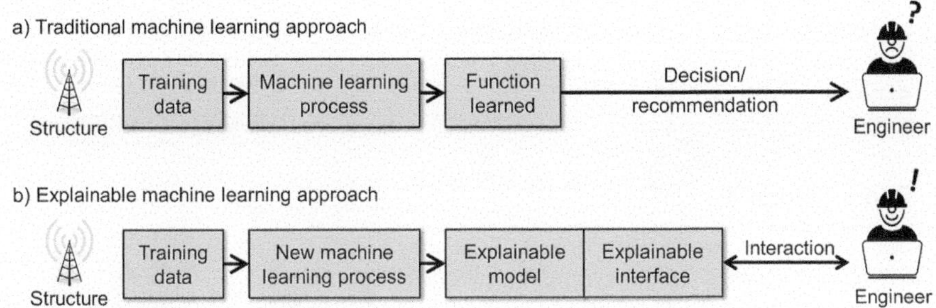

Fig. 9.9 **a** Traditional and **b** explainable machine learning approach in structural health monitoring (SHM)

9.8.3 An SHM Methodology

A minimalistic monitoring system would be quite desirable in the practicing structural engineering community, whereby two or three acceleration sensors with remote (i.e., wireless) connection to a central unit that would remain active at all times and transmit data in the form of time histories at intermittent time intervals. This data transmission could take place when certain pre-programmed markers, which may involve kinematic and/or stress variables, are exceeded. It is, of course, necessary that the monitoring system be calibrated rather carefully. There are two complementary ways of doing this: (a) By developing mechanical models for modal analysis of the structure and (b) by training neural networks to comprehend when a given signal cannot be assumed to fall within pre-established confidence limits. Within the former way, we again distinguish two possible paths which are again complimentary: (a) Finite element method (FEM) models which are detailed but time consuming and (b) simpler analytical models that are extremely efficient in terms of computing resources. This is important since the remote sensors will have limited computation capacity by design. Of course, structural models will have to be calibrated with response data coming from the structure in question under controlled loading scenarios. Once a robust modal analysis is established, it is possible to dispense with the structural models and replace them by trained neural networks. Figure 9.10 is a schematic representation of the proposed process.

9.8.4 The Mathematical Background

A standard choice for the proposed monitoring system architecture is a linear filter that simply creates a linear one-to-one mapping on the input and output data stream pairs. Explicit calculation of the filter coefficients can be done using simple matrix calculations and it is possible to do future modification of these coefficients using adaptive, least-mean-squares-error minimization. The filter operates in two modes, namely training and prediction. We note that although the linear filter is adequate for SHM purposes, there are more advanced

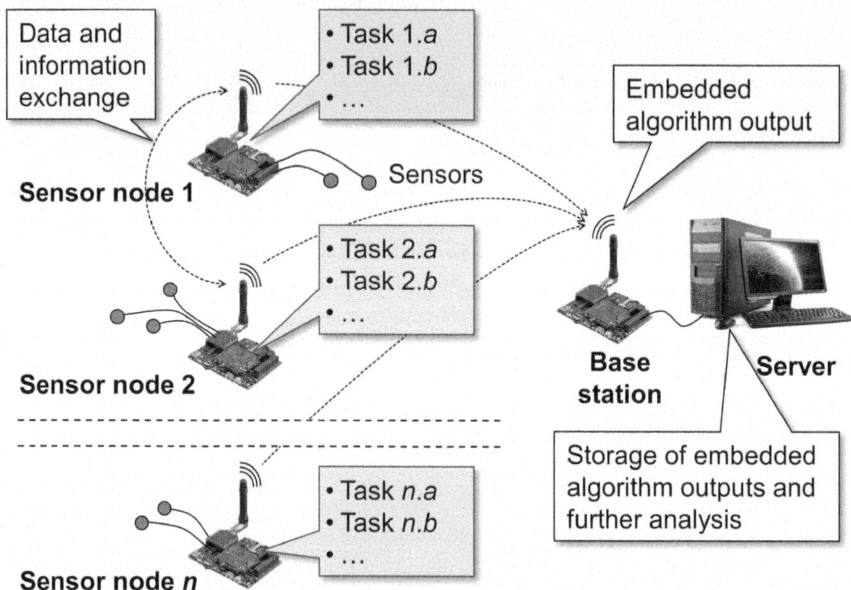

Fig. 9.10 **a** Traditional and **b** explainable machine learning approach in structural health monitoring (SHM)

systems such as neural networks that can replace them. In what follows, the temperature problem is described, although a similar venue is possible for traffic loads, ground-induced vibrations, and wing loads.

At first, we have the training regime that is described in terms of data streams comprising temperature variations for the structure (bridge or pylon) that are either functions of time or functions of space. The architecture of the linear filter takes a subset of these temporal and spatial temperature profiles as input and delivers a single output that represents the estimated, or predicted, fundamental frequency of vibration. We note that the same procedure can be repeated for the second mode frequency, although accuracy is lost at the higher frequency range. In this sense, the filter can be viewed as a multiple linear regression model but is more commonly termed a predictor or an estimator. Next, determining the appropriate subset of the available temperature profiles is termed the variable selection problem. Finally, the Least-Mean-Squares (LMS) error minimization technique is used to estimate the coefficients of the predictor. More specifically, the filter models the relationship between the selected structural temperature readings x, which is a column vector with r entries, and its measured frequency y at that temperature profile, by a linear function $y = w_0 + x^T w + \epsilon$, where w is a column vector of coefficients that weighs each temperature input, w_0 is the offset (or bias) term and ϵ is the filter error. Suppose now that n observations are available and let $x(i)$ and $y(i)$ denote the ith input/output pairs. The previous equation can be written in matrix form, simplified by absorbing the bias in the filter coefficients as $y = Xw + \epsilon$. LMS error minimization is now used for estimating the filter coefficients. This is done by computing values for the filter coefficient vector that minimizes the expected value of the square of the filter error created by the $i = 1, 2, \ldots n$ observations as $\min_w E\{\epsilon(i)^2\}$, where E is the expectation operator.

Once training is established, the prediction regime allows the adaptive filter to be used to predict the fundamental frequency of the structure. The predicted value is then used to discriminate between changes of this fundamental frequency caused by temperature effects versus changes caused by other environmental factors or by ageing or by damage in the structure. For example, let x_0 denote a new temperature reading vector. A point prediction $\hat{y}_0$ of the fundamental natural frequency at the new temperature profile becomes $\hat{y}_0 = x^T \hat{w}$, where $\hat{w}$ is a revised weight vector. It is not possible to expect a perfect match between the prediction and the measured modal parameters because of model incompleteness, insufficient training data sets, and uncertainties in the actual measurements. What is important, however, is that it is possible to compute a confidence interval around the point prediction $\hat{y}_0$ to account for the inherent uncertainties.

In sum, once the filter has been trained, a newly measured frequency can be compared against the confidence interval. If the fundamental frequency falls outside the confidence interval, then one possible explanation is that changes in the underlying structural characteristic are caused by damage or other effects.

9.9   Discovering the Governing Equations

A central challenge in several areas of engineering is the governing equations extraction from measured or known data. Considering the fact that physical systems of our interest are governed by differential equations of few terms we conclude on the sparsity of dynamic equations in a high-dimensional nonlinera function space. Based on that, recenlty the sparse identification of nonlinear dynamics (SINDy) algorithm has been investigated [9]. For the purpose of application of the algorithm we consider our dynamic system in its first order form given by,

$$\dot{x}(t) = f(x(t)) \tag{9.6}$$

with the vector $x(t) \in \mathcal{R}^n$ denoting the stae of a system at time t, and the function $d(x(t))$ representing the dyamic constraints that define the equations of motion. In order to determine f from data, time history $x(t)$ is collected, while $\dot{x}(t)$ is either measured or numerically approximated. Data are sampled at several discrete times $t_1, t_2, , t_m$ and arranged into two matrices:

$$X = \left\{ \begin{array}{c} x^T(t_1) \\ x^T(t_2) \\ \vdots \\ x^T(t_m) \end{array} \right\} = \begin{bmatrix} x_1(t_1) & x_2(t_1) & \cdots & x_n(t_1) \\ x_1(t_2) & x_2(t_2) & \cdots & x_n(t_2) \\ \vdots & \vdots & \ddots & \vdots \\ x_1(t_m) & x_2(t_m) & \cdots & x_n(t_m) \end{bmatrix} \tag{9.7}$$

and

$$\dot{X} = \begin{Bmatrix} \dot{x}^T(t_1) \\ \dot{x}^T(t_2) \\ \vdots \\ \dot{x}^T(t_m) \end{Bmatrix} = \begin{bmatrix} \dot{x}_1(t_1) & \dot{x}_2(t_1) & \cdots & \dot{x}_n(t_1) \\ \dot{x}_1(t_2) & \dot{x}_2(t_2) & \cdots & \dot{x}_n(t_2) \\ \vdots & \vdots & \ddots & \vdots \\ \dot{x}_1(t_m) & \dot{x}_2(t_m) & \cdots & \dot{x}_n(t_m) \end{bmatrix}. \tag{9.8}$$

Next, a library $\Theta(X)$ consisting of candidate non-linear functions of the columns of X is built. For example, $\Theta(X)$ may consist of constant, polynomial, and trigonometric terms,

$$\Theta(X) = \begin{bmatrix} | & | & | & | & & | & & | & \\ 1 & X & X^{P_2} & X^{P_3} & \cdots & \sin(X) & \cos(X) & \cdots \\ | & | & | & | & & | & & | & \end{bmatrix} \tag{9.9}$$

where higher polynomials are denoted as X^{P_j}. Each column of $\Theta(X)$ represents a canditate function for te right-hand side of equation (9.6). Since we expect that only a few of these non-linearities to be active in each row of f, we set up a sparse regression problem to determine the sparse vectors of coefficients $\Xi = [\xi_1 \, \xi_2 \, \cdots \, \xi_n]$ that determine which nonlinearities are active:

$$\dot{X} = \Theta(X)\Xi. \tag{9.10}$$

Each column ξ_k of Ξ is a sparse vector of coefficients that determines which terms are active in the right-hand side for each of the row equations $\dot{x}_k = f_k(x)$ in Eq. (9.6). After determining Ξ, a model of each row of the governing equations may be constructed as follows:

$$\dot{x}_k = f_k(x) = \Theta(x^T)\xi_k, \tag{9.11}$$

where $\Theta(x^T)$ is avector of symbolic functions of elements of x, as opposed to $\Theta(X)$, which is a data matrix. In what follows, we apply the SINDy algorithm for two well-known SDOF problems, namely that of the usual linear SDOF as well as for the case of the Duffing oscillator.

9.9.1   Discovering the SDOF Parameters and Equation

We first consider the SDOF case in order to illustrate the usage of SINDy. We transform the equation of motion to its first order form by choosing $x_1(t) = u(t)$ state function as the displacment function as well as $x_2(t) = v(t)$ state function.Therefore, we may re-write the equation as, the velocity:

$$\dot{x}_1 = x_2(t)$$

$$\dot{x}_2 = \frac{p(t)}{m} - \frac{k}{m}x_1(t) - \frac{c}{m}x_2(t) \tag{9.12}$$

We use the `sdof` entity in order to obtain the response of a SDOF oscillator under the step force function $p(t) = 1$ and we store in a variable labeled u. We also need the discrete time step of "measured" results dt. We are called to define the left hand side terms of equation (9.12). What we should define is the expected possible functions comprising library Θ. Here we will consider such library consisting of, up to a second order, polynomials,

$$\Theta(x) = [1 \quad x_1 \quad x_2 \quad x_1^2 \quad x_2^2 \quad x_1 x_2] \tag{9.13}$$

In our example $p(t)/m = 1$, $k/m = 1$ and $c/m = 0.5$. We also use only the displacement response as computed by `sdof` entity, while for the time derivative we use standard finite differences approximations. The system we are called upon to solve can be written as,

$$
\begin{bmatrix}
\dot{x}_1(t_1) & \dot{x}_2(t_1) \\
\dot{x}_1(t_2) & \dot{x}_2(t_2) \\
\vdots & \vdots \\
\dot{x}_1(t_m) & \dot{x}_2(t_m)
\end{bmatrix}
=
$$

$$
\begin{bmatrix}
1 & x_1(t_1) & x_2(t_1) & x_1^2(t_1) & x_2^2(t_1) & x_1(t_1)x_2(t_1) \\
1 & x_1(t_2) & x_2(t_2) & x_1^2(t_2) & x_2^2(t_2) & x_1(t_2)x_2(t_2) \\
\vdots & \vdots & \vdots & \vdots & \vdots & \vdots \\
1 & x_1(t_m) & x_2(t_m) & x_1^2(t_m) & x_2^2(t_m) & x_1(t_m)x_2(t_m)
\end{bmatrix}
\begin{bmatrix}
\xi_{1,1} & \xi_{1,2} \\
\xi_{2,1} & \xi_{2,2} \\
\xi_{3,1} & \xi_{3,2} \\
\xi_{4,1} & \xi_{4,2} \\
\xi_{5,1} & \xi_{5,2} \\
\xi_{6,1} & \xi_{6,2}
\end{bmatrix}
\tag{9.14}
$$

It would be instructive to mention certain details for the above formulation. We should mention that $\dot{x}_1(t_i)$ and $\dot{x}_2(t_i)$ in the left hand side matrix of Eq. (9.14) are provided by using measured data, this data refers to the given response of displacment with $\dot{x}_1$ the velocity and $\dot{x}_2$ the acceleration. Again the first matrix on the right hand side of equation (9.14) consists of selected possible functions that might be computed given the measured data. Therefore we may solve for the unkown coefficients $\xi_{i,j}$ with index i refering to the possible function and index j to the state variable. Yet, the system to be inverted is not square, and for that reason a least square solution as given by the SVD pseudo-inverse can be adopted. Finally we should point out that in order to use the some certain finite differences schemes in order to approximate derivatives of measured data we may drop out some first and last discrete values corresponding to specific initial and final time steps. These steps can be clearly seen in the code that follows below in Listing 9.4.

```groovy
import org.apache.commons.math3.linear.*
import courses.structuraldynamics.*
thesdof=new sdof(1.0,1.0,0.5)

thesdof.setRHS({1.0d} as DF)

dt=0.01
thesdof.solve(200.0,dt)

u=[]; thesdof.Disp().each{u.add(it)}

sz=u.size()
disp=[]; velc=[]; accl=[]
(1..<(sz-1)).each{
  disp.add(u[it])
  velc.add((u[it+1]-u[it-1])/(2.0*dt))
  accl.add((u[it+1]-2.0*u[it]+u[it-1])/(dt*dt))
}

functs=[]
functs.add({1.0})
functs.add({disp[it]})
functs.add({velc[it]})
functs.add({disp[it]*disp[it]})
functs.add({velc[it]*velc[it]})
functs.add({disp[it]*velc[it]})

M=disp.size()
N=functs.size()
da=new double[M][N]
db1=new double[M][1]
db2=new double[M][1]
(0..<M).each{
  i=it
  db1[i][0]=velc[i]
  db2[i][0]=accl[i]
  (0..<N).each{
    j=it
    da[i][j]=functs[j](i)
  }
}

A = new Array2DRowRealMatrix(da)
SVD = new SingularValueDecomposition(A)
solver = SVD.getSolver();

println solver.solve(new Array2DRowRealMatrix(db1))
println solver.solve(new Array2DRowRealMatrix(db2))
```

Listing 9.4 Discovering governing equations from data using SINDy

Table 9.5 Sparse coefficients of dynamics

Function	ξ_1 for $\dot{x}_1$	exact coefficient for $\dot{x}_1$	ξ_2 for $\dot{x}_2$	exact coefficient for $\dot{x}_2$
1	0	0	0.9999895833	1
x_1	0	0	-0.9999895833	-1
x_2	1	1	-0.4999989582	-0.5
x_1^2	0	0	0	0
x_2^2	0	0	0	0
$x_1 x_2$	0	0	0	0

The solution of the above example without any further iteration gave the results shown in Table 9.5.

9.9.2   Discovering the Duffing Nonlinear Oscillator Parameters and Equation

Based on results of response data given in example presented in Sect. 2.14.10 we proceed in discovering the Duffing's nonlinear SDOF equation. Assuming we have run script of Listing 2.15 with the response of the displacement stored in variable u, the discrete time step in dt and the frequency of the forcing term $\bar{\omega}$ in variable omega considering a possible nonlinear function library of fourteen terms given by,

$$\Theta(x) = [1 \quad x_1 \quad x_2 \quad x_1^2 \quad x_2^2 \quad x_1^3 \quad x_2^3 \quad x_1 x_2 \quad x_1^2 x_2 \quad x_1 x_2^2$$
$$x_1^3 x_2 \quad x_1 x_2^3 \quad \cos(\bar{\omega}t) \quad \sin(\bar{\omega}t)] \tag{9.15}$$

we use a similar structured script as the one listed in the previous Sect. 9.9.1, see Listing 9.5.

```
import org.apache.commons.math3.linear.*
sz=u.size()
disp=[]; velc=[]; accl=[]
(1..<(sz-1)).each{
    disp.add(u[it])
    velc.add((u[it+1]-u[it-1])/(2.0*dt))
    accl.add((u[it+1]-2.0*u[it]+u[it-1])/(dt*dt))
}

functs=[]
functs.add({1.0})
functs.add({disp[it]})
functs.add({velc[it]})
functs.add({disp[it]**2})
```

```
15  functs.add({velc[it]**2})
16  functs.add({disp[it]**3})
17  functs.add({velc[it]**3})
18  functs.add({disp[it]*velc[it]})
19  functs.add({disp[it]**2*velc[it]})
20  functs.add({disp[it]*velc[it]**2})
21  functs.add({disp[it]**3*velc[it]})
22  functs.add({disp[it]*velc[it]**3})
23  functs.add({cos(omega*it*dt)})
24  functs.add({sin(omega*it*dt)})
25
26
27  M=disp.size()
28  N=functs.size()
29  da=new double[M][N]
30  db1=new double[M][1]
31  db2=new double[M][1]
32  (0..<M).each{
33    i=it
34    db1[i][0]=velc[i]
35    db2[i][0]=accl[i]
36    (0..<N).each{
37      j=it
38      da[i][j]=functs[j](i)
39    }
40  }
41
42  A = new Array2DRowRealMatrix(da)
43  SVD = new SingularValueDecomposition(A)
44  solver = SVD.getSolver();
45
46  println solver.solve(new Array2DRowRealMatrix(db1))
47  println solver.solve(new Array2DRowRealMatrix(db2))
```

Listing 9.5 Discovering governing equations from data using SINDy for nonlinear Duffing's equation

The solution of the above example without any further iteration gave the results shown in Table 9.6.

9.10　Conclusions

In this chapter, the contemporary topic of Operational Modal Analysis (OMA) was presented and applied to the case of two coupled, single-story shear frames modelled as a MDOF system, whereby its dynamic properties are evaluated from its response to a known forcing function. Furthermore, the topic of Experimental Modal Analysis (EMA) was applied to a

Table 9.6 Sparse coefficients of dynamics for the Duffing's equation discovery

Function	ξ_1 for $\dot{x}_1$	Exact coefficient for $\dot{x}_1$	ξ_2 for $\dot{x}_2$	Exact coefficient for $\dot{x}_2$
1	0	0	−0.0000027579	0
x_1	0	0	−0.0004584262	0
x_2	1	1	−0.0499796934	−0.05
x_1^2	0	0	−0.000000778	0
x_2^2	0	0	0.0000001795	0
x_1^3	0	0	−0.9997830721	−1
x_2^3	0	0	−0.0000003431	0
$x_1 x_2$	0	0	0.000000955	0
$x_1^2 x_2$	0	0	−0.0000020377	0
$x_1 x_2^2$	0	0	−0.0000552119	0
$x_1^3 x_2$	0	0	−0.0000000828	0
$x_1 x_2^3$	0	0	−0.000000031	0
$\cos(\bar{\omega}t)$	0	0	7.4992604992	7.5
$\sin(\bar{\omega}t)$	0	0	−0.0749509169	0

simply supported, Euler-Bernoulli beam under impact. The first step in EMA is to extract the natural frequencies and modal shapes of this structure from its recorded dynamic response using two different methods, one defined in the time domain (PCA) and one in the frequency domain (FDD). The second step is to evaluate damping in the structure by applying the logarithmic decrement method to the free vibration part of the transient records. This completes the description of the dynamic properties of a structure from its recorded response.

In closing, it should be mentioned that an important question that arises in the field of structural dynamics is the degree to which the dynamic properties, including damping, of a structure are affected by the onset of damage. Damage is associated with the ageing of a structure and manifests itself by the presence of cracks, by the loss of support integrity, corrosion, residual thermal stresses, etc. Damage usually results in a shift of the natural frequencies to lower values, thus indicating structural degradation. It is also possible to see new harmonics appearing in the response spectrum of a structure, especially when external loads are cyclic in nature. Since damage detection impacts the built environment and its infrastructure, the techniques used in OMA are essential to the field of structural health monitoring, which in the case of damage detection, necessitates the implementation of repair and/or rehabilitation protocols.

Finally, some material is provided at the end of this chapter to illustrate the possibility of deriving the terms of a differential equation governing the dynamic response of both linear and nonlinear SDOF systems.

References

1. European Standard EN 1998-1 (2004) Eurocode 8: Design of structures for earthquake resistance—part 1: General rules, seismic actions and rules for buildings
2. Brincker R, Ventura C (2015) Introduction to operational modal analysis. Wiley
3. Zahid FB, Ong ZC, Khoo SY (2020) A review of operational modal analysis techniques for in-service modal identification. J Braz Soc Mech Sci Eng 42(8):398
4. Brandt A (2023) Noise and vibration analysis: signal analysis and experimental procedures. Wiley
5. Dadoulis GI, Manolis GD (2021) Model bridge span traversed by a heavy mass: analysis and experimental verification. Infrastructures 6(9):130
6. Inman DJ, Singh RC (1994) Engineering vibration, vol 3. Prentice Hall Englewood Cliffs, NJ
7. Peeters B, De Roeck G (2001) One-year monitoring of the z24-bridge: environmental effects versus damage events. Earthq Eng Struct Dyn 30(2):149–171
8. Sohn H, Dzwonczyk M, Straser EG, Kiremidjian AS, Law KH, Meng T (1999) An experimental study of temperature effect on modal parameters of the alamosa canyon bridge. Earthq Eng Struct Dyn 28(8):879–897
9. Brunton SL, Proctor JL, Nathan Kutz J (2016) Discovering governing equations from data by sparse identification of nonlinear dynamical systems. Proc Natl Acad Sci 113(15):3932–3937

10.1 Response Spectra

Response spectra are perhaps the most popular method used nowadays in earthquake engineering for computing the kinematic and stress fields imparted to a structure due to seismically induced ground motions [2]. It is used in the design stage of a structure and in post-evaluation of the state of integrity of that structure after it has been subjected to an earthquake, or more realistically, to a series of earthquake events [3]. Note that since the use of response spectra requires knowledge of the natural periods (eigenvalues) of a structure, it pre-supposes a modal analysis stage.

10.1.1 Designing for Earthquake-Induced Loads

The field of earthquake engineering, as we know it today, developed right after the El Centro 1940 earthquake in California that was probably the first earthquake for which reliable ground motion recordings became available. One ground motion recording became routine due to the advancement of sensors and their proliferation in earthquake-prone regions such as California, Japan, Southern Europe, at countries along the Pacific rim, etc., the development of response spectra become routine. Specifically, a response spectrum plots the maximum kinematic (displacements, velocities, accelerations) response or maximum forces (base shear) that develops in all single-degree-of freedom (SDOF) oscillators which are described by their natural period (or natural frequency) and their damping ratio. Response spectra axes are both regular as well as logarithmic, and the spectra often appear as tripartite, all in one plots of spectral accelerations (S_A), spectral velocities (S_V) and spectral displacements (S_D). This is possible because for low levels of damping (see the equation of motion of SDOF systems under free vibrations), the following relation (approximately) holds true:

© The Author(s), under exclusive license to Springer Nature Switzerland AG 2026   237
G. Manolis and C. Panagiotopoulos, *Vibrations of Structural Systems*, Synthesis Lectures on Mechanical Engineering, https://doi.org/10.1007/978-3-032-12279-7_10

$S_A = \omega S_V = \omega^2 S_D$. Note that most conventional, frame-type buildings, of either steel or concrete construction, exhibit a rather narrow range for their first fundamental frequency, which falls in the 0.3–3.0 Hz range. Similarly, there is an equally narrow range for their damping ratio, usually between 2 and 5%.

Prior to the use of response spectra, earthquake-induced loads were treated as extra lateral forces to be superimposed to the usual gravity loads for design, see Fig. 10.1a. For a simple

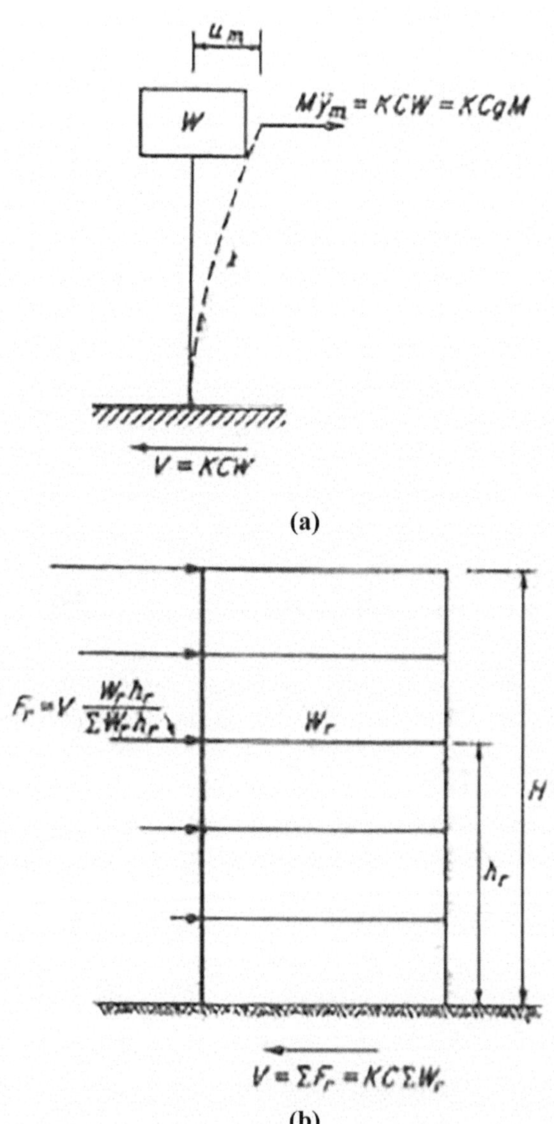

Fig. 10.1 a SDOF representation of a multi-story frame under a seismically induced horizontal force and **b** distribution of the base shear force V along the height of the frame

SDOF representation of a multi-story frame, the base shear is equal and opposite to the lateral force that develops the top (the location of the concentrated mass) because of ground motion. Specifically, in early earthquake design codes, this lateral force V was equal to a certain percentage of the structure's weight W as follows:

$$V = (ZIKSC)W$$

The dimensionless seismic coefficients in parenthesis comprise the seismic zone coefficient Z (ranging from 0 to 4), the importance coefficient I (ranging from 1.0 to 1.5), the type of structural system K (ranging from 1.0 to 2.5), the type of supporting soil S (ranging from 1.0 to 1.5), and finally a coefficient encompassing the basic dynamic property of the structure as $C = 1/15\sqrt{T}$, T being the fundamental period of the structure (in sec). Obviously, coefficient Z is read from a seismic intensity chart addressing the specific region in question, see Fig. 10.2. For the remaining coefficients, a high value of I is reserved for hospitals and power plants, a high value of K for water towers, and a high value of S for loose

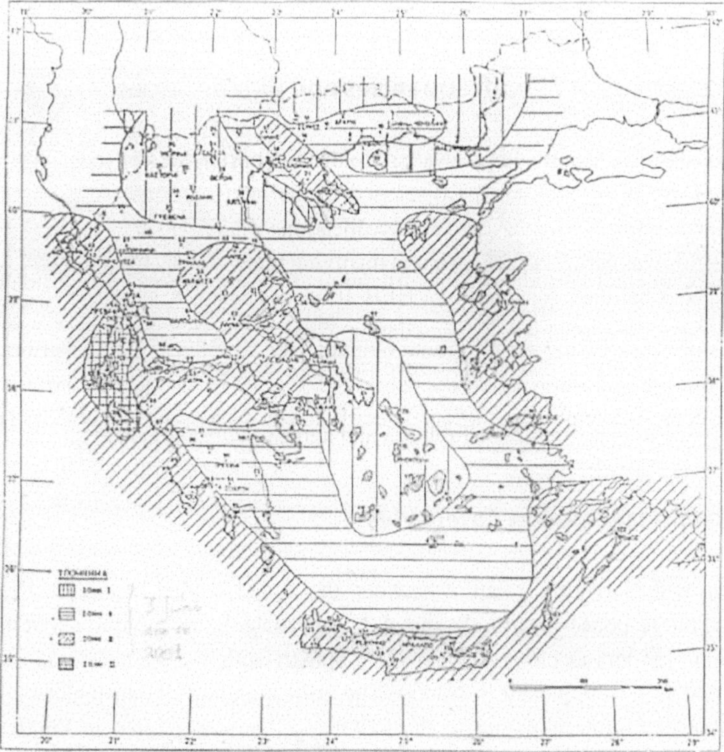

Fig. 10.2 The prior to 2001 Eurocode [3] division of the Greek domain into 4 seismic zones, with Zone 1 being aseismic and Zone 4 giving the largest ground acceleration value of 0.3 g

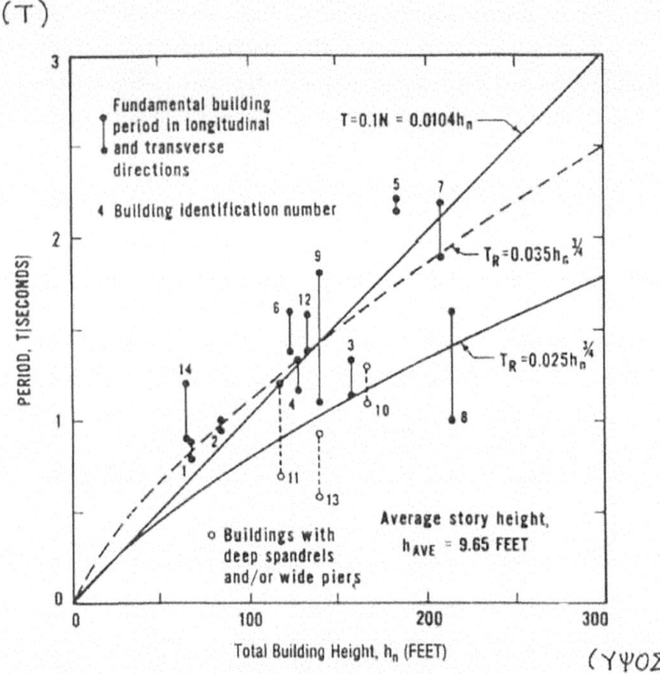

Fig. 10.3 The fundamental natural period of typical reinforced concrete frames as a function of their total height as determined by various approximate formulas

soil. The fundamental period T of the structure can be approximated as a function of the building's height, see Fig. 10.3. Otherwise, a dynamic analysis of the structure is warranted. For multi-story frames, the seismic force (equal to the base shear V) is distributed at each floor level in a linear fashion across the height, see Fig. 10.1b. For ordinary frame structures with a high degree of symmetry, the base shear V acts in each of the two principal directions.

10.1.2 Response Spectra Categories

At first, we will clarify the relation between the dynamic load factor (DLF) defined in Chap. 2 and the response spectra. As shown in Fig. 10.4a, b, the input to a SDOF oscillator is a harmonic ground acceleration $\ddot{y}(t) = \ddot{y}_{S_0} \sin \Omega t$, with $\ddot{y}_{S_0}$ the acceleration amplitude in (m/s^2) and Ω the frequency in (rad/s). The corresponding dimensionless DLF for the maximum accelerations imparted to the SDOF oscillator is plotted in Fig. 10.4c and is parametric in the dimensionless ratio (ω/Ω), where ω in is the natural frequency of a particular SDOF of interest. For a high damping ratio of 20% we observe that in the region of resonance, where $\omega \to \Omega$, the maximum acceleration experienced by the SDOF is 2.5

Fig. 10.4 Forcing function in the equivalent form of either harmonic ground **a** acceleration or **b** displacement and the resulting **c** acceleration dynamic load factor (DLF), **d** the relative displacement spectrum S_D and **e** the absolute acceleration spectrum S_A, all for a damping ratio of 20%

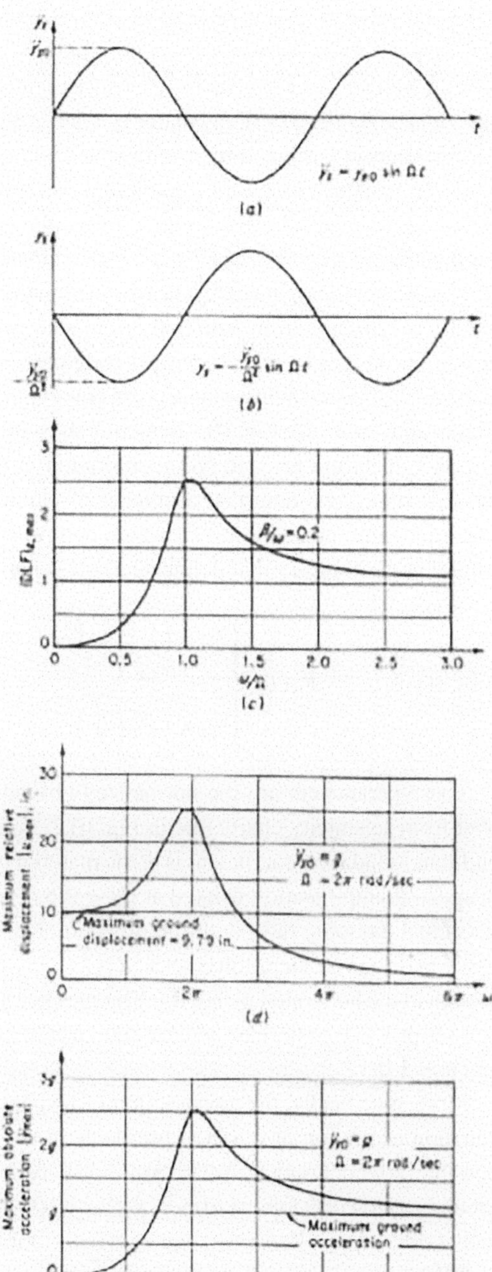

times the input ground acceleration. Both response spectra that appear in Fig. 10.4d, e are defined for a fixed ground motion amplitude $\ddot{y}_{S_0} = g = 9.84$ (m/s^2) and a fixed frequency $\omega = 2\pi$ (rad/s). As before, in the resonance region where $\omega = 2\pi$, the maximum acceleration experienced by the SDOF oscillator is again 2.5 g. Thus, the dynamic load factor is a more versatile concept as it is dimensionless and addresses a specific category of dynamic loads, while the response spectrum is particular to a given load.

By categorizing response spectra, we distinguish (a) response spectra for a given, recorded earthquake event as shown in Fig. 10.5; (b) smoothen response spectra for a given geographical region encompassing the contribution of many recorded seismic motions and/or response spectra constructed from artificially generated ground motions, all of which can be used for design purposes (see Fig. 10.6); (c) code-prescribed design spectra used for earthquake-resistant design and addressing a given country or a given federation of countries such as the European Union., see Fig. 10.7 depicting the national Greek earthquake design code (EAK) response spectrum prior to its incorporation in the EU codes [3] in 2001. In reference to this last figure, we note that a number of parameters enter the expression for the design spectrum symbolized as $R_d(T)$ which comprises three regions, depending on the marker natural periods T_1, T_2 that delineate a soft versus a stiff SDOF system, as follows:

$$R_d(T) = \begin{cases} \gamma_I A \left(1 + \frac{T}{T_1}\left(\frac{\eta\theta\beta_0}{q} - 1\right)\right) & \text{for } 0 \leq T \leq T_1 \\ \gamma_I A \frac{\eta\theta\beta_0}{q} & \text{for } T_1 \leq T \leq T_2 \\ \gamma_I A \frac{\eta\theta\beta_0}{q}\left(\frac{T_2}{T}\right)^{2/3} & \text{for } T_2 \leq T \end{cases} \tag{10.1}$$

These parameters are the normalized ground acceleration A as a fraction of g that is read from seismicity charts like in Fig. 10.2, the building importance coefficient is γ_I, the building foundation coefficient is θ, the material behavior coefficient is q and the coefficient of spectral amplification is fixed at $\beta_0 = 2.5$. The spectrum is set for a $\zeta = 5\%$ ratio, so if the actual damping ratio is different, coefficient $\eta = \sqrt{7/(2+\zeta)}$ is introduced. Finally, a baseline design acceleration is set as $R_d(T)/A\gamma_I \geq 0.25$. This design spectrum is for ground motion in the horizontal plane. If it becomes necessary to design for vertical ground motions (subscript V), as in the case of near-source earthquakes, then $A_V = 0.7A, q_V = 0.7q \geq 1.0$ and $\theta = 1.0$.

More specifically, the construction of response spectra hinges on the processing of the equation of motion of a SDOF oscillator in the case of ground motion. Two variants are possible, after defining the relative motion between the top $x(t)$ and the base $x_S(t)$ where the ground motion develops as $u(t) = x(t) - x_S(t)$, a relation that holds true for both velocities and accelerations. These two equations of motion are completely equivalent and are given as follows:

$$m\ddot{x}(t) + c\dot{x}(t) + kx(t) = kx_S(t) + c\dot{x}_S(t) \tag{10.2}$$

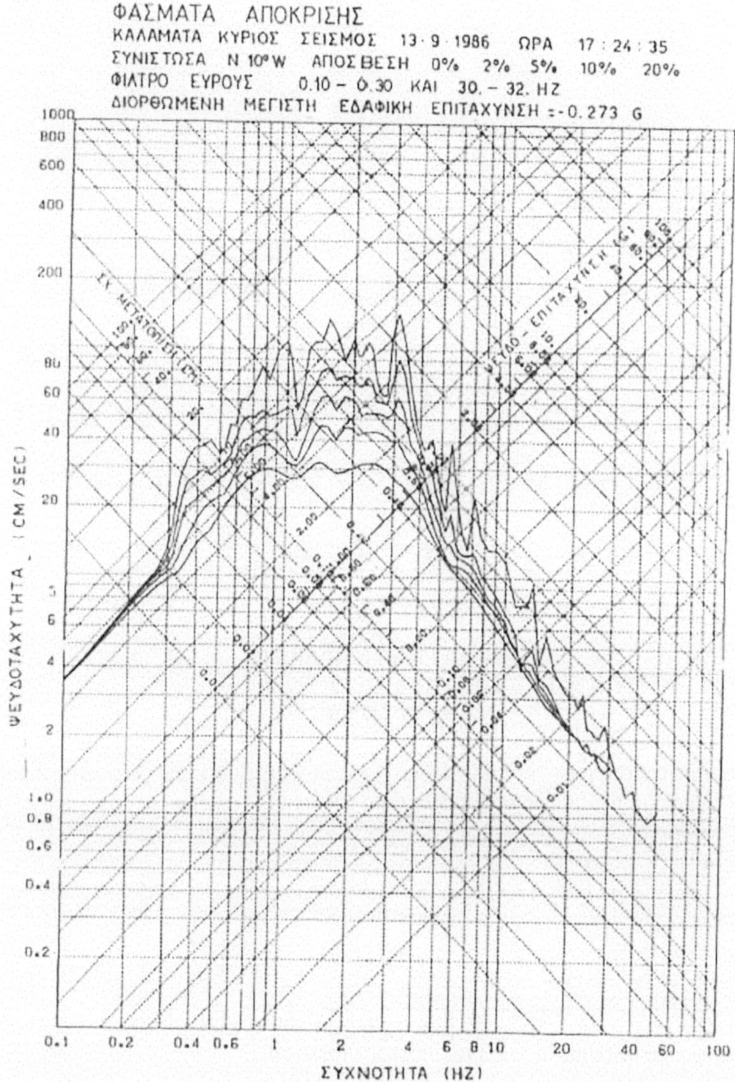

Fig. 10.5 The tripartite response spectrum as a function of frequency for the Kalamata, Greece, main shock of 13 September 1986 as developed by the Institute of Engineering Seismology and Earthquake Engineering (ITSAK) in Thessaloniki, Greece. https://www.itsak.gr/en for a reference damping ratio of 5%

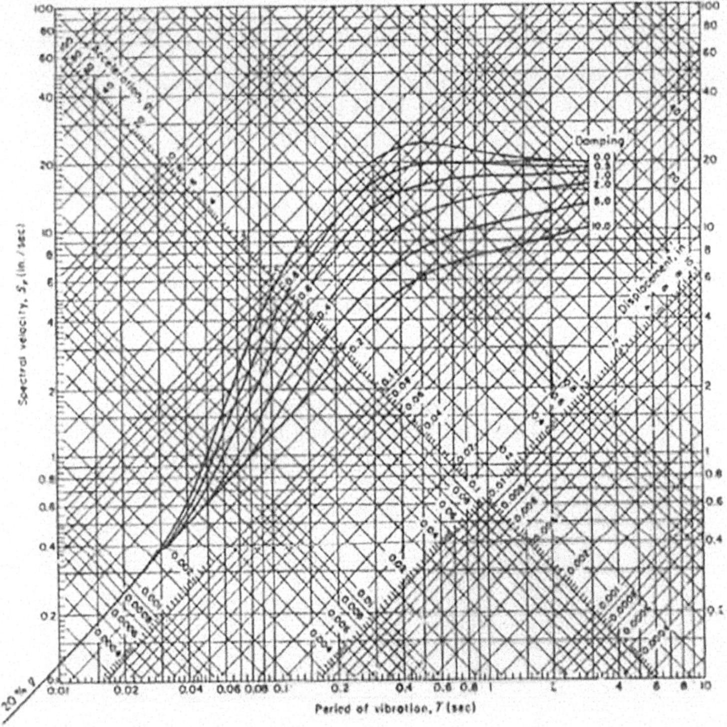

Fig. 10.6 A tripartite response spectrum defined in terms of period T and parametric in the damping ratio, as developed from artificial ground motions

and

$$m\ddot{u}(t) + c\dot{u}(t) + ku(t) = -m\ddot{x}_S(t) \tag{10.3}$$

As before, m, c, k are the SDOF oscillator mass, damper and stiffness. It is simple to see that in the absence of damping and under free vibrations,

$$m\ddot{x}(t) + ku(t) = 0 \rightarrow |\ddot{x}(t)| = \sqrt{k/m}\,|y(t)|,$$

so that the spectral relation is $S_A = {}^2 S_D$, provided the absolute accelerations and the relative displacements are involved. Note that in the presence of damping, this relation becomes approximate and that the spectral velocities are the relative ones.

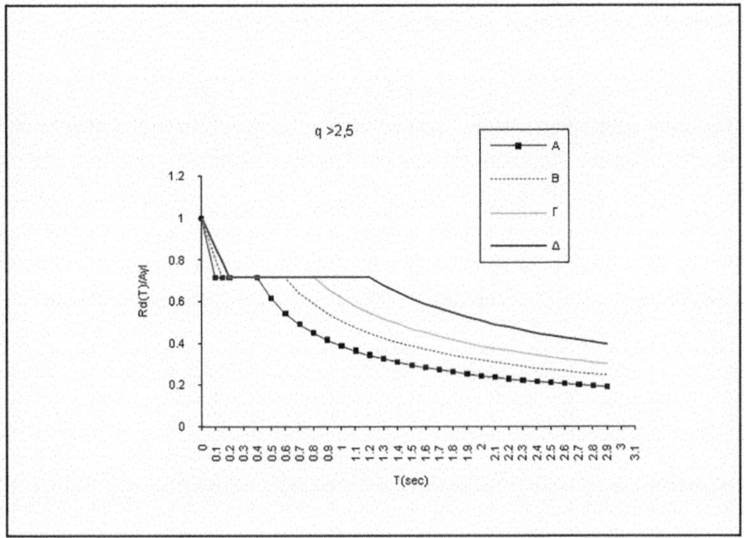

Fig. 10.7 The design spectrum $R_d(T)$ specified by the national Greek Earthquake Code (EAK) prior to its inclusion in the EU-wide earthquake design code [3]. Note that markers A, B, Γ, Δ refer to soil of progressively poorer quality, starting from competent rock to alluvial soil

10.2   Class `specalc`

Description:
The `specalc` appearing in the software package `courses.structuraldynamics` implements standard method for calculating repsonse spectra for a given ground motion.

Instance

- specalc(double[] acc, double z, double dt)

Methods

- void getSpectrums(double[] periods)
- void computeMotion()
- double max(ArrayList<Double> arr)
- double min(ArrayList<Double> arr).

10.2.1 Numerical Example of Response Spectrum

Listing 10.1 below describes how to construct response spectra for a giving earthquake record (real or artificially generated). Here we consider a recorded motion, shown in Fig. 10.8, during the earthquake at Lefkada of Greece on August of 2003. We use the `specalc` class to calculate response spectrum (method `getSpectrums`) over a specific periods range from 0.2 to 3.0 s and for some specific damping ratio. Then we make the plot of S_V shown in Fig. 10.9. We then use the same class to calculate the displacement and velocity records as result from integration of acceleration. The resulted displacement record is shown in Fig. 10.10.

```
// acceleration record can be downloaded from:
// https://symplegma.civil.auth.gr/docs/LEF10302-L.
   txt
import courses.structuraldynamics.specalc

plot()
thePlot.setAutoColor(true)
dt=0.01
a=[]
path2accel=\.. path to record ..\

new File(path2accel).eachLine { line ->
    a.add(Double.parseDouble(line))
}
plot(dt,a)
periods=linspace(0.2,3.0,1000)
z=1.0/100.0

gmp = new  specalc(a as double[],z,dt)

gmp.getSpectrums(periods)

Specplot = new PlotFrame()
Specplot.setAutoColor(true)
Specplot.addFunction(new plotfunction(periods,gmp.SV)
    )
Specplot.show()

gmp.computeMotion()
plot()
plot(dt,gmp.dsp)
```

Listing 10.1 Response Spectra generation

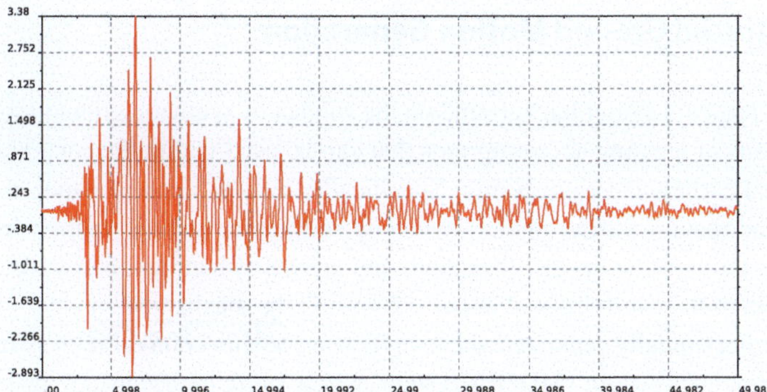

Fig. 10.8 Acceleration record of Lefkada's earthquake on August 2003, https://www.itsak.gr/en

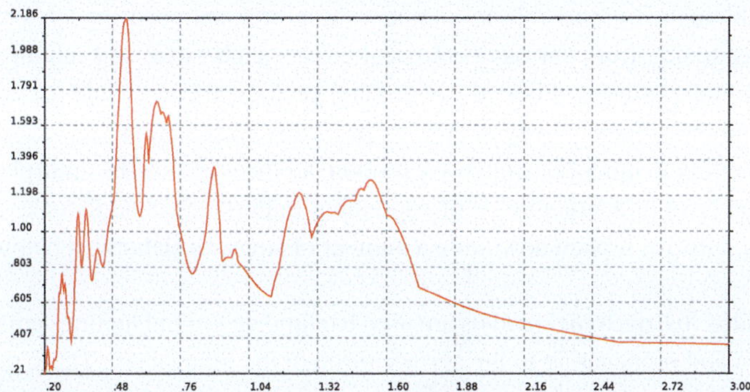

Fig. 10.9 Velocity spectrum for the record of Lefkada's earthquake on August 2003

Fig. 10.10 Displacement for the recorded Lefkada's earthquake resulting after double integration in time of the respective acceleration record

10.3　Artificial Ground Motion Generation

This section presents a brief background on the subject of synthetic ground motion generation as a means of creating accelerograms that can be used within the context of structural dynamics and earthquake engineering. Specifically, we start from a standard seismological model (the specific barrier model), whereby earthquake induced motion from the fault plane travels through the upper lithosphere and arrives at bedrock. Next, we consider the filtering effect of the overlying soil deposits by using the mechanical model of uniform soil layers under horizontally polarized shear (SH) waves. This configuration is modeled as a multiple degree-of freedom (MDOF) system, whose complex frequency response matrix is transformed to the time domain and convoluted with artificial time histories (accelerations/velocities/displacements) recorded at bedrock. Thus, motions at the free surface are produced. Although it is possible to augment the aforementioned model with additional complications arising from various geological effects such as (a) soil inhomogeneity and anisotropy, (b) the presence of buried cavities and geological cracks, and (c) surface topography, these modifications fall beyond the scope of this presentation.

Thus, the ability to model seismic wave motion, all the way from its fault-plane source to the foundation level of a structure, is of paramount importance in earthquake engineering. The modeling effort is broken down into a number of steps, as elaborated below and shown in Fig. 10.11.

The so-called 'barrier-type' models are used to simulate ground motions emanating from a fault system and propagating to the surface through the lithosphere. These models incorporate the simulation of fault geometry and the relation between seismic moment and stress drop to estimate the number of cracks appearing in the fault. The simulated peak ground

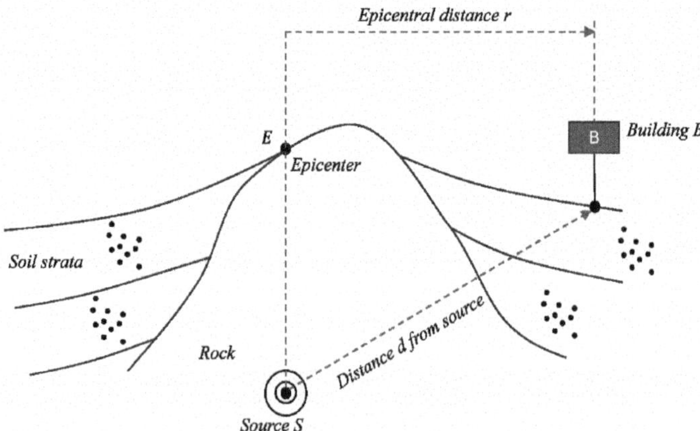

Fig. 10.11 Transmission of earthquake-induced waves from source to surface

accelerations (PGA) are then calibrated against strong motion data. In terms of applications, we mention here the works [4, 5], who constructed acceleration-source spectra that show good agreement with recorded data, except at high frequencies where spectral amplitudes are over-predicted. Since the variability in the simulated motions is generally less than that manifested in recorded motions, randomness is added to key parameters in these models [6], a process well understood since the 1960s [7]. Over the last thirty or so years, many researchers conducted studies for establishing ground seismicity models for sites around the world, including not only seismicity resulting from tectonic plate movement, but also seismicity because of volcanic action.

Local site effects essentially deal with the influence of the upper soil layers, which may reach few hundred meters in depth [8], on seismic signals that are either computed or measured at bedrock. Again, many researchers have examined the seismic response, predominantly of two-dimensional soil deposits with or without lateral inhomogeneity, by breaking up the problem in its anti-plane (SH) and in-plane (P and S) wave fields and using spectral methods. Also, attempts have been made to bridge the gap between earthquake source mechanism, path attenuation and local site conditions using integrated approaches.

Wave propagation techniques for elastic waves generated in solid media are an indispensable computational tools for site response. Usually, advanced numerical methods are used for this purpose, such as finite differences, hybrid boundary element - discrete wave number methods and finite elements. The complexity that needs to be accounted for comes primarily from the presence of surface topography, e.g., valleys and canyons, see for instance [9]. An efficient, yet approximate, technique results through the use of complex transfer functions, defined as the ratio of the Fourier transfer of accelerations at the free-surface to the Fourier transfer of input accelerations defined at bedrock. These transfer functions correspond to systems of coupled oscillators, whose properties comprise a deterministic plus a random part, mirroring the composition of actual soil layers. For instance, these techniques were used by [BAR] for constructing artificial ground motion time-histories in soft soil deposits.

10.3.1 Generation of Synthetic Ground Motions

There are, of course, numerous applications for synthetic ground motions in earthquake engineering, given the paucity of detailed recorded motions at particular geographic locations. It should be noted however that in many instances response spectra (either processed from actual motion records, or synthetic, or code-prescribed) are used in many studies involving dynamic response of structures in lieu of either actual or synthetic ground motions. Furthermore, design response spectra can be used to scale the magnitude of synthetic ground motions, although some empiricism is involved in this process since the scaling factor itself is a function of frequency. Among numerous applications of artificially constructed accelerograms we mention their use in dam analysis under 1D and 2D conditions [10], in Newmark's displacement method for assessing earthquake-induced landslides [11] and in long structures with multiple supports [12].

Ground motion signals can be further processed for conducting inverse-type analyses, structural response computations, etc. For instance, [13] processed a collection of over one-thousand ground motion records from Taiwan's SMI program to verify models that were developed for various Taiwanese earthquake-prone regions. Also, [14] calculated site coefficients and constructed displacement response spectra for California sites taking into account non-linear soil behavior and validated these computations with data from the Northridge 1994 earthquake. Based on extensive ground motion simulations, [15] constructed uniform hazard spectra for Hong Kong by using probabalistic hazard assessment methods. Finally, using stochastic neural network techniques, [16] generated multiple accelerograms for regions for which a single response spectrum is available.

The most prominent feature of the upper soil strata is layering, which modifies bedrock signals because it acts as a filter, amplifying motion in the frequency range near the soli layers' eigenfrequencies and de-amplifying motions past that range. Also, surface topography serves to differentiate seismically induced motion across the free surface of the ground, because of wave scattering phenomena. This is quite the same to what happens when upward moving seismic waves encounter tunnels, cavities and other buried structure [9]. In addition, it is important to model anisotropy and inhomogeneity in the soil, since this type of material structure leads to continuous scattering of the traveling seismic waves. In general, there has been a surge of work in the last three decades regarding numerical work on elastic wave propagation in various types of naturally occurring media. Summarizing, we mention the following cases: (a) Anisotropic Media: Mostly boundary element methods have been used for steady-state vibrations of 2D and 3D surface foundations resting on orthotropic or general anisotropic sub-bases, for transient vibrations of underground structures in 2D orthotropic media and for the time-harmonic response of 2D basins and 3D cavities in anisotropic media, see for instance [17]. (b) Inhomogeneous Media: A number of configurations in the otherwise homogeneous material come under the banner of inhomogeneity. For instance, the presence of discontinuities of various types such as geological cracks are a form of inhomogeneity. Also, functionally graded materials are a special class of inhomogeneous media, mostly man-made but possibly naturally-occurring [18]. Their dynamic behavior has been the focus of much recent work on developing mechanical models for wave motion in the general inhomogeneous continuum [19]. In closing, there is a relatively small amount of work dealing with the combined inhomogeneous-anisotropic material.

The practical side of synthesizing theoretical seismograms for design purposes that account for important, but often neglected effects such as inhomogeneity and anisotropy in the soil material, should be stressed here. The presence of large geological discontinuities (cracks, cavities) in the upper soil strata and their dynamic interaction must also be studied in conjunction with the previous effects. If these effects are included, then parametric studies conducted for realistic geological deposits are directly applicable to the important question of seismic protection of structures and infrastructure.

10.3.2 The Kanai-Tajimi Filters

Given the importance of response spectra for structural analysis and design purposes, it is essential that ground motion records are readily available. This can be done by searching seismic databases, but it is most convenient to be able to produce synthetic seismic motions upon demand, which require as input information on the surrounding geological profile and a measure of the magnitude of the expected earthquake. This type of software interfaces perfectly with entity `specalc`, of Sect. 10.2.1, for continuing on with the construction of earthquake spectra.

An elementary, yet basic model is that of horizontally polarized shear (SH) waves. As can be seen in Fig. 10.12, first we have a seismic source at large depth which emits a white-noise signal, i.e., one which remains constant over the frequency range of interest. Next, the signal passes through three filters, the High-Pass (H-P), the Kanaji-Tajimi (K-T) and the Low-Pass (L-P) filters, which respectively model a stiff halfspace and the overlying soil deposits, ranging from firm to loose [20].

The power spectral density function (PSDF) for ground surface acceleration S_a is as follows:

Fig. 10.12 Seismic motion simulation on ground considering a triple filter system

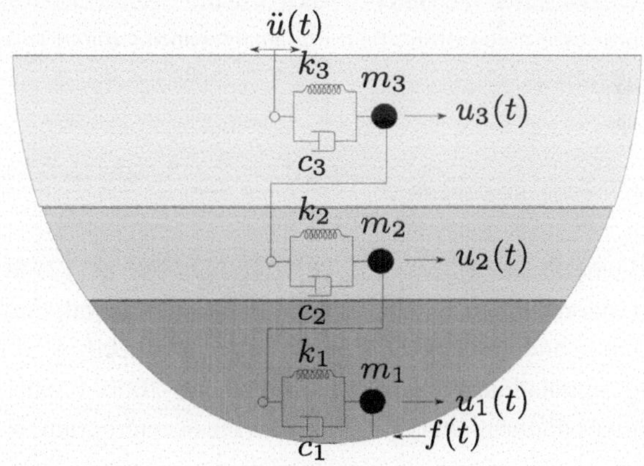

$$S_a(\Omega) = \underbrace{\frac{(\Omega/\omega_1)^4}{\left(1 - (\Omega/\omega_1)^2\right)^2 + 4\zeta_1^2(\Omega/\omega_1)^2}}_{\substack{\text{high-pass} \\ \text{filter}}}$$

$$\times \underbrace{\frac{1 + 4\zeta_2(\Omega/\omega_2)^2}{\left(1 - (\Omega/\omega_2)^2\right)^2 + 4\zeta_2^2(\Omega/\omega_2)^2}}_{\text{Kanai-Tajimi filter}}$$

$$\times \underbrace{\frac{1}{\left(1 - (\Omega/\omega_3)^2\right)^2 + 4\zeta_3^2(\Omega/\omega_3)^2}}_{\text{low-pass filter}} \times S_0. \tag{10.4}$$

The above equation represents the triple filtering of the white noise power spectrum density $2\pi S_0$ from the source through H-P (index 1), K-T (index-2) and L-P (index 3) filters. The corresponding equations of dynamic equilibrium for the individual filters assume the following normalized form:

$$\ddot{u}_1(t) + 2\zeta_1\omega_1\dot{u}_1(t) + \omega_1^2 u_1(t) = f(t), \tag{10.5}$$

$$\ddot{u}_2(t) + 2\zeta_2\omega_2\left(\dot{u}_2(t) - \dot{u}_1(t)\right) + \omega_1^2\left(u_2(t) - u_1(t)\right) = 0, \tag{10.6}$$

$$\ddot{u}_3(t) + 2\zeta_3\omega_3\left(\dot{u}_3(t) - \dot{u}_2(t)\right) + \omega_3^2\left(u_3(t) - u_2(t)\right) = 0. \tag{10.7}$$

In the numerical implementation it is possible for the user to choose which of these filters will be active. The next step is to use the free surface PSDF S_a from Eq. (10.4) as the startimg point for generation of ground accelerations under (pseudo) non-stationary conditions [21] as follows:

$$\ddot{u}(t) = \sum_{n=1}^{N} A_n \sin\left(\Omega_n t + \phi_n\right) \psi(t). \tag{10.8}$$

In the above, $A_n = \sqrt{2S_a(\Omega_n)\Delta\Omega}$, $\Omega_n = (n - 1/2)\Delta\Omega$, $\Delta\Omega = (\Omega_\beta - \Omega_\alpha)/N$ and ϕ_n is the random phase angle defined in the $[0, 2\pi]$ interval. Furthemore, Ω_β and Ω_α respectively are upper and lower bounds in the frequency spectrum of interest. Also, index n is the number of simulations, values in the range 500–1000. Finally, the envelope function $\psi(t)$, which gives non-stationary properties to the accelerogram, is given below as:

$$\psi(t) = \begin{cases} \left(\frac{t}{t_1}\right)^2, & 0 < t < t_1 \\ 1, & t_1 \leq t \leq t_2 \\ \exp\left(-b(t - t_2)\right), & t_2 < t \end{cases} \tag{10.9}$$

where time markers t_1 and t_2 range from 10 to 90% of the total time (duration) of motion T, and $b = 2/(t_2 - t_1)$ is a constant (in 1/s).

10.4   Class `synthaccel`

Description: The `synthaccel` appearing in the software package `courses.structuraldynamics` implements the genration of synthetic accelerograms as described in Sect. 10.3.2.

Instance

- synthaccel()
- synthaccel(double So, double om1, double xsi1, double om2, double xsi2, double om3, double xsi3)

Attributes

Class `synthaccel` attributes include the double `So` variable appearing in Eq. (10.4), variables (as double) om_i and xsi_i which for $i = 1$ correspond to frequency and damping of the high-pass filter, for $i = 2$ for that of the Kanai-Tajimi filter and finally for $i = 3$ for the low-pass filter. To be mentioned that boolean variables `H_P`, `K_T` and `L_P`, with default valuew being `true`, are responsible for considering or not the high-pass, the Kanai-Tajimi and low-pass filters. Furthermore, integer `N` as the number of terms appeared in Eq. (10.8), while double parameters `Om_a` and `Om_b` stand for Ω_α and Ω_β. Finally the total duration of the record is considered through the double variable `duration` with t_1 and t_2 values of the envelope function of Eq. (10.9) given by the double valued parameters `t1` and `t2`.

Methods

- void Calc(int steps)
- double[] getSynthAcc()
- double[] getSynthDis()
- double[] getSynthVel().

10.4.1 Synthesis of Seismic Motions and Numerical Example

A complete model behind the construction of artificial seismic motions (displacements, velocities, accelerations) based on spectra that goes beyond the usual 'seismic waves on bedrock model' must encompass correction factors accounting for (a) soil layering (b) surface geometry (c) cavities (d) cracks (e) material inhomogeneity and (e) material anisotropy, for the basic cases of P, SV and SH wave types. These effects can be summarized as frequency dependent, dimensionless functions constructed through comparisons with the equivalent homogeneous case and modify the PSDF input defined at the bottom of the soil deposit (i.e., bedrock). Synthetic ground motions, in the absence of recorded data, find extensive use in the analysis of a wide range of structures and/or individual structural systems for examining their earthquake-resistant behavior.

A compact software module was developed for generating synthetic motions as time-dependent graphs and in digital form, along with the corresponding input spectra and the correction functions used, based the material previously discussed. The first version of program 'Infoseismo',[1] now absorbed within the Symplegma Development Environment (SDE), places three filters in sequence to model the passage of fault-generated seismic motion through the upper lithosphere, see [22]. Thus, white noise (with energy equally distributed along the frequency range of interest) is released from the earthquake fault source is subsequently filtered through a Kanai-Tajimi (KT) filter, a high-pass (HP) filter, and a low-pass (LP) filter before it reaches the free surface. Essentially, what these filters do is to modify the energy transmission of the seismic disturbance contained within a fixed frequency band 0.2–20 Hz across the lithosphere and bring it up to rock outcrop. The mechanical interpretation of these filters is three single-degree-of freedom oscillators in series, all of them highly damped, and whose natural frequencies are strategically placed at the low, mid and high end of the aforementioned frequency spectrum. No interaction phenomena between the incoming wave and the structure of the upper soil deposits (e.g., layering, cavities, cracks, free surface and surface topography) is taken into account, and neither is the possible inhomogeneous structure of the said deposits considered in any way. A typical range of parameters entering the three-filter model is taken from [4] and listed in Table 10.1 for three earthquakes of increasing magnitude used in the numerical example presented here.

More specifically, the 'barrier-type' model is modified through the presence of overburden in the form of layered soil, which is modeled as an MDOF system, whereby each individual oscillator represents a typical type of layer. Given that all DOF correspond to horizontal motions, the model describes SH wave propagation through layers of constant thickness, parallel to the free surface of the ground [8]. Soil properties for near-surface layers can be inferred from comparing vertical to horizontal spectra resulting from microtremor recordings at the surface of a soil deposit, known as the quasi-transfer spectra technique. In Table 10.1, ω is the natural frequency of the specific soil layer and ξ its damping, while S_0 is the value of the power spectral density function, an energy measure, directly related to the earthquake magnitude M as measured in the Richter scale. Listing 10.2 provides base code in order to generate synthetic accelerograms reslting records similar to plots of Fig. 10.13.

---

[1] http://infoseismo.civil.auth.gr.

Table 10.1 Input parameters for synthetic ground motion generation for three categories of earthquakes classified according to their energy release.

Magnitude $M = 6.0$ Event

PSDF	KT-filter	KT-filter	HP-filter	HP-filter	LP-filter	LP-filter
$S_0(\text{cm}^2/\text{s}^3)$	$\omega_g(Hz)$	ξ_g	$\omega_f(Hz)$	ξ_f	$\omega_m(Hz)$	ξ_m
150.0	5.0	0.7	0.8	1.0	10.0	1.4

Magnitude $M = 6.5$ Event

PSDF	KT-filter	KT-filter	HP-filter	HP-filter	LP-filter	LP-filter
$S_0(\text{cm}^2/\text{s}^3)$	$\omega_g(Hz)$	ξ_g	$\omega_f(Hz)$	ξ_f	$\omega_m(Hz)$	ξ_m
300.0	6.0	3.0	0.3	1.0	5.0	0.7

Magnitude $M = 7.0$ Event

PSDF	KT-filter	KT-filter	HP-filter	HP-filter	LP-filter	LP-filter
$S_0(\text{cm}^2/\text{s}^3)$	$\omega_g(Hz)$	ξ_g	$\omega_f(Hz)$	ξ_f	$\omega_m(Hz)$	ξ_m
800.0	4.0	3.5	0.2	1.0	3.5	0.7

```
import courses.structuraldynamics.synthaccel
SynthPlot = new PlotFrame()
SynthPlot.setAutoColor(true)
steps=1000
data=[
[0.015,0.8,1.0,5.0,0.7,10.0,1.4],
[0.03,0.3,1.0,6.0,3.0,5.0,0.7],
[0.08,0.2,1.0,4.0,3.5,3.5,0.7]
]
data.each{
   synthacc= new synthaccel(it[0], it[1], it[2], it
      [3], it[4], it[5], it[6])
   synthacc.duration=60.0
   synthacc.t1=0.1*synthacc.duration
   synthacc.t2=0.6*synthacc.duration
   synthacc.Om_a=0.2
   synthacc.Om_b=25.0
   synthacc.Calc(steps)
   acc=synthacc.getSynthAcc()
   pf=new plotfunction(synthacc.duration/steps,acc)
   pf.Name="S0="+it[0]*10000.0+"cm^2/s^3"
   SynthPlot.addFunction(pf)
}
SynthPlot.makeLegend()
SynthPlot.show()
```

Listing 10.2 Base code to reproduce examples given in Table 10.1

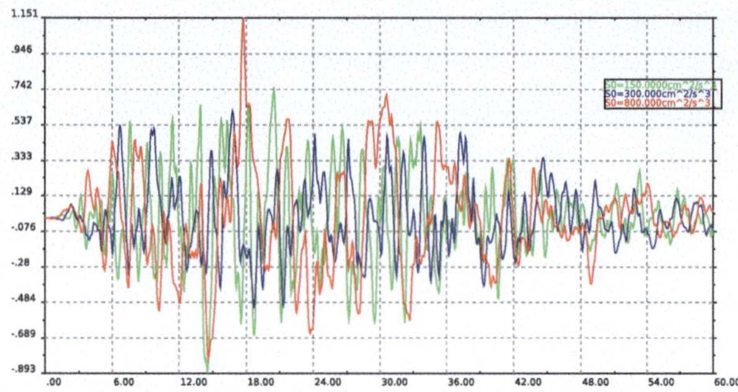

Fig. 10.13 Synthetic accelation records (in m/s^2) for the three scenarios of Table 10.1 as produced by code of Listing 10.2

10.5 Soil-Structure Interaction

Soil-structure-interaction (SSI) is a topic intimately related with structural dynamics, since in the majority of civil engineering applications the structure in question is founded on compliant soil. This implies that it is not possible to consider the structure as either simply supported or fixed on a rigid base. In fact, part of the surrounding soil acts as a secondary system that modifies the properties of the primary structure. Furthermore, ground motions are filtered through the surrounding soil as they propagate upwards from bedrock and reach the base of the structure. Finally, the type of foundation that is used to support the superstructure also plays a role in the overall dynamic response of this composite structural system. Generally speaking, the presence of soil results in a more flexible system than would otherwise exist if the structure rested on bedrock, which magnifies the kinematic response but reduces the forces that are transmitted to the superstructure.

10.5.1 Overview of SSI

Soil-structure-interaction is a discipline comprising structural mechanics, geotechnical and earthquake engineering. As such, it requires specialized methods of analysis that fall within two major groups: (a) sub-structuring techniques, whereby the problem is broken into its constituent parts, with each part solved by different methods. The complete solution is then assembled through use of continuity and equilibrium conditions at the common interfaces; (b) domain techniques, which nowadays invariably means use of the finite element method to model the problem in its entirety including superstructure, foundation and surrounding soil. There is an additional complexity that arises in SSI studies and involves the genesis and transmission of the seismic signals from their source upwards to the ground surface where the structural system is located. This step involves record selection for use in SSI analysis, which is a process with open questions on what constitutes a 'correct' selection set. It contents can be it recorded time histories, artificially generated signals or code-prescribed spectra [3]. More specifically, the effect of SSI on the dynamic response of structures has been a subject of research for well over five decades [23]. Among the various techniques that have been developed, the sub-structuring method is of practical interest, since it is most commonly used in civil engineering analysis and design. Following this approach, the problem is decomposed into the kinematic and the inertial interaction parts, in order to determine the foundation input motion and the response of the structure, respectively [24]. Inertial interaction analysis involves the determination of the foundation's impedance functions, which represent the stiffness and damping coefficients of the soil continuum, defined at the soil-foundation interface. In general, the derivation of impedance functions for the soil continuum forms a subclass of elastic wave propagation phenomena in geological media [2]. These problems involve a number of important engineering details such as layering, material inhomogeneity, the presence of a liquid phase in the soil, the presence of discontinuities, free-surface relief, etc. Numerical simulations and measurements regarding elastic wave propagation phenomena have clearly demonstrated the importance of all the above factors on the resulting scattered elastic wave fields and the generation of localized stress concentration phenomena.

The literature on dynamic SSI is vast and can be traced back to the seminal work of Eric Reissner in the 1930s. However, it can be argued that its origins lie on the recovery of fundamental solutions dating the 19th century, as pointed out in a review of the early history of SSI [23]. Another early state-of-the-art review [25] gives the essence of this problem: soil flexibility results in natural period lengthening and damping modification of the combined soil-structure system. This fact has been the subject of numerous studies utilizing various methodologies [26]. Furthermore, SSI effects can be studied even if the soil impedances are in the form of simple elastic springs. The fact that the problem is time-dependent, however, requires the use of complex-valued impedance functions, where the real part corresponds to a spring, while the imaginary part to a dashpot. In addition, each part must be frequency dependent, since the soil continuum responds differently at low versus high frequencies of vibration. Moreover, there is always a soil mass around the structure that vibrates in tandem with the foundation, giving rise to the concept of added (or virtual) mass. Finally, given the complex composition of the ground formations, the impedance functions should be able to account for soil inhomogeneity, anisotropy, the presence of geological cracks and other surface or subsurface discontinuities, plus the possibility for soil nonlinearities and for liquefaction if the level of ground shaking is high. Thus, the problem is practically intractable due to the high complexity manifested by the soil continuum, but this has not prevented the development and use of impedance functions as a means of producing acceptable engineering solutions.

10.5.2 Impedance Functions for SSI

Consider the simple, one-story space frame of Fig. 10.14 comprising four columns supporting a rectangular plate as the roof. The frame may rest on a spread foundation or alternatively

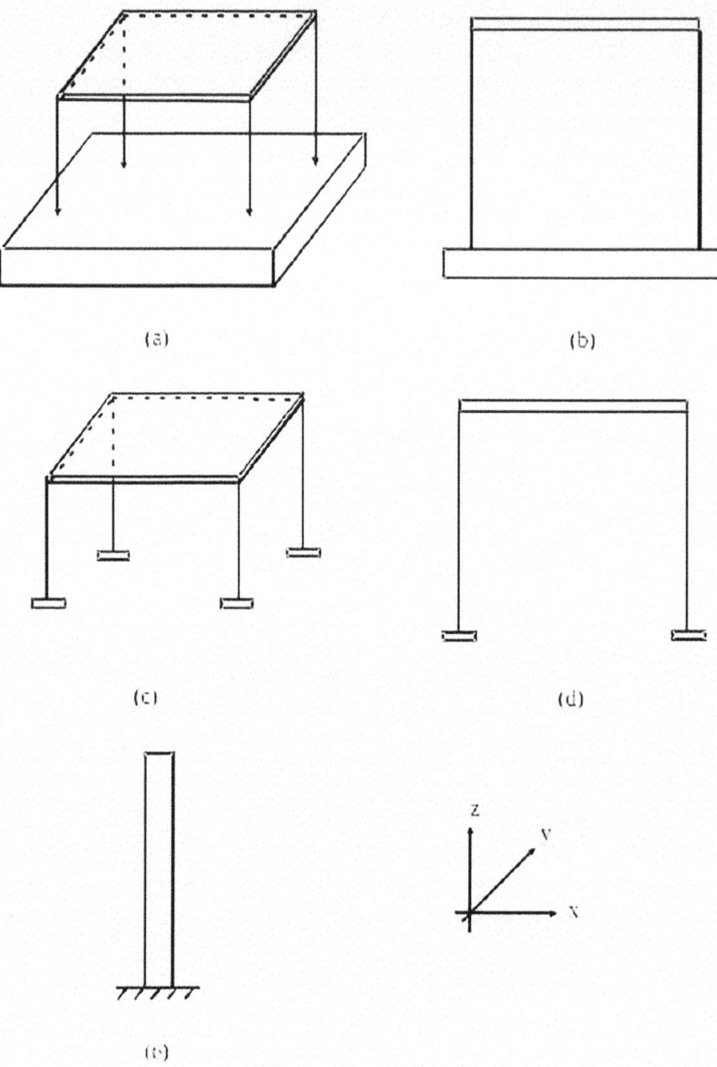

Fig. 10.14 Simplified representations of a single-story frame: **a** 3D representation resting on a mat foundation: **b** 2D representation resting on a mat foundation; **c** 3D representation resting on footings **d** 2D representation resting on footings; **e** simplified 'stick' model.

each column may rest on an individual footing. Its projection on the X-Z plane corresponds to a plane frame in the strong direction, and similarly the projection on the Y-Z plane is a plane frame the weak direction. Further simplification results if the structural model is a cantilever beam fixed to the ground, i.e., a stick model. Assume the structure is founded on compliant soil with a shear modulus G (kN/m^2) and a Poisson's ratio ν. The key idea behind a dynamic SSI analysis is to replace the surrounding soil with a series of springs, dashpots and virtual masses, although in practice only the first representation is considered. Furthermore, these springs are defined for the six components of the kinematic field in 3D, namely three translational and three rotational springs. Although these equivalent springs are frequency-dependent, it is usual practice to consider them as constants. Thus, for the simple stick model, the base fixity is relaxed and a maximum of six springs are introduced, yielding a MDOF system with one axial, two transverse, two rotational and one torsional DOF.

The foundation type supporting the aforementioned structure is a crucial parameter in the determination of the spring constants. At first, it can be classified as either a surface or an embedded foundation. Furthermore, it is possible to have a mat foundation or individual footings that may or may not be linked by connector beams. Finally, for sub-standard soil it is possible to use piles supporting the foundation cap, but this case is beyond the scope of this chapter. Much information has been generated regarding the impedance functions representing the supporting soil, which in their general form are complex functions with the real part corresponding to a stiffness and the imaginary part corresponding to a dashpot [27]. These functions result from the solution of boundary value problems involving continuous media of semi-infinite extent and are frequency dependent. Their complete form is a 6×6 matrix involving the DOF previously mentioned and corresponds to their corresponding generalized forces. Note that the inverse of an impedance matrix is the compliance matrix corresponding to the generalized displacements.

Some basic results are collected in Table 10.2 regarding the static stiffness of a rigid circular plate or radius R resting on the elastic half-space and are due to Reissner, see [28]. A list of static stiffness values for a rigid, rectangular surface foundation of dimensions $2L \times 2$, $L \geq B$ resting on the elastic half-space is compiled in Table 10.3, see [29, 30]. More specifically, the table lists three translational springs, two along the horizontal plane parallel to the foundation sides, and one vertical spring, namely K_x, K_y, K_y in (kN/m). Also listed are two rotational and one torsional spring, namely K_{rx}, K_{ry}, K_t in (kNm). Note that I_{bx}, I_{by} are the moments of inertia of the flat foundation about its center and $J_b = I_{bx} + I_{by}$ is its polar moment of inertia. Obviously, the square foundation is a special case of the rectangular one.

Table 10.2 Stiffness Coefficients for a Circular Rigid Foundation on the Surface of a Homogeneous Half-space.

Horizontal	Verical	Rocking	Torsional
$K_x = \frac{8.0GR}{2-\nu}$	$K_z = \frac{4GR}{1-\nu}$	$K_{rx} = K_{ry} = \frac{8GR^3}{3(1-\nu)}$	$K_t = \frac{16GR^3}{3(1-\nu)}$

Notation: R is the radius of circular foundation; G, ν are the soil's shear modulusand Poisson ratio

Table 10.3 Stiffness coefficients for a rectangular rigid foundations on the surface of a homogeneous half-space

Vibration mode	Stiffness coefficients K for a foundation soil contact surface area A_b circumscribed by a rectangle $2L \times 2B$, with $L > B$ Also, G, ν are the soil's shear modulus and Poisson ratio	Square Shape $L = B$
Vertical, z	$K_z = \frac{2GL}{1-\nu}(0.73 + 1.54\chi^{0.75})$ with $\chi = \frac{A_b}{4L^2}$	$K_z = \frac{4.54GB}{1-\nu}$
Horizontal, y (in the lateral direction)	$K_y = \frac{2GL}{2-\nu}(2 + 2.5\chi^{0.85})$	$K_y = \frac{9GB}{2-\nu}$
Horizontal, x (in the longitudinal direction)	$K_x = K_y - \frac{0.2}{0.75-\nu}GL\left(1 - \frac{B}{L}\right)$	$K_x = K_y$
Rocking, r_x (about the X-axis) Rocking, r_y (about the Y-axis)	$K_{rx} =$ $\frac{G}{1-\nu}I_{bx}^{0.75}\left(\frac{L}{B}\right)^{0.25}\left(2.4 + 0.5\frac{B}{L}\right)$ $K_{ry} = 3\frac{G}{1-\nu}I_{by}^{0.75}\left(\frac{L}{B}\right)^{0.15}$ I_{bx}, I_{by} are the area moments of inertia of the foundation contact surface around X, Y axes	$K_{rx} = \frac{3.6GB^3}{1-\nu}$ $K_{ry} = K_{rx}$
Torsional, t	$K_t =$ $GJ_B^{0.75}\left(4 + 11\left(1 - \frac{B}{L}\right)^{10}\right)$ $J_b = I_{bx} + I_{by}$ is the polar moment of inertia of the foundation contact surface	$K_t = 8.3GB^3$

Note: As $L/B \to \infty$ (a strip footing), the values of K_z and $K_y \to \infty$ can be computed from the formulas corresponding to a footing with $L/B \approx 20.0$

10.5.3 SSI Modelling Details

Some comments regarding Tables 10.2 and 10.3 are as follows:

1. For a mat foundation with dimensions $2L \times 2B$ that cover the surface area projected by the supported 3D frame, the spring constants are defined at the center of mass of the rigid foundation. In the case where plane frames are examined, the same assumption can be made, save for the fact that the spring coefficients computed must be divided by the number of frames in that particular direction. For instance, for the 3D single story frame of Fig. 10.14, there are two plane frames in each principal direction.
2. Analysis of the coupled structure-soil spring system is usually carried out using the finite element method, which as a first step solves the eigenvalue problem and yields the natural periods of vibration T_i, $i = 1, 2, \ldots n$ plus the corresponding modal shapes.
3. For individual footings of dimensions $2L \times 2B$, the spring constants are computed at the base of the column that rests on the footing. Following that, a space frame can be analyzed. In the case of a plane frame, the computed spring stiffness for frames along the perpendicular direction must be added. As example, in Fig. 10.14 for the X-Z plane frame, there are two columns in the Y- direction.
4. Regarding Table 10.2 for the circular surface foundation, of radius R, the static spring coefficient values pertain to the horizontal and vertical directions K_x, K_y, plus the rotation and the twist K_{rx}, K_t. Note that there is angular symmetry, as these values are defined at the center of the foundation. Thus, irrespective of the type of foundation, the spring constants are defined at its center.
5. The 'stick' model, i.e., the equivalent cantilever beam representation of a 3D frame, is composed by combining the plane frames in either principal direction. This yields the properties of a beam whose height is defined along the Z- direction with principal axes in the X- and Y- directions. As shown in Fig. 10.14, the original 3D space frame is supported by a total of four columns, implying that the flexural stiffness is the sum of the constituent four stiffness, for both principal directions. The same holds true for the computation of the soil spring stiffness at the base of the cantilevered beam. Given that this mechanical representation is quite simple, the eigenvalue analysis can be treated analytically.
6. As an alternative way for computing stiffness values for the supporting soil, one may use the allowable stresses which range from $\sigma_{al} = 10\text{N/cm}^2$ to $\sigma_{al} = 50\text{N/cm}^2$ for most categories of soils within the context of approximate, strength of materials formulas.

10.5.4 The Influence of SSI on the Structural Response

In using Tables 10.2 and 10.3 for circular and rectangular foundation analysis, the following points must be considered:

1. The flexibility of the horizontal members (i.e., beams and floor elements) of the frame in question must also be taken into account, even if the final analysis involves only lateral motions. The computation of the relevant flexibilities necessitates a separate static analysis, unless diaphragm action is assumed, in which case the only stiffness is due to the supporting columns.
2. It is important that an eigenvalue analysis of the soil-structure system takes place before a full-fledged dynamic analysis, so that a preliminary assessment of the influence of compliant soil can be established.
3. There are a number of ways for normalizing the eigenvectors resulting from the eigenvalue analysis of the SSI problem. If the first entry of all eigenvectors is set equal to unity and the remaining are adjusted correspondingly, the resulting participation factors are all positive and smaller than unity. If the last entry of all eigenvectors is set equal to unity, then the resulting participation factors are either positive or negative and are either greater or smaller than unity.
4. Generally speaking, SSI phenomena decrease the value of the base shears but increase the value of the roof displacements. The computation of these forces and displacements following an eigenvalue analysis requires use of the SRSS (square root of the sum of squares) superposition rule, which does not seem to give much different values compared to a simple summation of modal values for these quantities.
5. Under SSI effects, the fundamental natural period of typical frame-type structures falls within the range 0.5–3.0 s.
6. The most crucial parameter affecting SSI seems to be the height of the structure in question.
7. The impedance functions representing the soil continuum are in reality frequency ω dependent and are given as function of the dimensionless ratio $\omega l/c_s$, where c_s is the velocity of the propagating shear waves in the soil and l is a representative dimension of the foundation, They are complex numbers with their real part corresponding to a stiffness and the imaginary part to a damper. At the same time, there is a soil mass that participates in the SSI phenomenon that should be accounted for as a virtual mass element.

10.5.5 SSI Example

We now present an elementary example involving an SDOF representation of a structure, to which the presence of a rigid foundation resting on complaint soil is added in order

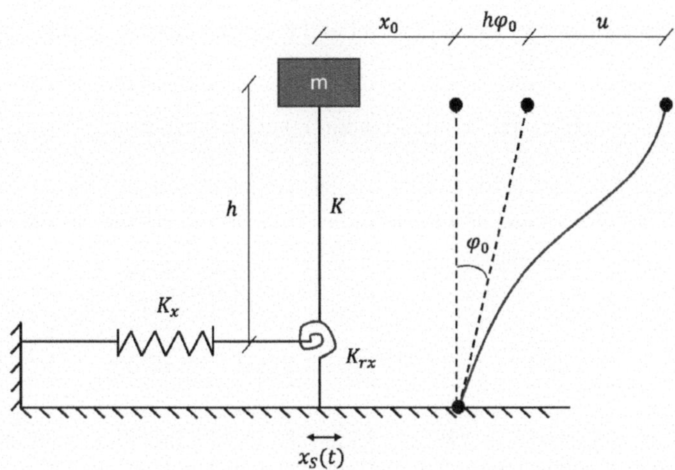

Fig. 10.15 SDOF system augmented by the addition of a foundation with horizontal and rocking motion

to study soil-structure interaction (SSI) effects. More specifically, as shown in Fig. 10.15, the structural SDOF model has a height of $h =15$ m and is described by a lumped mass $m =10^5$ tn, a stiffness $k =10^8$ kN and an effective damping $d = 0.07$, giving an undamped fundamental frequency $\omega =31.6$ rad/s (or $f =5.03$ Hz) and a natural period $T =0.2$ s. The kinematics of this SDOF model are described by the horizontal displacement $x(t)$ of the lumped mass.

The introduction of a rigid, massless foundation resting on soil with a shear wave speed $c_s =200$ m/s, a shear modulus $G = 8 \cdot 10^8$ kN/m^2 and a Poisson's ratio of $\nu = 0.25$ introduces two kinematic variables corresponding to the horizontal translation of the center of mas of the foundation $x_0(t)$ and its rotation $\phi_0(t)$ about its center of mass. Thus, the SSI problem now involves an MDOF system with three degrees of freedom for lateral vibrations. Corresponding to the two new kinematic variables, we have a translational stiffness K_x and a rotational stiffness K_{rx}. Values for these mechanical parameters can be found in tables such as Table 10.3 appearing in this chapter.

A number of comments are now due: (i) There are also corresponding dampers C_x and C_{rx}, plus a virtual soil mass (of volume comparable to that of the foundation) m_x and a mass moment of inertia I_{rx} participating in the combined system vibrations; (ii) the soil springs and dampers are in their most general case frequency dependent. The forcing function is assumed to be a ground displacement $x_s(t)$ caused by the seismic action of (i) the Imperial Valley 1979 earthquake, a standard seismic record used in earthquake engineering and (ii) the Kobe 1995 earthquake, a low frequency yet destructive seismic record. As shown in

Fig. 10.15, the total displacement at the level of the lumped mass of the SSI system is given as

$$x(t) = x_0(t) + h\phi_0(t) + u(t) + x_s(t)$$

where $u(t)[= x(t) - x_s(t)$ is the relative displacement between the top and the base of the structure. The SDOF equation of motion for the SSI system, following condensation of the extra two DOF of the foundation, assumes the generic form $M\ddot{x} + Kx = F$. Specifically for the case of the relative motion u(t) of the structure we have

$$m(1 + \frac{k}{K_x} + \frac{kh^2}{K_{rx}})\ddot{u}(t) + ku(t) = -m\ddot{x}_s(t)$$

The presence of SSI leads to a number of changes in the original structure's mechanical parameters. Thus, the new system natural frequency and damping respectively are

$$\omega_s = \frac{\omega}{\sqrt{1 + \frac{k}{K_x} + \frac{kh^2}{K_{rx}}}}$$

$$d_s = d(\frac{\omega_s}{\omega})^2 + d_{soil}(1 - (\frac{\omega_s}{\omega})^2) + (\frac{\omega_s}{\omega})^2 \left(\frac{k}{K_x}\frac{\omega_s C_x}{2K_x} + \frac{kh^2}{K_{rx}}\frac{\omega_s C_{rx}}{2K_{rx}} \right)$$

If we assume that values for the structural and soil effective damping are comparable, and that dampers C_x and C_{rx} are neglected, then $d_s = d = d_{soil}$. Consulting Table 10.3 for the soil stiffness coefficients, we have

$$K_x = 8GR/(1 - \nu) = 2.56 \cdot 10^7 (\text{kN}),$$
$$K_{rx} = 8GR^3/3(1 - \nu) = 7.68 \cdot 10^9 (\text{kN-m})$$

Values for the ratios are computed as $k/K_x = 3.91$ and $kh^/K_{rx} = 2.93$, yielding a value for the SSI system natural frequency of $\omega_s = 11.28$ rad/s (or $f_s = 1.79$ Hz) and a natural period $T_s = 0.56$s. We therefore observe that when SSI effects are present, the system becomes more compliant, resulting in larger kinematic values (e.g., the top displacement) and smaller stress values (e.g., the base shear force). In the numerical implementation using the SDE that follows, Fig. 10.16 depicts the time evolution of the top displacement in the absence and presence of SSI for the Imperial Valley 1979 earthquake, and similarly in Fig. 10.17 for the Kobe 1995 earthquake. Since damping in the system was neglected, the amplification of motion at the top of the structure is impressive, almost an order of magnitude more. One additional fact is that the Kobe 1995 earthquake is known to be unusual for having its energy concentrated at the low frequency range, which is exactly where the combine SSI system natural frequency falls.

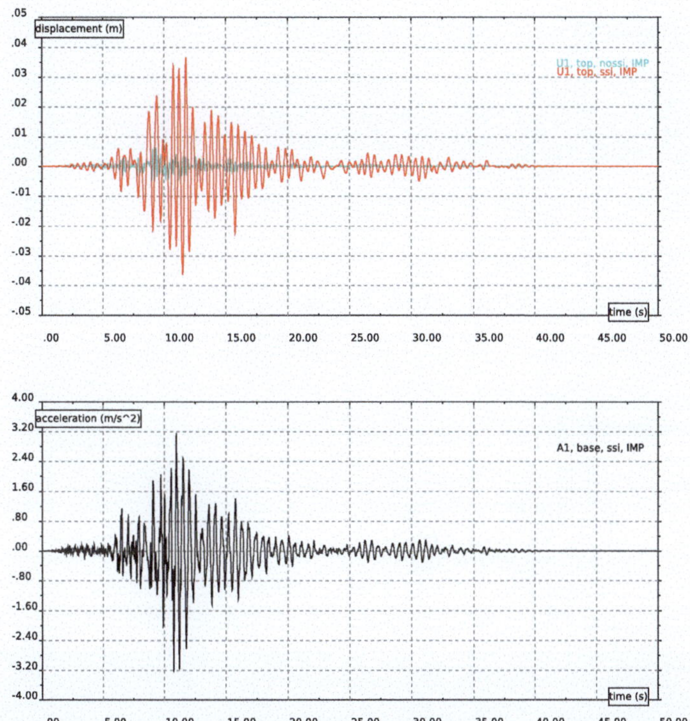

Fig. 10.16 **a** Displacement at the top of the SDOF system in the absence/presence of SSI phenomena due to **b** the Imperial Valley 1979 acceleration record at the base

10.6   Elements of Poroelasticity

The design of vibro–acoustic systems incorporating poroelastic materials, primarily aimed at noise reduction or cancellation, requires knowledge of several material parameters. These material parameters are usually possible to be defined using some standard testing in an acoustic laboratory. Here, we study an inverse technique for the characterization of poroelastic materials based on Biot's theory of poroelasticity. Experimental setups for such a procedure typically consist of a configuration of several sequentially positioned layers, each of which could be a poroelastic medium, a fluid, or an elastic solid. As an indicative example we mention a configuration formed by a fluid, two porous layers saturated by a fluid, a solid layer and a second fluid. The configurations considered here result in a one-dimensional problem defined in the longitudinal direction, free of external forces per unit volume, stationary, and in the frequency domain ω. The problem represents the ideal conditions of the Kundt's (or impedance) tube. In our work we take advantage of analytical solutions for the one-dimensional case of the poroelasticity's boundary value problem. The macroscopic model is valid from low to high frequencies [31]. Therefore, the analytical solutions are also

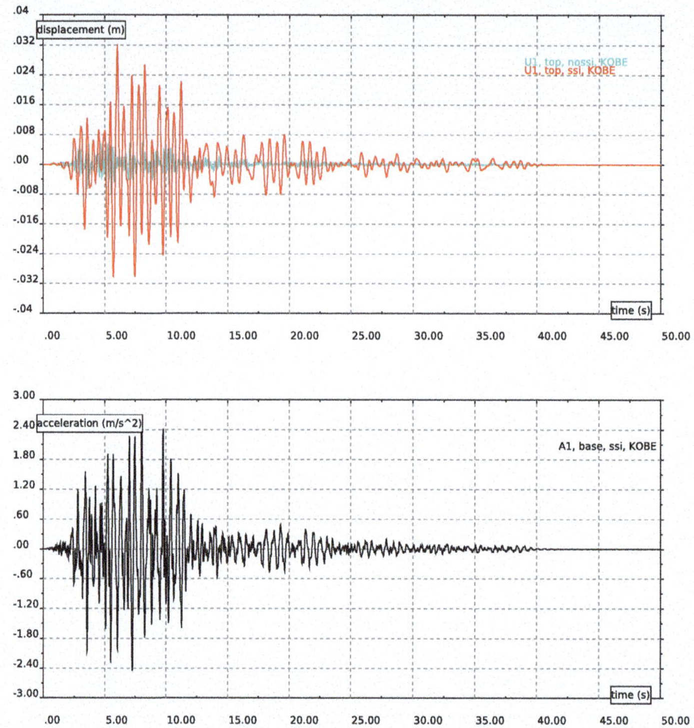

Fig. 10.17 **a** Displacement at the top of the SDOF system in the absence/presence of SSI phenomena due to **b** the Kobe 1995 acceleration record at the base

valid from low to high frequencies. Utilizing these solutions, one of the authors (CGP), during his collaboration with Pythmen R&D,[2] developed a java-based toolkit that solves general case of an arbitrary finite number of layers multidomain problem and calculates acoustical indicators, e.g. the surface impedance, the reflection coefficient or the absorption coefficient. These parameters are important for poroelastic material characterization in vibro-acoustic applications. Similar material has been published in conference proceedings [32].

10.6.1 Acoustic Multilayer Domains

It is a common procedure to treat acoustic problems in poroelastic media saturated by air by neglecting the two waves that propagate mainly through the solid skeleton. Therefore, models that account only for the wave propagating through the fluid interacting with the solid skeleton are typically adopted. One of the most common such models is that of Biot, Johnson and Allard [33], which accounts for the air compressibility and the temperature

---

[2] https://pythmen.com/.

effects. Nevertheless, such an approximation would not be acceptable if the saturated fluid is a liquid. That is even more pronounced when inverse characterization of materials is sought, because in such cases we need the full volume of information that is embedded into the response of such mediums. A typical example are porous metals immersed in noisy liquid environments.

In the present work we trace back to Biot's more general theory of propagation of elastic waves in a fluid-saturated porous solid, since the waves that propagate through the solid skeleton can not be considered to be negligible when the saturated fluid is a liquid. Following previous publications on analytical solutions [34] for the general Biot's theory in the range from low to high frequency content when the medium can be considered to be one dimensional. Having faced situations where the sequence of successive material parts lead to extremely intricate and lengthy analytical solutions we proceed by developing a numerical framework that deploys analytical solution for single one dimensional domains arriving at a toolkit capable to handle effectively a arbitrary number of sequential successive materials that constitute a coherent wave propagation medium.

10.6.2 Developed Module for 1D Poroelasticity

Using well-established analytical solutions from the literature [34], we an appropriate module developed by (CGP) at Pythmen R&D. The module can effectively address a configuration of successive sequentially positioning material layers, as shown in Fig. 10.18, interconnected by appropriate interface conditions. The first and the last layer can be considered to extend in the infinite.

Each layer of length l_i is supposed to be an one dimensional domain of specific type of fluid, solid or poroelastic medium.

For the fluid domain in the absence of body forces, with compresibility modulus λ_f and density ρ_f, the governing equation is the standard wave equation:

$$\lambda_f w_{xx} = \rho_f \ddot{w}. \tag{10.10}$$

where subscripts denote spatial derivatives and overdots indicate time derivatives. Furthermore under the assumption of harmonic condition we have the analytic solution of the displacements w, given as,

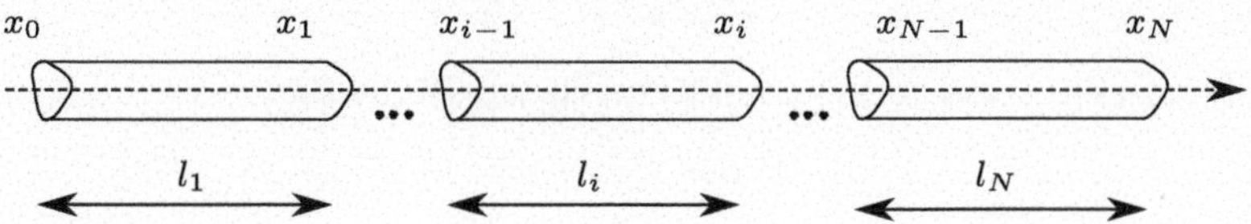

Fig. 10.18 Configuration of successive material layers

$$w(x, t, \omega) = W_1 e^{i\omega\sqrt{\frac{\rho_f}{\lambda_f}}x} e^{i\omega t} + W_2 e^{-i\omega\sqrt{\frac{\rho_f}{\lambda_f}}x} e^{i\omega t} \tag{10.11}$$

Boundary conditions referred to a fluid layer are expressed in terms of the displacements w as well as the derivative of displacement in respect to x from which the pressure is derived as,

$$p = -\lambda_f w_x \rightarrow p = -\lambda_f \frac{\partial w}{\partial x}. \tag{10.12}$$

Explicitly Eq. (10.12) can be written after introducing the solution for the displacement as follows:

$$p(x, t, \omega) = -W_1 \left(i\omega\sqrt{\rho_f \lambda_f}\right) e^{i\omega\sqrt{\frac{\rho_f}{\lambda_f}}x} e^{i\omega t} + W_2 \left(i\omega\sqrt{\rho_f \lambda_f}\right) e^{-i\omega\sqrt{\frac{\rho_f}{\lambda_f}}x} e^{i\omega t}. \tag{10.13}$$

For compatibility conditions between two successive fluid layers we consider the continuity of displacement and equilibrium of pressure, that is,

$$w^i = w^{i+1}, \quad \lambda_f^i \frac{\partial w^i}{\partial x} = -\lambda_f^{i+1} \frac{\partial w^{i+1}}{\partial x}. \tag{10.14}$$

An elastodynamic solid under the assumption of plane wave in the longitudinal direction e_x free of external forces per unit volume, has the longitudinal displacement as d, and is governed by,

$$(\lambda_s + 2\mu_s)d_{xx} = \rho_s \ddot{d}, \tag{10.15}$$

The solution of the above equation under the assumption of harmonic conditions in frequency ω, can be given as,

$$d(x, t, \omega) = D_1 e^{i\omega\sqrt{\frac{\rho_s}{\lambda_s + 2\mu_s}}x} e^{i\omega t} + D_2 e^{-i\omega\sqrt{\frac{\rho_s}{\lambda_s + 2\mu_s}}x} e^{i\omega t} \tag{10.16}$$

which is quite similar to Eq. (10.11), while here λ_s and μ_s the Lame's coefficients for the material of the solid and ρ_s the mass density. Returning to the boundary conditions, when these are present on some solid layer, they expressed in terms of displacements d and the stress component τ which is derived from the displacement's derivatives,

$$\tau = (\lambda_s + 2\mu_s)\frac{\partial d}{\partial x}. \tag{10.17}$$

Explicitly Eq. (10.17) can be written after introducing the strain as,

$$\tau(x, t, \omega) = -D_1 \left(i\omega\sqrt{\rho_f \lambda_f}\right) e^{i\omega\sqrt{\frac{\rho_f}{\lambda_f}}x} e^{i\omega t} + D_2 \left(i\omega\sqrt{\rho_f \lambda_f}\right) e^{-i\omega\sqrt{\frac{\rho_f}{\lambda_f}}x} e^{i\omega t}. \tag{10.18}$$

For the case of two successive solid layers' compatibility we need to consider continuity of displacements and stress equilibrium,

$$d^i = d^{i+1}, \quad (\lambda_s^i + 2\mu_s^i)\frac{\partial d^i}{\partial x} = -(\lambda_s^{i+1} + 2\mu_s^{i+1})\frac{\partial d^{i+1}}{\partial x}. \tag{10.19}$$

Finally, we mention that poroelastic medium is the most general one, which under certain assumptions can simulate the behaviour of a solid or a fluid and is therefore capable to degenerate into these types. The analytic solution for this case is the one already given in Eq. (10.32). The boundary conditions in this case are four in number and more specific are expressed in terms of solid skeleton displacements, pore fluid displacements and their derivatives. Similarly, we should address a two–porous layers, in the case that we assume that are glued together. In that case, the displacement of solid skeleton of the ith poroelastic layer u^i and that of the $i + 1$ layer must be equal across that interface. In order to guarantee the equality of interstitial fluid at the interface, the displacement of the fluid at macroscopic level of the two adjacent layers must be inversely proportional to their open porosities:

$$u^i = u^{i+1}, \quad \beta^i v^i = \beta^{i+1} v^{i+1}. \tag{10.20}$$

Furthermore, we assume that the interstitial pressures of adjacent porous layers are equal and similarly for their total stresses, then we have

$$\beta^{i+1} s^i = \beta^i s^{i+1}, \quad \sigma^i + s^i = \sigma^{i+1} + s^{i+1}. \tag{10.21}$$

For the case of different physics layers' compatibility, appropriate interface conditions are taken into account depending on the type of the two individual neighbouring fields. For the case of fluid to poroelastic interface conditions we consider the set given below,

$$w = (1 - \beta)u + \beta v. \tag{10.22}$$

Furthermore, the interface conditions state that the stress of the solid skeleton σ and the stress of the the fluid s of the porous layer are related with the pressure of the fluid p under the equations,

$$\sigma = -(1 - \beta)p, \quad s = -\beta p, \tag{10.23}$$

As regards a poroelastic to solid layer compatibility we take into account the conditions that follow,

$$u = d, \quad v = d. \tag{10.24}$$

Besides, the stress of the solid τ must be equal to the sum of the macroscopic stress of the solid skeleton of porous layer σ,

$$\sigma + s = \tau \tag{10.25}$$

10.6.3 Implementation of 1D Poroelasticity

The kinematic equations in Biot's theory express the relation between the macroscopic displacement of the solid u_i and that of the fluid v_i, with the macroscopic deformations of the solid ϵ_{ij} and that of the fluid e_{ij}, respectively,

$$\epsilon_{ij} = \frac{1}{2}(u_{i,j} + u_{j,i}),$$
$$e_{ij} = \frac{1}{2}(v_{i,j} + v_{j,i}). \tag{10.26}$$

Furthermore, Biot's constitutive model of poroelasticity establishes the behaviour of a fluid-filled porous solid,

$$\sigma_{ij} = \delta_{ij}(\lambda' \epsilon_{kk} + Be) + 2\mu\epsilon_{ij},$$
$$s = B\epsilon_{kk} + Ae, \tag{10.27}$$

where, σ_{ij} the macroscopic stress of the solid skeleton, while s is the negative value of the hydrostatic pressure of the fluid and e is the cubic dilatation of the fluid, both at the macroscopic level. Then, λ' and μ are constants connected to Lame's parameters, that relate the stress and strain of the solid skeleton. The elastic constant, A, represents the modulus of elasticity that relates the spherical waves components of the macroscopic strain and stress of the fluid. Finally, B serves for the static coupling of fluid and the solid skeleton. Equations of motion for external forces per unit volume over the solid skeleton f_i and the fluid ϕ_i can be given as,

$$\sigma_{ij} + f_i = b_{ij}(\dot{u}_j - \dot{v}_j) + \rho_{11}\ddot{u}_i + \rho_{12}\ddot{v}_i,$$
$$s_j + \phi_i = b_{ij}(\dot{v}_j - \dot{u}_j) + \rho_{12}\ddot{u}_i + \rho_{22}\ddot{v}_i. \tag{10.28}$$

where, ρ_{11}, ρ_{12} and ρ_{22} are directly related with the open porosity β, the density of the base material of the solid skeleton ρ_s, the density of the fluid ρ_f and the apparent density of the dynamics coupling ρ_α. For isotropic permeability of the solid skeleton, the flow resistivity b_{ij} reduces to a single constant $\delta_{ij}b$ with δ_{ij} the Kronecker delta. After replacement of the stresses by the strains in the equilibrium equations of Eq. (10.28) and using the constitutive relations of Eq. (10.27), we arrive at,

$$(\lambda' + \mu)u_{j,ji} + Bv_{j,ji} + \mu u_{i,jj} + f_i = b(\dot{u}_j - \dot{v}_j) + \rho_{11}\ddot{u}_i + \rho_{12}\ddot{v}_i,$$
$$Bu_{j,ji} + Av_{j,ji} + \phi_i = b(\dot{v}_j - \dot{u}_j) + \rho_{12}\ddot{u}_i + \rho_{22}\ddot{v}_i \tag{10.29}$$

s can be related with the interstitial fluid pressure p the single stress variable of interest in fluid,

$$s = \frac{s_{kk}}{3} = -\beta p, \quad s_{ij} = \delta_{ij}s. \tag{10.30}$$

Under the assumption of one-dimensional problem in the longitudinal direction e_x free of external forces per unit volume, stationary in harmonic condition of frequency ω, the solutions of the fluid and the skeleton displacements have the form,

$$u(x, t, \omega) = \tilde{u}e^{i\lambda x}e^{i\omega t}, \quad v(x, t, \omega) = \tilde{v}e^{i\lambda x}e^{i\omega t}. \tag{10.31}$$

In the above, $\tilde{u}$ and $\tilde{v}$ constant complex numbers, x the spatial variable in the longitudinal direction, t the time and λ the complex wave number. Expression for the above solutions, deployed in the current work, are given in the literature, e.g. in [35]. Here, we limit ourselves to state that solution is composed by the superimposition of two type of waves the one propagating mainly by the solid skeleton and another one that propagates mainly by the fluid, come to the solution of the form,

$$v(x, t, \omega) = \left(C_1^I e^{i\lambda_1 x} + C_2^I e^{-i\lambda_1 x_3} + C_3^{II} e^{i\lambda_3 x} + C_4^{II} e^{-i\lambda_3 x}\right), \tag{10.32}$$

where v stands for u or v the solid skeleton and fluid displacements. Finally, to write the explicit expressions of stresses in respect to displacements we introduce the kinematics relations of Eq. (10.26) into the constitutive ones of Eq. (10.27) and considering the case of one–dimensional problems, we recover

$$\sigma = (\lambda' + 2\mu)u_x + Bv_x, \quad s = Bu_x + Av_x. \tag{10.33}$$

10.6.4 Loading an External Module for 1D Poroelasticity

We intetionaly left out the module for poroelasticity in order to show how the user can load an external java package and smoothly incorporate it in the environment of the SDE.

For simplicity, we consider a configuration for a Kundt's tube experiment of two layers, the first one being a fluid (air) while the second is foam given as a poroelastic material for which we want to compute the acoustic parameters. The code for the solution of that problem using the `jpythmen` external package is given in Appendix app:LEMP. This current example is similar to some test case presented in the literature [36] (Fig. 10.19).

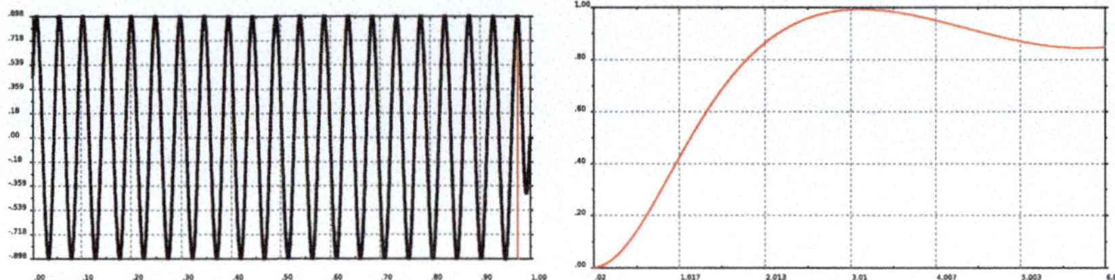

Fig. 10.19 (Left) Solution for the kinematic field over the length of the tube x for the test problem. With a red vertical line it is shown the location where the poroelastic medium rests. (Right) Sound absorption coefficient over the considered frequency range

References

1. Biggs JM (1964) Introduction to structural dynamics. McGraw-Hill
2. Villaverde R (2009) Fundamental concepts of earthquake engineering. CRC Press
3. European Standard EN 1998-1 (2004) Eurocode 8: Design of structures for earthquake resistance—part 1: General rules, seismic actions and rules for buildings
4. Apageorgiou AS, Aki K (1983) A specific barrier model for the quantitative description of inhomogeneous faulting and the prediction of strong ground motion. i. description of the model. Bull Seismol Soc Am 73(3):693–722
5. Papageorgiou AS, Aki K (1985) Sealing law of far-field spectra based on observed parameters of the specific barrier model. Pure Appl Geophys 123:353–374
6. Tsai C-CP (1998) Engineering ground motion modeling in the near-source regime using the specific barrier model for probabilistic seismic hazard analysis. Pure Appl Geophys 152:107–123
7. Shinozuka M, Sato Y (1967) Simulation of nonstationary random process. J Eng Mech Div 93(1):11–40
8. Hadjian AH (2002) Fundamental period and mode shape of layered soil profiles. Soil Dyn Earthq Eng 22(9–12):885–891
9. Berg P, If F, Nielsen P, Skovgaard O, Helbig K (1994) Modeling the earth for oil exploration
10. Cascone E, Rampello S (2003) Decoupled seismic analysis of an earth dam. Soil Dyn Earthq Eng 23(5):349–365
11. Miles SB, Ho CL (1999) Rigorous landslide hazard zonation using newmark's method and stochastic ground motion simulation. Soil Dyn Earthq Eng 18(4):305–323
12. Trifunac MD (1995) Pseudo relative velocity spectra of earthquake ground motion at long periods. Soil Dyn Earthq Eng 14(5):331–346
13. Sokolov V, Loh C-H, Wen K-L (2000) Empirical model for estimating fourier amplitude spectra of ground acceleration in Taiwan region. Earthq Eng Struct Dyn 29(3):339–357
14. Tsang H-H, Chandler AM, Lam NTK (2006) Estimating non-linear site response by single period approximation. Earthq Eng Struct Dyn 35(9):1053–1076
15. Megawati K, Chandler AM (2006) Distant-earthquake simulations considering source rupture propagation: refining the seismic hazard of hong kong. Earthq Eng Struct Dyn 35(5):613–635
16. Lin C-CJ, Ghaboussi J (2001) Generating multiple spectrum compatible accelerograms using stochastic neural networks. Earthq Eng Struct Dyn 30(7):1021–1042

17. Dravinski M, Wilson MS (2001) Scattering of elastic waves by a general anisotropic basin. part 1: a 2d model. Earthq Eng Struct Dyn 30(5):675–689
18. Manolis GD, Shaw RP, Pavlou S (1999) Elastic waves in nonhomogeneous media under 2d conditions: I. fundamental solutions. Soil Dyn Earthq Eng 18(1):19–30
19. Nayfeh AH (1995) Wave propagation in layered anisotropic media: with application to composites. Elsevier
20. Manolis GD, Koliopoulos PK (2001) Stochastic structural dynamics in earthquake engineering. (No Title)
21. Shinozuka M, Jan C-M (1972) Digital simulation of random processes and its applications. J Sound Vib 25(1):111–128
22. Hanyga A (2016) Seismic wave propagation in the earth. Elsevier
23. Kausel E (2010) Early history of soil-structure interaction. Soil Dyn Earthq Eng 30(9):822–832
24. Wolf J (1985) Dynamic soil-structure interaction. Prentice Hall, Inc
25. Johnson JJ (1981) Soil structure interaction: the status of current analysis methods and research: seismic safety margins research program, vol 88. Office of Nuclear Regulatory Research, US Nuclear Regulatory Commission
26. Kausel E (2017) *Advanced structural dynamics*. Cambridge University Press
27. NIST GCR 12-917-21. Soil-structure interaction for building structures (2012) Prepared by NEHRP consultants joint venture (a partnership of the Applied Technology Council and the Consortium of Universities for Research in Earthquake Engineering)
28. Veletsos AS, Wei YT (1971) Lateral and rocking vibration of footings. J Soil Mech Found Div 97(9):1227–1248
29. Gazetas G (1991) Formulas and charts for impedances of surface and embedded foundations. J Geotech Eng 117(9):1363–1381
30. Pais A, Kausel E (1988) Approximate formulas for dynamic stiffnesses of rigid foundations. Soil Dyn Earthq Eng 7(4):213–227
31. Sanchez-Ricart L, Garcia-Pelaez J (2014) Viscous dissipative power density function for poroelastic saturated materials at high frequencies. Mech Mater 73:11–27
32. Panagiotopoulos C, Sanchez-Ricart L (2024) Inverse characterization of sound absorbing media using one dimensional analytical biot's poroelasticity theory solutions. In of Acoustics [74], pp 572–579
33. Allard J, Atalla N (2009) Propagation of sound in porous media: modelling sound absorbing materials. Wiley
34. Sanchez-Ricart L, García-Peláez J (2014) Analytical solutions of a covering formed by two porous layers: reflected and transmitted energy reductions. Acta Acust Acust 100(6):1001–1012
35. Cheng AH-D, Badmus T, Beskos DE (1991) Integral equation for dynamic poroelasticity in frequency domain with bem solution. J Eng Mech 117(5):1136–1157
36. Sanchez-Ricart L, García-Peláez J (2013) Fluid-solid static coupling for saturated porous fiber metals in the biot's acoustic theory. Acta Acust Acust 99(5):716–728

The Symplegma Development Environment (SDE) with the Groovy Programming Language   A

The most appropriate programming language underlying the Symplegma Development Environment SDE developed in conjunction with this textbook, which allows for best utilization of the routines that are included in the Climax library module, is Groovy. Part of the Climax library is also the package courses.structuraldynamics mainly utilised and presented in the book. Furthermore, in order to fully utilize either Groovy or the full capabilities of the entire library Climax, one can run problems using the environment SDE, which can be freely downloaded, for example from book's webpage.[1] SDE can be launched online using a mechanism such is the Java Web Start or OpenWebStart.[2] In order to separate the intrinsic methods of Groovy, noting that this programming language is essentially an extension of the better-known commercial programming language Java from those which the authors have compiled as a combination of packages that comprise Climax and/or SDE, the sections appearing in this Appendix will be marked with an asterix (*) in the latter case. We will not discuss in here the possibilities for utilizing the computational methods contained in Climax library, but instead we will focus on the use of Groovy as an alternative programming language for scientific and educational purposes. A complete description can be found in the official documentation at its website.[3]

A.1   Variables

Variables can be defined by using uppercase or lowercase characters in combination with numbers. Acceptable names may have the following form:

---

[1] http://symplegma.civil.auth.gr/.

[2] https://openwebstart.com/.

[3] http://groovy-lang.org/documentation.html.

G. Manolis and C. Panagiotopoulos, *Vibrations of Structural Systems*, Synthesis Lectures
on Mechanical Engineering, https://doi.org/10.1007/978-3-032-12279-7

```
NetCost, Left2Pay, x3, X3, z25c5
```

It is not allowed to assign names which contain special characters or variable or start with a number. Examples of this are:

```
Net-Cost, 2pay, %x, *sign
```

Furthermore, names already in use by either Groovy of the environment SDE), such as PI=3.14159 … $\simeq \pi$, are also unacceptable.

```
1  x=13;  y=5*x
2  z=x**2+y
3  println x
4  println "y= "+y
```

In order for a numerical value of a variable to appear, the command `print` or the command `println` must be given. The hyphen mark (;) is used for separating parts appearing in the same line.

A.2   Arrays and Tables

An often used object for arranging and manipulating variables are either the array or the ArrayList.

```
1  x=[]
2  x.add(1.0);  x.add(4.5);  x<<3.5
3  println x.size()
4  x[0]=2.5;  x.remove(1)
5  println x
```

Alternatively, in order to recover a variable one may use the command getClass().

A.3   Intrinsic Functions

Among the intrinsic functions available in Groovy are the trigonometric functions sin, cos, often used functions such assqrt, log, exp, etc. In order to make these functions available through Groovy, as well as in Java, an activation instruction must be given using the command import static java.lang.Math.*) to the Math library of Java. This has already been set up in SDE so that all functions and often used variables, such as PI, E which pertain to $\pi \approx 3.14$ and $e \approx 2.718$, are available. A typical examples would be

```
1  x=PI**2
2  println sqrt(x)
```

In contrast to the Matlab/Octave software, intrinsic functions cannot be utilized within arrays and matrices.

A.4   Control Statements

Control statements are pieces of code which will or will not execute depending if a certain condition or a set of conditions holds true. Use of intrinsic variables in the languages Java/-Groovy is the logical boolean variable which assumes the self-evident values true or false. If at a certain point we have prescribed a value to variable x, then we can run a number of checks:

- x == 2 is x equal to 2?
- x ! =2 is not x equal to 2?
- x > 2 is x greater than 2?
- x < 2 is x lesser than 2?
- x >=2 is x greater of equal to 2?
- x <=2 is x smaller or equal to 2?

Special attention must be paid to the equality symbols ==. In contrast to Matlab/Octave, the intrinsic functions cannot be applied within the context of arrays and matrices.

A.4.1   Control Statements If/Else

Control statement if decides if a proposition is true or not and acts accordingly. This statement can be accompanied by the follow-up statement else which prescribes what happens if the proposition under examination is not true. In the absence of else no further action tales place if the proposition is not true. Given below is an example.

```
// Initializing a local variable
int a = 2

//Check for the boolean condition
if (a<100) {
    //If the condition is true print the following
    statement
    println("The value is less than 100");
} else {
    //If the condition is false print the following
    statement
    println("The value is greater than 100");
}
```

A.4.2   Control Statement switch

Another control and decision statement in Java/Groovy is the switch, through which a number of cases are examined before a command/proposition is made. An example follows.

```
//initializing a local variable
a = 2

//Evaluating the expression value
switch(a) {
    //There is case statement defined for 4 cases
    // Each case statement section has a break
   condition to exit the loop
    case 1:
        println("The value of a is One");
        break;
    case 2:
        println("The value of a is Two");
        break;
    default:
        println("The value is neither One or Two");
        break;
}
```

A.5   Repetitive Statements

These statements repeat a procedure, and the number of repetitions depends on the truth/strength of a control statement. Groovy is rich in these types of statements, and in what follows we will present some of the more commonly used ones.

A.5.1   Repetitive Statement for

This statement will executed the action given in block five times, for integer values of i going from 0 to 4.

```
for(int i = 0;i<5;i++){println(i)}
// or equivalent
for(int i in 0..<5){println(i)}
```

A.5.2   Repetitive Statement while

As long as the relevant control statement holds true, the actions specified in block of statement while will execute. The result given below follows the example given above in for.

```
int i = 0;
while(i<5) {
    println(i);
    i++;
}
```

A.5.3   Repetitive Statement in a range

A most useful module in Groovy is the (range) for ordering integers in the interval (start..end) that includes numbers starting from start reaching to end. In combination with the function each and by defining the function object (closure) (to be discussed in the next section) allows us to define a repetitive structure. This is shown in the example that builds on the immediately previous one.

```
(0..4).each{println(it)}
```

A.6   Functions

A method (or function) in Groovy is define in such a way as to return a value for a specified definition. The object returned can even be an empty one (void) or an undefined object (def). Once the function completes the procedures specified, it exits using the command return. In case the functions has been defined as (void) or (def), this last command return can be omitted. In the example that follows, we will define and use the following function:

$$f(x, y) = \sin(2\pi x)\sin(2\pi y)$$

```
double f(double x, double y){
  return Math.sin(2.0*Math.PI*x)*Math.sin(2.0*Math.PI*y)
}
println f(0.85,0.2)
```

The definition of the types of parameters as input to a function is voluntary. Within the SDE, the class name Math can be omitted. Thus, we can write the above statement in a compact way as

```
double f(x, y){
  return sin(2.0*PI*x)*sin(2.0*PI*y)
}
println f(0.85,0.2)
```

A.7   Function Objects (closures)

For the purposes of the present book, we prefer to use function objects (closure) instead of functions (method).A simple way to understand a function object closure, is to view it as a function that is the input to another function. These objects are basic ingredients, which together with the dynamic programming capabilities and the high degree of compatibility with Java, that make Groovy extremely useful. A function object is defined within brackets with input parameters (definitions) and is separated from the main programming body with the symbol($->$). Furthermore a functional object can end with a return statement (return) or without one. In the second case, the return is defined by a computed entity and by the last process that took place within the object. We will present the definition of the trigonometric function presented previously as a function object.

```
f={x, y->
  sin(2.0*PI*x)*sin(2.0*PI*y)
}
println f(0.85,0.2)
```

It is worth noting that even if the input parameters are missing, a functional object in Groovy has a predetermined variable labelled as it. This is shown in the example that follows.

```
f={sin(2.0*PI*it[0])*sin(2.0*PI*it[1])}
println f([0.85,0.2])
```

A.8   Input/Output of Data

Groovy furnishes a group of ancillary routines for communication and overall management of data, to and from files. A first step is to define such a file.

```
SomeFile = new File("somedirectory/Example.txt")
```

A.8.1   Data Storage

In order to store data in a file, e.g., the file SomeFile created previously, the commands given below can be used.

```
1  SomeFile.write "This is the first line\n"
2  SomeFile << "This is the second line\n"
3  SomeFile.leftShift "This is the third line\n"
4  SomeFile.append "This is the fourth line\n"
5  println SomeFile.text
```

Command write erases past data and inserts alphanumeric characters for the definitions given. Command leftShift and append, which in Groovy are symbolized as $<<$, inserts the specified content without easing what was previously there.

A.8.2   Data Recovery

In order to read data from a file, the simplest way is to use the command text that returns the file's content defined by an alphanumeric String variable.

```
1  filetext=SomeFile.text
2  println filetext
```

If we wish to read the file line by line we insert it in a list using the command readLines

```
1  lines = SomeFile.readLines()
2  lines.each{println it}
```

A.9   Script Files (Scripts) *

Groovy provides scripting programming language capabilities. The term script, is a a file with a bundle of commands (cluster command file) that contains a group of commands and/or instructions written according to the protocol of a specific programming language. The purpose is execution of specific processes by a computing device. Groovy can read and manage such file clusters, which could be defined by any ending. Within the context of SDE, the convention is that these file clusters are specified by the ending .climax. This way, the naming is consistent with the library of files available in Java and allows for their managing within the Climax environment. Essentially, the mechanism responsible for executing the commands script are files contained in object GroovyShell. In what follows, we give some script files from climax which are relevant in solving the types of problems treated in this book.

A.10   Vectors ad Matrices *

When using SDE, the pre-set type used for algebraic is the one offered in the library of JAMA : A Java Matrix Package, see http://math.nist.gov/javanumerics/jama/. Vectors are considered as special cases of matrices. At the same time, there are intrinsic routines for solving linear systems of equations and for eigenvalue analysis. As example, we list the following group of commands:

```
order = 3
// define an array double[][]
da=new double[order][order]

for(i in 0..<order){
    for(j in 0..<order){
        da[i][j]=random()
    }
}

// use the above defined double array to equip a
    matrix M
M=Matrix(da)

// update array da for another use
for(i in 0..<order){
    for(j in 0..<order){
        da[i][j]=random()
    }
}

// use the above defined double array to equip a
    matrix K
K=Matrix(da)

M.print()
K.print()

(K+M).print()
(K-M).print()
(K*M).print()
M.inverse().print()
M.transpose().print()
```

Next is an example of coding for solving the generalized eigenvalues problem.

$$(K - \lambda M)x = 0$$

```
Eigen= EigenDescomposition(M.inverse()*K)
EigenValues=Eigen.getRealEigenvalues()
EigenVectors=Eigen.getV()

(1..order).each{println EigenValues[it-1]}
EigenVectors.print()
```

We note that the method getRealEigenvalues() returns a double array with entries the eigenvalues λ_i, while getV() returns a matrix where every column contains an eigenvector x_i.

A.11 Complex Numbers *

In the SDE, the formula used to represent complex numbers is the one used in the library of Apache Commons Math, see http://commons.apache.org/proper/commons-math/. An example of this format now follows.

```
c1= new Complex(1.0,2.0)
c2= new Complex(3.0,2.4)
println c1+c2
println c1-c2
println c1*c2
println c1/c2
println c1.conj() // c1.conjugate()
println c1.re() // c1.getReal()
println c1.im() // c1.getImaginary()
println c1.abs() // absolute value of c1
```

A.12 Figures (Plotting)*

The plotting in the form of a diagram of various objects is accomplished using plotfunction.The construction of an object for plotting requires the declaration of discrete values of the function (ordinate) through the ordering of its numbers, as well as of the argument (abscissa), although this is optional since these will be assumed to increase with unit value starting from zero. Alternatively, only the increment can be defined for the horizontal axis, and the abscissa will start from zero. For the construction of a figure, there exists a standard object in SDE defined by the variable name thePlot. The object that creates plots is labelled as PlotFrame and can handle any number of plotfunction. In the example that follows, we sketch diagrams for the trigonometric functions sine and cosine. Furthermore, execution of this code will result in diagrams such as those shown in Fig. A.1 which follows. Finally, by augmenting an if necessary, modifying the aforementioned code, we can enrich the diagrams

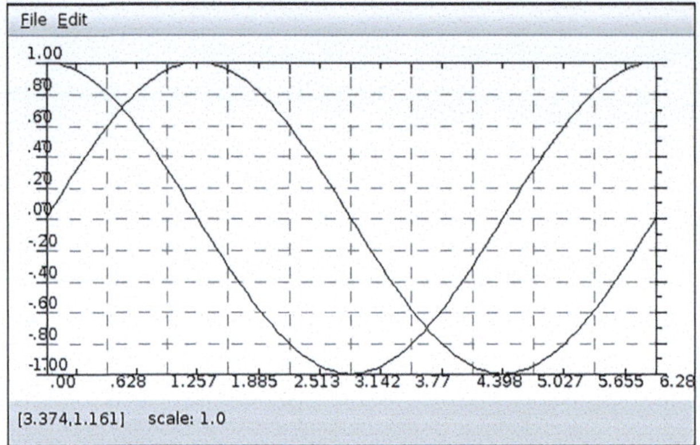

Fig. A.1 Trigonometric functions

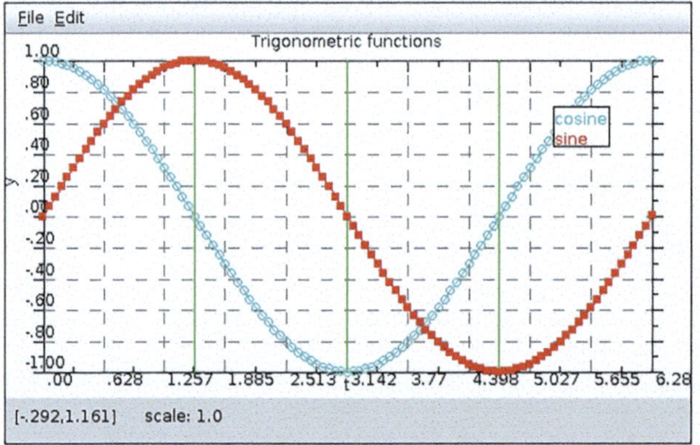

Fig. A.2 Trigonometric functions from Fig. A.1 enriched with colors and an inclusion

with further external information and also alter their original appearance. The final result will be quite similar to the one appearing in Fig. A.2, but with coloring and the insertion of a box with additional labels.

```
trigPlot = new PlotFrame()
n=100; dt=2.0*PI/n
t=new double[n+1]
s=new double[n+1]
c=new double[n+1]
(0..n).each{
    td=dt*it
    t[it]=td
    s[it]=sin(td)
    c[it]=cos(td)
}
```

```
12  trigPlot.addFunction(new plotfunction(dt,s))
13  trigPlot.addFunction(new plotfunction(t,c))
14  trigPlot.show()
```

```
1   trigPlot = new PlotFrame()
2   n=100; dt=2.0*PI/n
3   s=[];c=[]
4   (0..n).each{s[it]=sin(dt*it);c[it]=cos(dt*it)}
5   fp=new plotfunction(dt,s); fp.setMarker(true)
6   fp.setMarkerFill(true); fp.setName("sine")
7   trigPlot.addFunction(fp)
8   fp=new plotfunction(dt,c); fp.setMarker(true)
9   fp.setMarkerStyle(1);   fp.setName("cosine")
10  trigPlot.addFunction(fp)
11
12  trigPlot.setAutoColor(true)
13  trigPlot.Title("Trigonometric functions")
14  trigPlot.makeLegend()
15  trigPlot.xLabel("t")
16  trigPlot.yLabel("y")
17  trigPlot.vline(PI/2,java.awt.Color.green)
18  trigPlot.vline(PI,java.awt.Color.green)
19  trigPlot.vline(3*PI/2,java.awt.Color.green)
20  trigPlot.show()
```

The user defined groovy class `oneStorey` for the three dimensional single storey entity is given below. Obviously one could take that code and further extend it in order to consider more details and capabilities for the one storey three dimensional `oneStorey` building entiy.

```
import climax.contraption
import java.awt.Graphics2D
import java.awt.BasicStroke;
import java.awt.Stroke;

class oneStorey extends courses.structuraldynamics.
    mdof implements contraption{
    int id=0
    Color SelfColor = Color.blue
    Color MotionColor = Color.yellow
    Color ColSelfColor = Color.red
    Color COlMotionColor = Color.green
    double Elast,Poiss=0.25
    double Lx,Ly,mass
    double x0=0.0, y0=0.0, h=1.0;
    boolean stif=false
    ArrayList columns

    void setmdof(){
        NDOFS=3;
        K=new double[NDOFS][NDOFS];
        double G=E/(2.0*(1.0+Poiss))
        double kx,ky,kz,Ix,Iy,Id
        columns.each{
            if(stif){
                kx=it[2];
```


G. Manolis and C. Panagiotopoulos, *Vibrations of Structural Systems*, Synthesis Lectures on Mechanical Engineering, https://doi.org/10.1007/978-3-032-12279-7

```
                ky=kx; if(it[3])ky=it[3]
                kz=0.0; if(it[4])kz=it[4]
            }else{
                Ix=it[2];
                Iy=Ix; if(it[3])Iy=it[3]
                Id=0.0; if(it[4])Id=it[4]
                kx=12*Elast*Ix/h**3
                ky=12*Elast*Iy/h**3
                kz=G*Id/h
            }
            double x=it[0]; double y=it[1]

        K[0][0]+=kx; K[1][1]+=ky
        K[0][2]+=-y*kx; K[2][0]=K[0][2]
        K[1][2]+=x*ky; K[2][1]=K[1][2]
        K[2][2]+=kz+y*y*kx+x*x*ky
      }
    M=new double[NDOFS][NDOFS];
    M[0][0]=mass; M[1][1]=mass; M[2][2]=mass*(Lx*
Lx+Ly*Ly)/12.0
    C=new double[NDOFS][NDOFS];
    this.u0=new double[NDOFS];
    this.v0=new double[NDOFS];
    this.df=new DoubleFunction[NDOFS];
    this.diagonalMass=true
    this.u0=new double[NDOFS];
    this.v0=new double[NDOFS];
  }

  public double[] getElasticCenter(){
      double xE,yE,rE
      xE=(K[0][0]*K[1][2]-K[0][2]*K[1][0])/(K
[0][0]*K[1][1]-K[0][1]*K[1][0])
      yE=(K[0][1]*K[1][2]-K[1][1]*K[0][2])/(K
[0][0]*K[1][1]-K[0][1]*K[1][0])
      rE=sqrt(xE**2+yE**2)
      return ([xE,yE] as double[])
  }


  @Override
  public int getID() {
      return id
  }


  @Override
  public void SelfPortrait(Graphics2D g2, double
ymin, double ymax, int ye, int ys, double xmin,
double xmax, int xe, int xs) {
      def transX = {double x ->
```

```
            return ((int) ((int) ((-xmin * xe + x * (
 xe - xs) + xmax * xs) / (xmax - xmin))))
        }

    def transY = {double y ->
            return ((int) ((int) ((-ymin * ye + y * (
 ye - ys) + ymax * ys) / (ymax - ymin))))
        }

    Color origColor=g2.getColor()
    g2.setColor(SelfColor)

    // top vertical dashpot and spring
    ArrayList nods=[[-Lx/2.0,-Ly/2.0],[Lx/2.0,-Ly
/2.0],[Lx/2.0,Ly/2.0],[-Lx/2.0,Ly/2.0]]

    (-1..2).each{
        double x1=nods[it][0]; double y1=nods[it
][1]
        double x2=nods[it+1][0]; double y2=nods[
it+1][1]
        g2.drawLine( transX(x0+x1), transY(y0+y1)
, transX(x0+x2),  transY(y0+y2) )
    }

    g2.setColor(ColSelfColor)
    double dx=min(0.025*Lx,0.025*Ly);
    double dy=dx; int num=20;
    def drawColumn={double x, double y->
        int[] xPoints = new int[num]; int[]
yPoints = new int[num]
        (0..<num).each{
            xPoints[it]=transX(x0+x+dx*cos(2.0*pi
*it/num))
            yPoints[it]=transY(y0+y+dy*sin(2.0*pi
*it/num))
        }
        g2.fillPolygon(xPoints, yPoints, num)
    }

    columns.each{
        drawColumn(it[0],it[1])
    }


    g2.setColor(origColor);
  }
```

```
@Override
public void Motion(Graphics2D g2, double ymin,
double ymax, int ye, int ys, double xmin, double
xmax, int xe, int xs, int step, double scale) {
    def transX = {double x ->
        return ((int) ((int) ((-xmin * xe + x * (
xe - xs) + xmax * xs) / (xmax - xmin))))
    }

    def transY = {double y ->
        return ((int) ((int) ((-ymin * ye + y * (
ye - ys) + ymax * ys) / (ymax - ymin))))
    }

    Color origColor=g2.getColor()
    g2.setColor(MotionColor)

    // top vertical dashpot and spring
    ArrayList nods=[[-Lx/2.0,-Ly/2.0],[Lx/2.0,-Ly
/2.0],[Lx/2.0,Ly/2.0],[-Lx/2.0,Ly/2.0]]
    double ux=u[0][step]; double uy=u[1][step];
double th=u[2][step];
    (-1..2).each{
        double x1=nods[it][0]+scale*(ux-nods[it
][1]*th ); double y1=nods[it][1]+scale*(uy+nods[
it][0]*th )
        double x2=nods[it+1][0]+scale*(ux-nods[it
+1][1]*th ); double y2=nods[it+1][1]+scale*(uy+
nods[it+1][0]*th )
        g2.drawLine( transX(x0+x1), transY(y0+y1)
, transX(x0+x2),  transY(y0+y2) )
    }

    g2.setColor(C01MotionColor)
    double dx=min(0.025*Lx,0.025*Ly);
    double dy=dx; int num=20;
    def drawColumn={double x, double y->
        int[] xPoints = new int[num]; int[]
yPoints = new int[num]
        (0..<num).each{
            double x1=(x+dx*cos(2.0*pi*it/num))+
scale*(ux-(y+dy*sin(2.0*pi*it/num))*th)
            double y1=(y+dy*sin(2.0*pi*it/num))+
scale*(uy+(x+dx*cos(2.0*pi*it/num))*th)
            xPoints[it]=transX(x0+x1)
            yPoints[it]=transY(y0+y1)
        }
        g2.fillPolygon(xPoints, yPoints, num)
```

```
145        }
146
147        columns.each{
148            drawColumn(it[0],it[1])
149        }
150
151
152        g2.setColor(origColor);
153    }
154
155    @Override
156    public double maximum_X_coordinate() {
157        return x0+Lx/2.0
158    }
159
160    @Override
161    public double minimum_X_coordinate() {
162        return x0-Lx/2.0
163    }
164
165    @Override
166    public double maximum_Y_coordinate() {
167        return y0+Ly/2.0
168    }
169
170    @Override
171    public double minimum_Y_coordinate() {
172        return y0-Ly/2.0
173    }
174
175    @Override
176    public double maximum_Z_coordinate() {
177        return 0.5
178    }
179
180    @Override
181    public double minimum_Z_coordinate() {
182        return -0.5
183    }
184 }
```

Loading an External Module for 1D Poroelasticity

The java package for one dimensional poroelasticity can be downloaded from book's webpage.[4] In order to load the external java package `jpythmen`, from the main menu of the SDE window we choose the `Load` button and then `Add jar` (or by simply using the shortcut `Ctrl+L`. We locate the `jpythmen` on some directory of our pc and then we may use the code of the following Listing to produced results shown in Sect. 1DPoroElastExample.

```
1  // external module jpythmen can be downloaded at:
2  // \url{https://symplegma.civil.auth.gr/docs/jpythmen
      .jar}
3  import OneDimTube.*
4
5  expTube=new tube()
6
7  r0=1.9e-5
8  lp=0.02331
9  asection= new section(2.0*r0)
10
11 pf=1.2
12 lamf=1.01d*1.0e5
13 muf=1.8d*1.0e-5
14
15 ps=2691.0d
16 lams=5.035d*1.0e10
17 mus=2.594*1.0e10
18
19 mat1=new material(id:1, lamF:lamf,muF:muf)
20 mat1.rhoS=ps
21 mat1.rhoF=pf
22
```

---

[4] https://symplegma.civil.auth.gr/docs/jpythmen.jar.

© The Editor(s) (if applicable) and The Author(s), under exclusive license to Springer Nature Switzerland AG 2026
G. Manolis and C. Panagiotopoulos, *Vibrations of Structural Systems*, Synthesis Lectures on Mechanical Engineering, https://doi.org/10.1007/978-3-032-12279-7

```
23  fluid1= new fluid(mat:mat1, id:1)
24  fluid1.name="layer-"+1;
25  fluid1.length=1.0-lp
26  fluid1.sec=asection
27  expTube.putPart(fluid1)
28
29  poro=0.9094;
30  mat2=new material(2, lams, mus, poro,lamf,muf)
31  mat2.rhoS=ps; mat2.rhoF=pf; mat2.br=18980.0;
32  mat2.cnl=0.9; mat2.ae=0.75; mat2.cd=0.25
33
34  probeta = new poroelastic(mat:mat2, id:2)
35  probeta.name="layer-"+2
36  probeta.length=lp
37  probeta.sec=asection
38  expTube.putPart(probeta)
39
40  F0=new Complex(10.0e6,0.0)
41  expTube.bc_l([1] as int[], [F0] as Complex[])
42  expTube.bc_r([0,0] as int[], [re*0.0,re*0.0] as
        Complex[])
43
44  omega=linspace(2.0*pi*20.0d,2.0*pi*6000.0d,1000*8)
45  expTube.set(omega as double[])
46
47  sac_anl=[]
48  (0..<omega.size()).each{
49      sac_anl.add(expTube.parts[1].
        AbsorptionCoefficient(1.0, it))
50  }
51
52  plot()
53  thePlot.setAutoColor(true)
54  thePlot.setMarker(true)
55  plot(omega.collect{it/(2.0*pi*1000.0)}, sac_anl )
56
57  thePlot.getPlotPanel().numx=6
58  thePlot.getPlotPanel().numy=5
59  thePlot.show()
60
61  respPlot=new PlotFrame()
62  data=expTube.getKinematicField2Plot(4000,omega.length
        -1)
63  respPlot.addFunction(new plotfunction(data[0],data
        [1].collect{it.re()}))
64  respPlot.vline(fluid1.length,Color.red)
65  respPlot.setMarker(true)
66  respPlot.show()
```

The manufacturer's authorised representative in the EU is Springer
Nature Customer Service Centre GmbH, Europaplatz 3, 69115 Heidelberg,
Germany. If you have any concerns regarding our products, please
contact ProductSafety@springernature.com

Printed and bound by CPI Group (UK) Ltd, Croydon, CR0 4YY

11/05/2026

02107715-0011